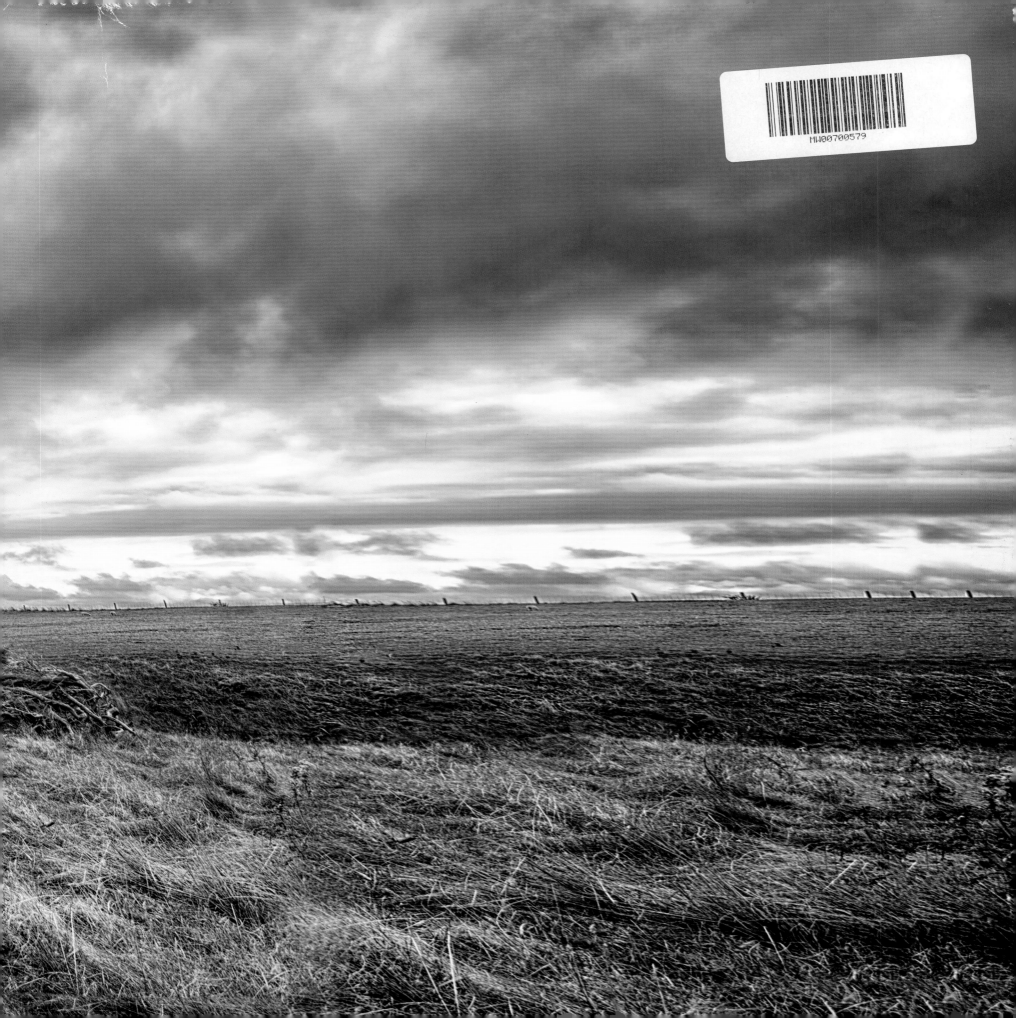

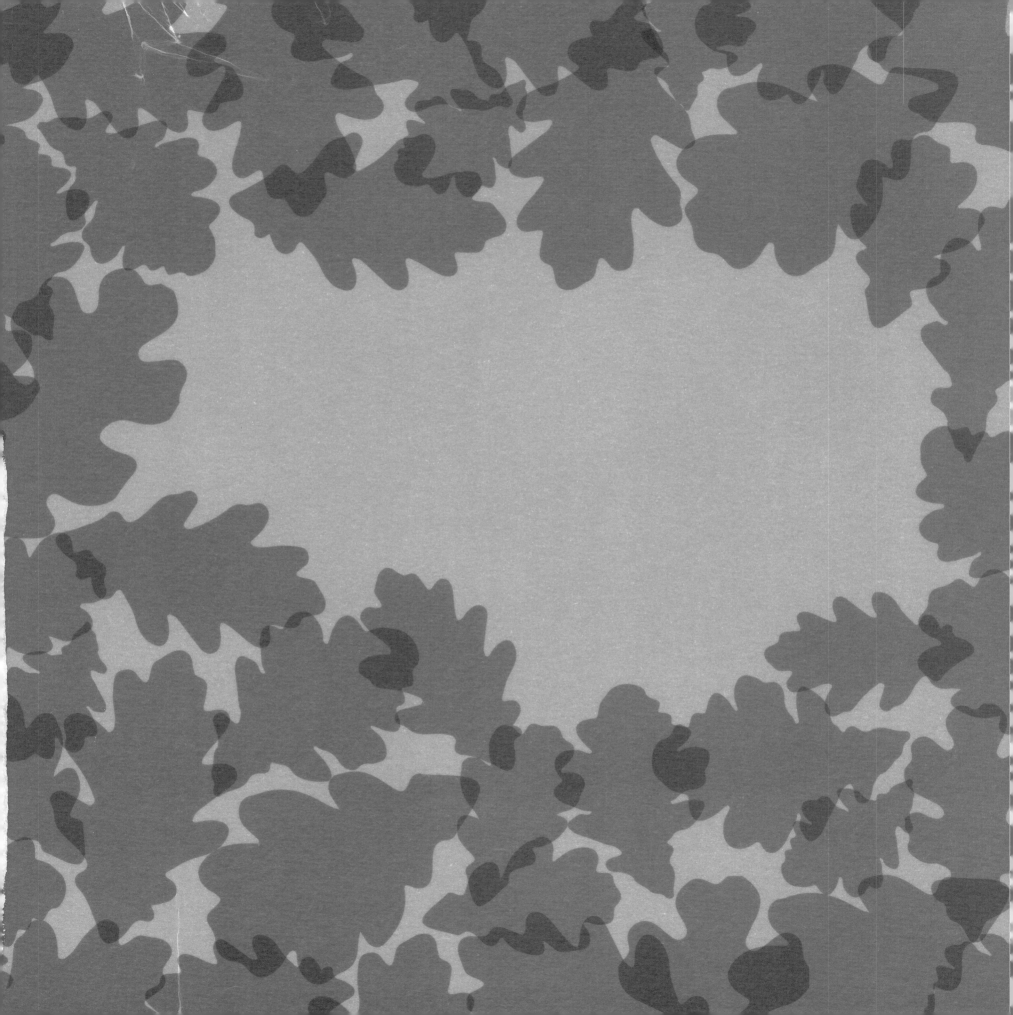

TreeGirl

Intimate Encounters with Wild Nature

TreeGirl | Intimate Encounters with Wild Nature

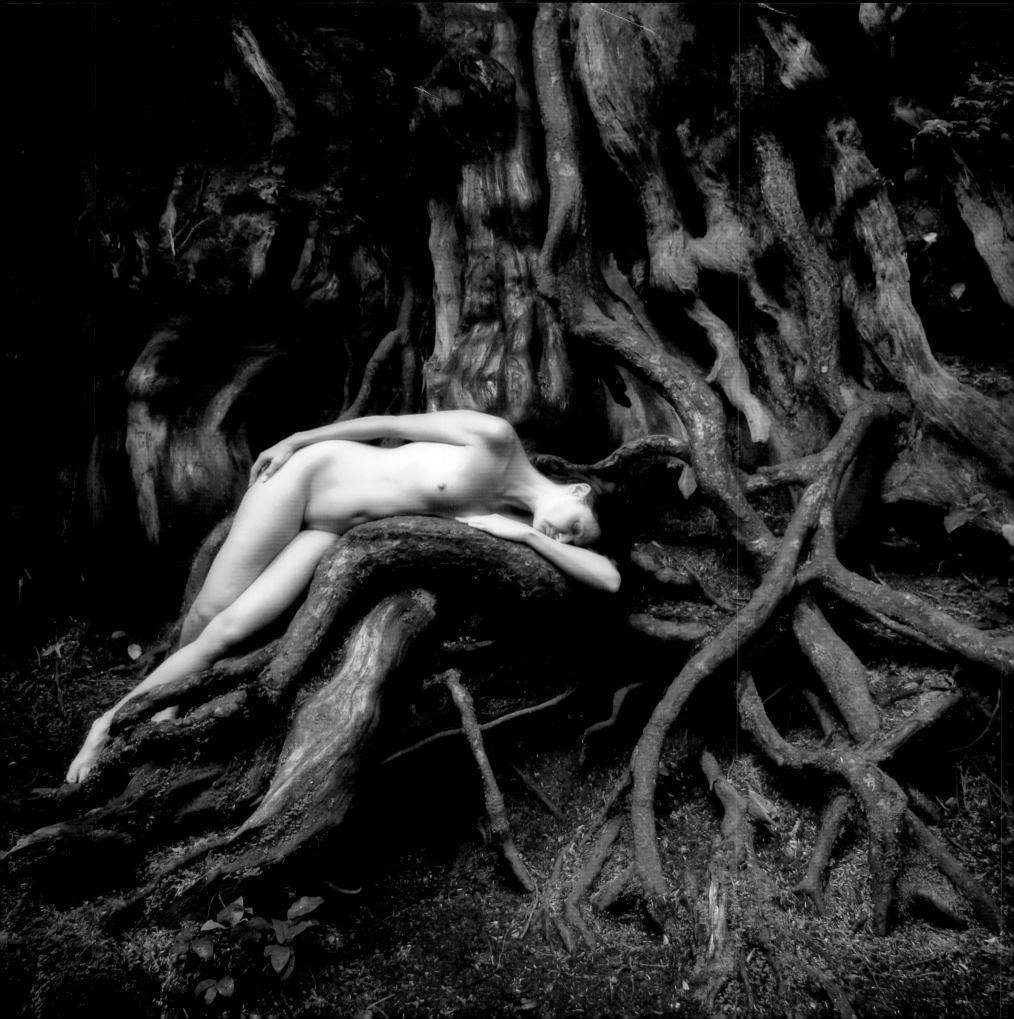

For the trees and their lovers

In memory of Diane Osborne

Lily Draped Over Redcedar Roots (2012)

Kalaloch Redcedar, Olympic National Park, Washington, US. See also p. 159.

Published by TreeGirl Studios LLC
3205 Dutton Ave. Suite 56
Santa Rosa, CA 95407
www.treegirlstudios.com

First Edition
ISBN 978-0-692-72604-4

Book Design and Layout by Annett Börner
www.dn.com.au/annett-boerner.html
Book Cover Design by Julianne Skai Arbor

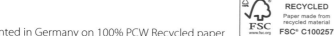

RECYCLED
Paper made from recycled material
FSC
www.fsc.org   FSC® C100257

Printed in Germany on 100% PCW Recycled paper

A portion of the funding of this book has been donated to TreeSisters.org – www.treesisters.org for tropical reforestation.

PHOTO CREDITS:
All photographs ©Julianne Skai Arbor/TreeGirl®, with the exception of:

p. 29   Blackbutt Leaves and Flower Buds © Dean Nicolle
p. 29   Blackbutt Trunk © Dean Nicolle
p. 37   Roasted California Bay Nuts © Kristen Rasmussen/www.rootedfood.com
p. 41   Buckeye Recline © Caiyloirch Rupert Marques and TreeGirl®
p. 66   Sequoia Meditation IV © Devin Fleurdujon and TreeGirl®
p. 6   Sequoia Meditation V © Devin Fleurdujon and TreeGirl®
p. 59   Büchtel von Eibenzweigen (*Taxus baccata*) © unpict/Fotolia by Adobe Stock Photo
p. 73   Huon Pine Wood © Rozenn Leard/Dreamstime Stock Photo
p. 89   Mangrove Seeds © Annett Börner
p. 91   Bees on Mesquite Flowers © Marilyn McFarlin
p. 107   *Quercus garryana* Leaves and Acorns © Martin Fletcher
p. 115   Pacific Dogwood in Bloom © randimal Fotolia by Adobe Stock Photo
p. 125   Silver Gimlet Leaves and Flower Buds © Dean Nicolle
p. 129   Australian Native Small Leaf Fig Tree – *Ficus oblique* © Sheryl Caston | Dreamstime.com
p. 131   Spotted Gum Bark © Dean Nicolle
p. 133   Frutti autunnali (Sweet Chestnut) © beppenob/Fotolia by Adobe Stock Photo
p. 152   Caterpillar Pinyon © Caiyloirch Rupert Marques and TreeGirl®
p. 153   Cliff Chipmunk on Two-Needle Pinyon Pine @ Hal Brindley/www.travelforwildlife.com
p. 157   Western Hemlock Needles and Cones © Dave Ingram
p. 191   Author Photo © Aura May

Front Cover:
Seven Sisters Full (2008)

Kalahari Desert, on the Ntwetwe Salt Pan, Botswana. See p. 17.

Front end image:
Monterey Cypress Wind (2015)

Monterey Cypresses (see p. 92) make excellent windbreaks, so New Zealanders planted them all over the islands. These windswept Monterey Cypresses were some of the most challenging, but aesthetically fulfilling, trees I've ever shot. I had driven all the way to the southern tip of the South Island of New Zealand to find them. Others driving the coast line pilgrimaged to the lookout point, but I crossed the fence facing the other direction, into the wild world of sheep. It was a mere 40˚F (4˚C) and windy. I was freezing. Inside the little forest of Cypresses was a whole other world of shelter: twisted trees, curious sheep, sheep bones, dilapidated sheds, and privacy to change clothes. It was also a few precious degrees warmer. The shoot took about an hour. I should have stopped earlier, but I couldn't help myself. That's part of being a treegirl: photographing can sometimes be so invigorating, and usually an once-in-a-lifetime opportunity, that I have to stay as long as I can bear it. You can find these trees along the road at Slope Point, but you have to ask the sheep permission if you trespass.

Page 192:
Panasos Olive Wave (2015)

Panasos, Crete, Greece. See p. 102

Back end image:
Pafuri Baobab Big (2008)

Just before sunset, I climbed 20 feet with a boost from my guide, to the top of this giant African Baobab—a world unto itself, full of nooks and hollows—where I took refuge in a giant arm that was perfectly shaped to fit my body curled into fetal position. I felt as if I had re-entered the womb and didn't want to leave. This Baobab resides near Pafuri, at the edge of Kruger National Park, near the border where South Africa, Zimbabwe, and Mozambique come together. See p. 19.

Back Cover:
Wanaka Water Willow (2015)

White Willow, on Lake Wanaka, North Island of New Zealand. See p. 166.

Contents

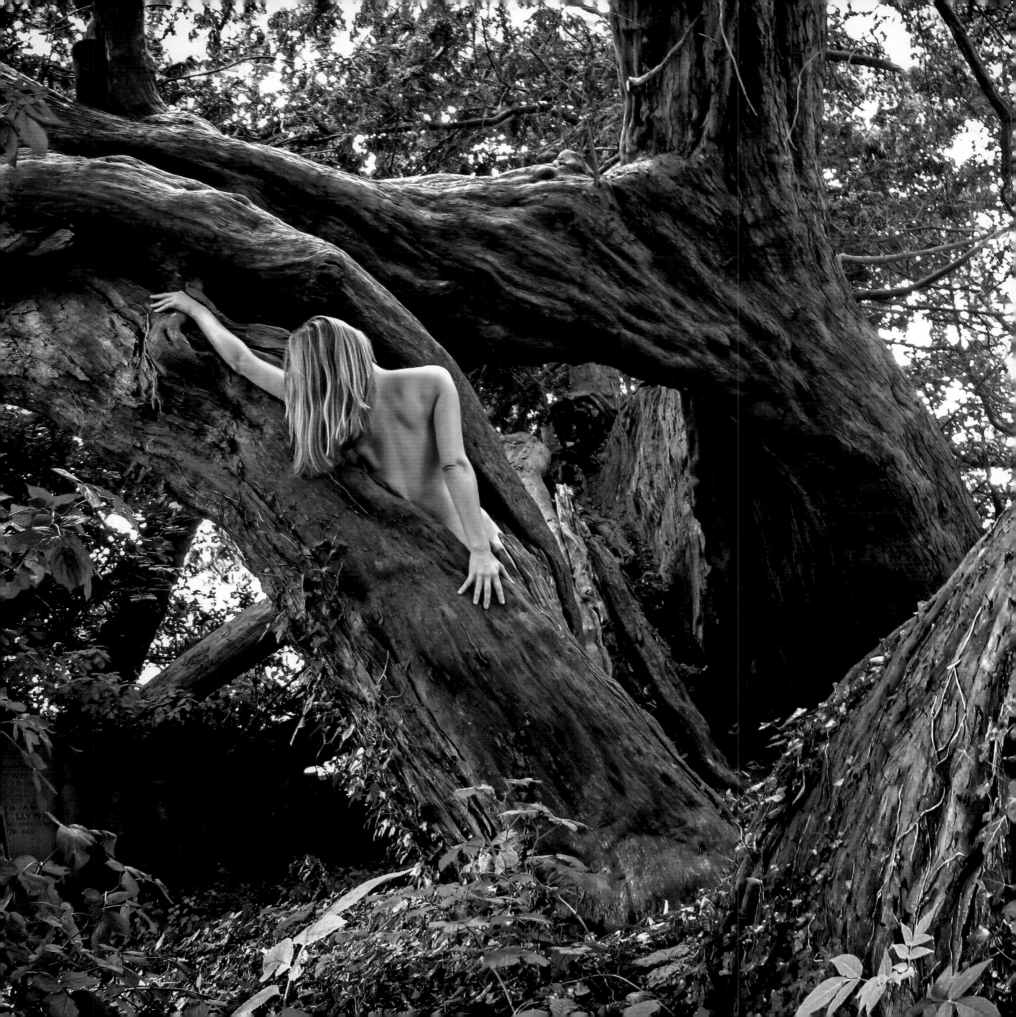

This book is not just about trees; it is also about responsibility and freedom. Responsibility because TreeGirl dares to talk about the things that really matter, by which I mean our part in Nature, and the imminent necessity to attune again to the life rhythms of planet Earth instead of believing stoically that everything is just there to be consumed by our species. For millennia, cultures have cared about the environment, and the secret of their success was a bond with Nature—personal, emotional, mythical, and experiential. The modern age has tried to replace that with "reason" alone. But being presented graphs of carbon dioxide emissions just doesn't do the trick of getting anyone moving, does it? It just disempowers us; "oh, let them do it! I am no expert." Fact is, "they" don't know either just how complex the interconnections of the elements of life are, and "they" don't really change our self-destructive course either. It takes all of us to do something, and what stimulates us to get going is beauty, emotion, and personal experience.

As for freedom, our media keep telling us that we live in the free Western world and that nobody ever was luckier. We have credit cards and smartphones, don't we? Just how "free" we really are you can tell in the moment we lose them. There is a new term for this dawning among sociologists and psychologists: it's addiction. What freedom really means I understood when I read Wilfred Pelletier's book *No Foreign Land*. Pelletier (1927–2000) was a Native American who tried to live in the white man's world for a few years. When returning to the Odawa reservation he fully recognized what freedom means to the Natives; he could wander in the wilderness with no need to "return" somewhere, he had everything he needed—his survival skills and a knife—and could be home everywhere in Nature. No dependence on gadgets, social status, an income, or bureaucracy. Life can be simple. And joyful.

And joy is something that TreeGirl really can teach! It is amazing to see someone moving in the wildwood with such trust and openness. I think she has found a way to revive that same joy and gratitude that people in ancient times all around the planet felt in their traditional sacred groves. Because in early history, humankind contemplated the eternal questions, meditated and prayed, and made offerings of gratitude to the life force mostly beneath trees. Trees have always been regarded as kind bridges between the human world and Spirit. Even the white marble splendour of the Acropolis was only there to honour Athena's sacred Olive tree which stood in the centre of her sanctuary. In fact, all deities of ancient Greece had their associated tree species, and ethnologists deem it highly probable that the Greek pantheon evolved from tree spirits in the first place. Rome itself had just as many sacred groves as the Celtic and Germanic woodland tribes. And so did the ancient Hebrews—Christian anthropologists of the 19th century were surprised to see just how much tree worship there was in the Holy Land, and how peacefully it co-existed with early monotheism. And Jesus was a teacher of the ancient Olive grove of Gethsemane—after the last supper "he went out and made his way *as usual* to the Mount of Olives, accompanied by the disciples" (Lk. 22. 39).

The mythical symbol of the World Tree was known on all continents (except Antarctica). In Asia Minor it was dominantly referred to as the Tree of Life, standing at the heart of paradise. The Quran adapted it from the Old Testament, and it can be traced back to ancient Sumerian cuneiform tablets. The World Tree is perhaps the most ecological of mythical images because it denotes the unity and interconnectedness of all life, and describes the biosphere as one huge organism. As the English poet, painter and visionary William Blake (1757–1827) says: "For everything that lives is holy. Life delights in life."

The notion that the whole Earth is alive and that it has a spirit did not, as most people think, disappear with Christianity. Rather, the change came with the "Enlightenment", the 'age of reason' and its mechanistic science which declared the material world as void of spirit. As the biologist Rupert Sheldrake says: "Until the seventeenth century, university scholars and Christian theologians taught that the universe was alive, pervaded by the Spirit of God, the divine breath of life. All plants, animals and people had souls. The stars, the planets and the earth were living beings, guided by angelic intelligences." [1] Seeing Nature as a mechanical, inanimate system may offer many comforts; it grants the delusion that we are in control and that we, the "crown of creation", have risen above the animistic, superstitious ways of primitive peoples. But by omitting the "mother" in Mother Nature we only manage to force our recognition of the mother principle into the subconscious. And, ironically, "matter" (Latin *materia*) stems from "mother" (*mater*).

Through outdoor education, psychology has recently begun to recognize that all living beings are a part of us; a child discovers a new part of herself every time she encounters an animal or a plant. And often children express this in play, by wanting to be a tiger or a horse, a bird or a crocodile. Soon after "mama" and "papa", animal names form part of the earliest vocabulary of toddlers. And although animals are closer to us emotionally, plants and trees too create unique individual sensations of self in a child (and in grown-ups too, if we care to recognize them). In the presence of an Oak we discover new variations of our sense of identity. There is another world waiting for us at the Redwood, just as rooted in our ancient soul, just as unawakened. Elm, Beech and Maple too are gates to other dimensions of our greater self. Beneath each tree there is an experience of self-discovery awaiting us—are you ready to explore it? TreeGirl surely lives by example. The Earth is not just physical; it is also a landscape of the soul. The segregation of inner and outer is imaginary.

So if inside and outside are a reflection of each other, what does nudity mean in these pictures from the wildwood? In the sacred space of the paradisiacal grove untouched by human greed, nakedness is an expression of trust, of respect, of love, of innocence. "See? I bring no weapons, no secret agenda. And I make myself vulnerable because I really want to see the true You." Or in the words of TreeGirl herself: "To be alive on the planet is to surrender to being her partner. To be in service of the Earth is to be enraptured by her."

I also give TreeGirl ten out of ten for photography. Light, contrast, composition are all highly enjoyable. But most noteworthy is the fact that—while trees in Western art have for centuries merely provided backgrounds for anthropocentric scenes—here trees are never reduced to serve the human story. I delight how in each photo the human figure gracefully finds their place within the tree, adding a new note to the tree's song, with a heart always beating to the tree's drum.

Enjoy TreeGirl's journeys, and enjoy your own!

Fred Hageneder, August 2016
www.themeaningoftrees.com

◀ Llanerfyl Emerging (2009)

St. Erfyl, in Llanerfyl, Powyrs, Wales. See also p. 58.

[1] Rupert Sheldrake, *The Science Delusion* (Coronet, London, 2013) p. 21.

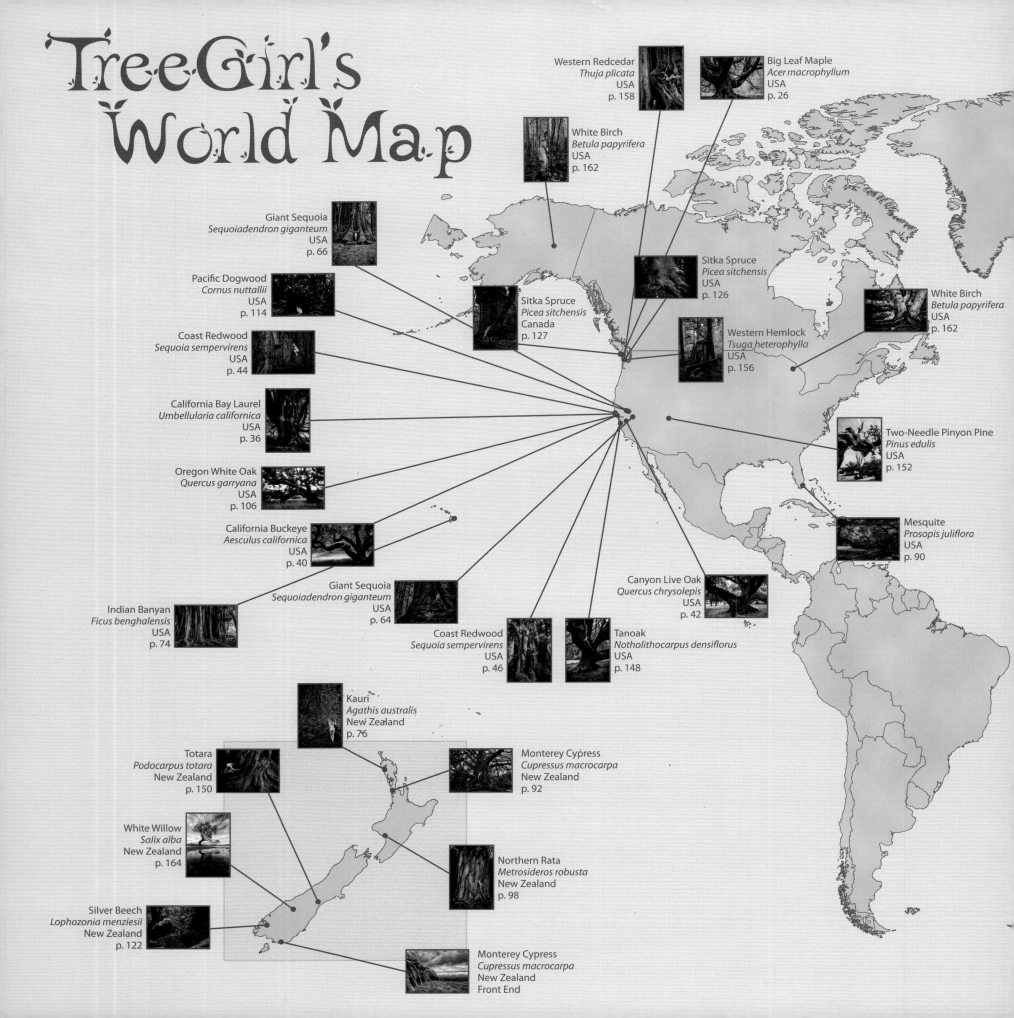

TreeGirl's World Map

Western Redcedar
Thuja plicata
USA
p. 158

Big Leaf Maple
Acer macrophyllum
USA
p. 26

White Birch
Betula papyrifera
USA
p. 162

Giant Sequoia
Sequoiadendron giganteum
USA
p. 66

Sitka Spruce
Picea sitchensis
USA
p. 126

White Birch
Betula papyrifera
USA
p. 162

Pacific Dogwood
Cornus nuttallii
USA
p. 114

Sitka Spruce
Picea sitchensis
Canada
p. 127

Western Hemlock
Tsuga heterophylla
USA
p. 156

Coast Redwood
Sequoia sempervirens
USA
p. 44

Two-Needle Pinyon Pine
Pinus edulis
USA
p. 152

California Bay Laurel
Umbellularia californica
USA
p. 36

Oregon White Oak
Quercus garryana
USA
p. 106

Mesquite
Prosopis juliflora
USA
p. 90

California Buckeye
Aesculus californica
USA
p. 40

Indian Banyan
Ficus benghalensis
USA
p. 74

Giant Sequoia
Sequoiadendron giganteum
USA
p. 64

Canyon Live Oak
Quercus chrysolepis
USA
p. 42

Coast Redwood
Sequoia sempervirens
USA
p. 46

Tanoak
Notholithocarpus densiflorus
USA
p. 148

Kauri
Agathis australis
New Zealand
p. 76

Totara
Podocarpus totara
New Zealand
p. 150

Monterey Cypress
Cupressus macrocarpa
New Zealand
p. 92

White Willow
Salix alba
New Zealand
p. 164

Northern Rata
Metrosideros robusta
New Zealand
p. 98

Silver Beech
Lophozonia menziesii
New Zealand
p. 122

Monterey Cypress
Cupressus macrocarpa
New Zealand
Front End

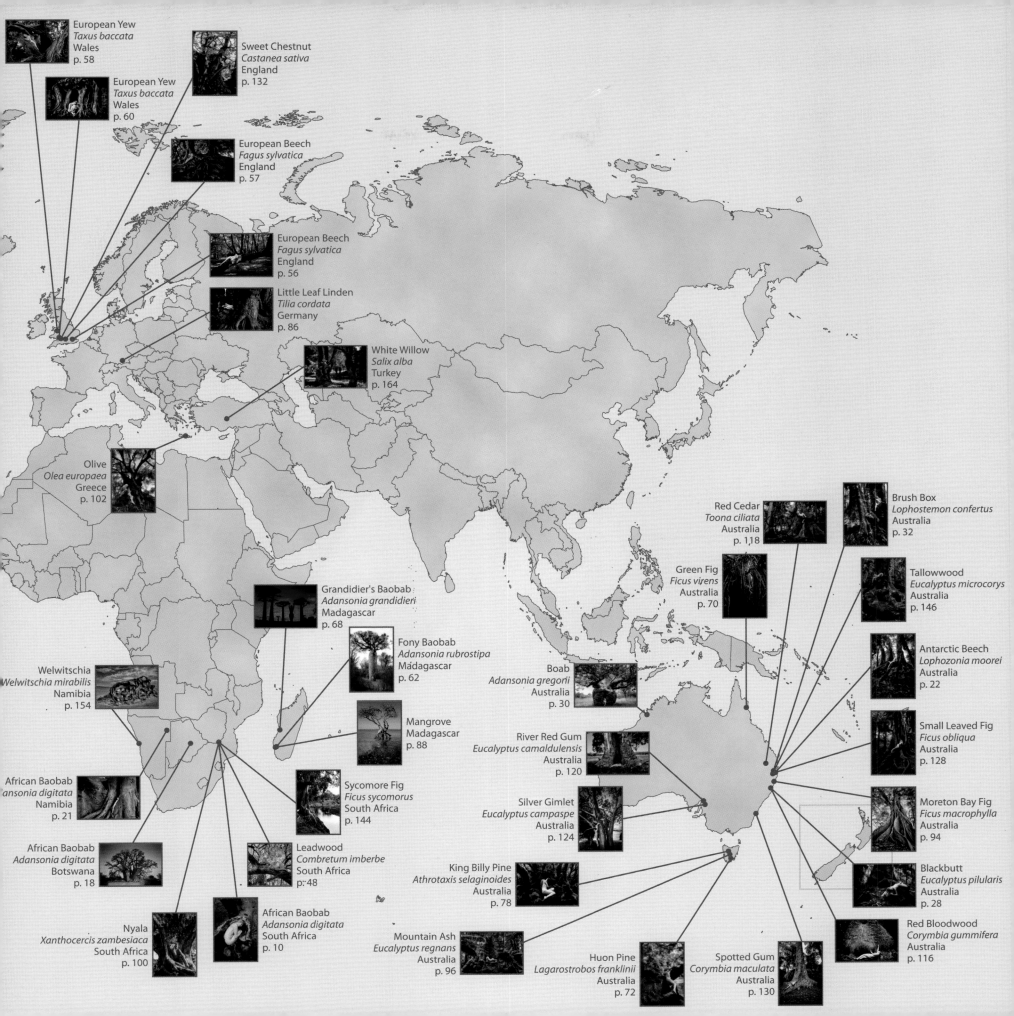

European Yew
Taxus baccata
Wales
p. 58

European Yew
Taxus baccata
Wales
p. 60

Sweet Chestnut
Castanea sativa
England
p. 132

European Beech
Fagus sylvatica
England
p. 57

European Beech
Fagus sylvatica
England
p. 56

Little Leaf Linden
Tilia cordata
Germany
p. 86

White Willow
Salix alba
Turkey
p. 164

Olive
Olea europaea
Greece
p. 102

Grandidier's Baobab
Adansonia grandidieri
Madagascar
p. 68

Fony Baobab
Adansonia rubrostipa
Madagascar
p. 62

Welwitschia
Welwitschia mirabilis
Namibia
p. 154

Mangrove
Madagascar
p. 88

Boab
Adansonia gregorii
Australia
p. 30

Green Fig
Ficus virens
Australia
p. 70

Red Cedar
Toona ciliata
Australia
p. 118

Brush Box
Lophostemon confertus
Australia
p. 32

Tallowwood
Eucalyptus microcorys
Australia
p. 146

Antarctic Beech
Lophozonia moorei
Australia
p. 22

African Baobab
Adansonia digitata
Namibia
p. 21

African Baobab
Adansonia digitata
Botswana
p. 18

Sycomore Fig
Ficus sycomorus
South Africa
p. 144

Leadwood
Combretum imberbe
South Africa
p. 48

River Red Gum
Eucalyptus camaldulensis
Australia
p. 120

Silver Gimlet
Eucalyptus campaspe
Australia
p. 124

Small Leaved Fig
Ficus obliqua
Australia
p. 128

Moreton Bay Fig
Ficus macrophylla
Australia
p. 94

Nyala
Xanthocercis zambesiaca
South Africa
p. 100

African Baobab
Adansonia digitata
South Africa
p. 10

King Billy Pine
Athrotaxis selaginoides
Australia
p. 78

Blackbutt
Eucalyptus pilularis
Australia
p. 28

Mountain Ash
Eucalyptus regnans
Australia
p. 96

Huon Pine
Lagarostrobos franklinii
Australia
p. 72

Spotted Gum
Corymbia maculata
Australia
p. 130

Red Bloodwood
Corymbia gummifera
Australia
p. 116

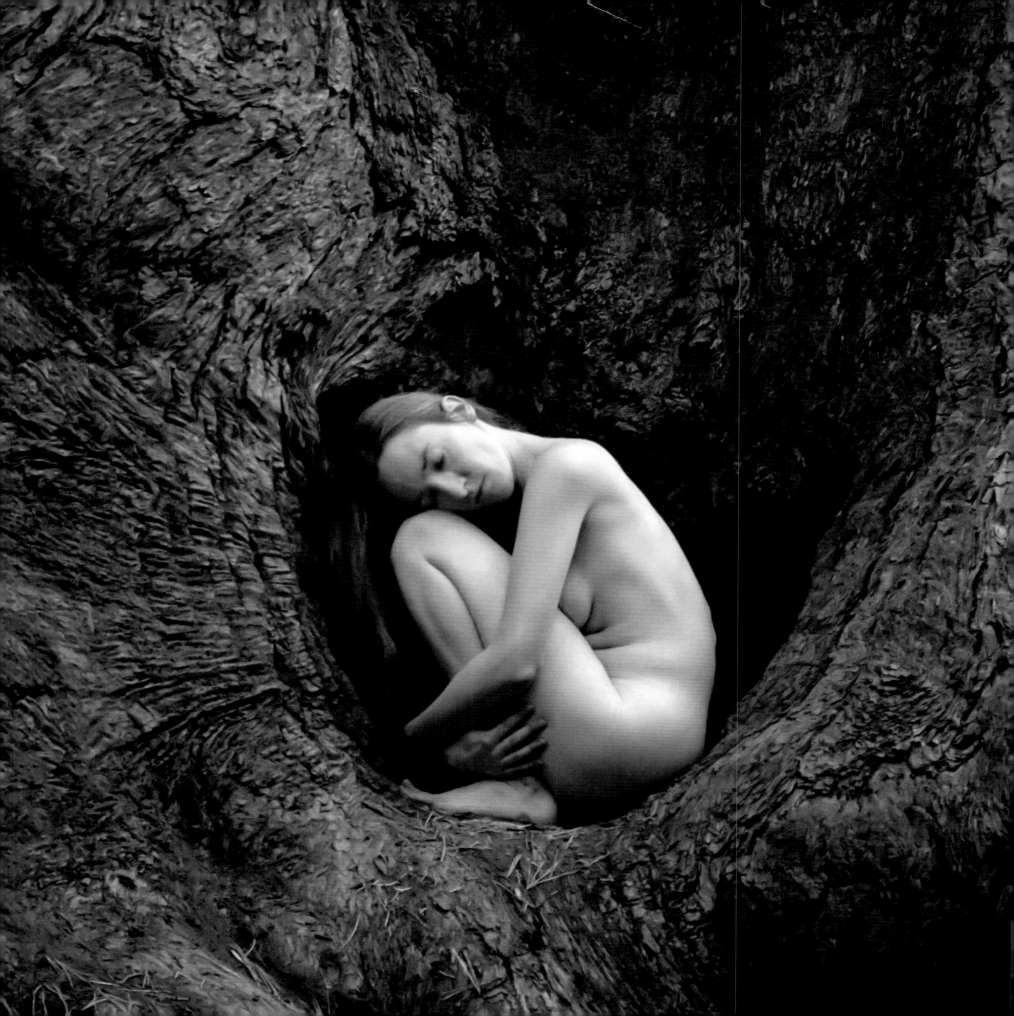

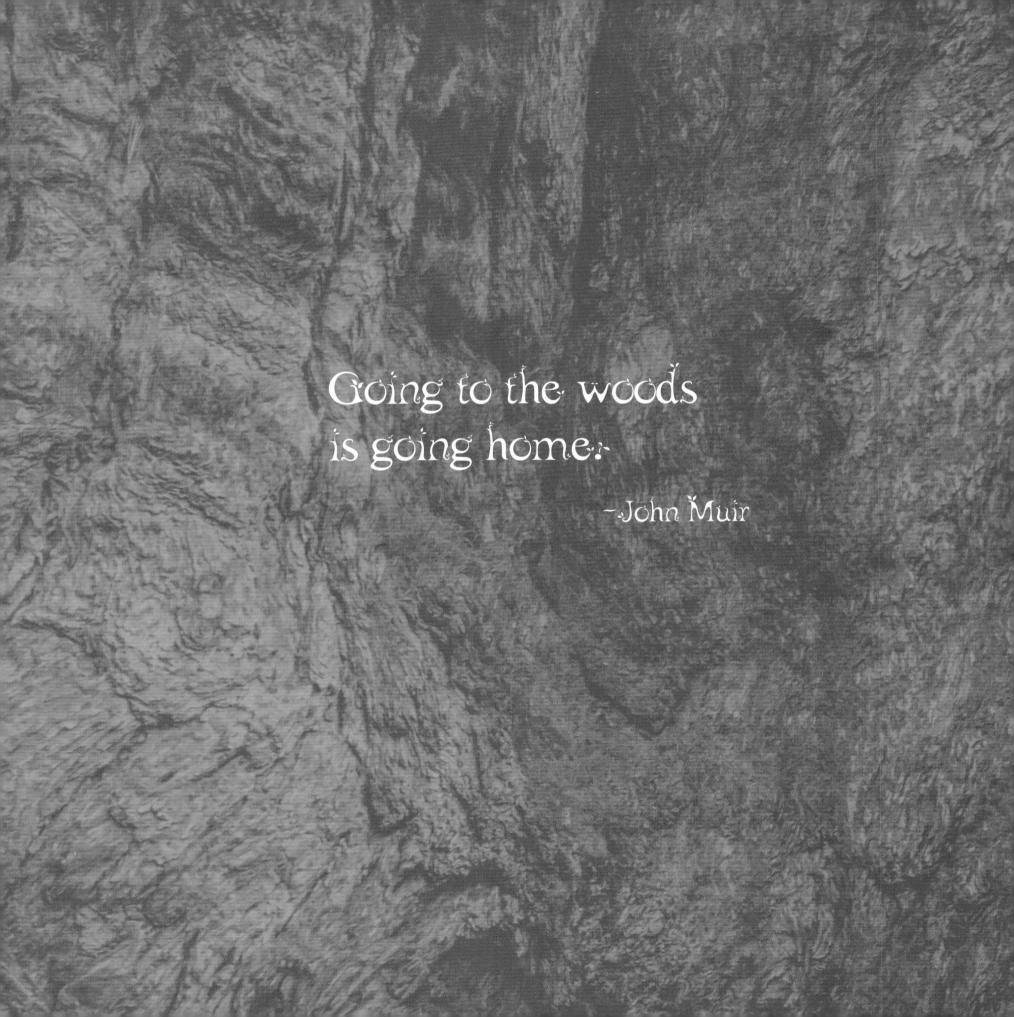

Going to the woods
is going home.

~John Muir

Introduction

This book is an invitation for you to recover your ancient bond with wild Nature—to rediscover trees as a source of intimate comfort, sensorial pleasure, and unexpected companionship, as a sanctuary from the madness of our modern, civilized world. It is an invitation to rewild your body and consciousness, to reforest your soul.

Trees

Trees are my secret love. I can gaze endlessly at their beautiful complexity, marvel at their simplicity of being, and envy their tenacious rootedness in one place. Every day, I long to be with trees, to sit under them, climb high up in them, collect their abundant gifts, learn about them, and listen to their ethereal songs in collaboration with the wind.

Instead, I spend too many days sitting in front of a screen, zooming behind a wheel, accumulating manufactured objects, and interacting with other two-leggeds and their noisy, handheld gadgets. But when I can, I steal away from the scheduled, maintained world to be alone with the trees, whose strong arms exalt the blue sky, whose gnarled roots kiss the rich earth, and whose generous, cascading foliage grants me peaceful shelter and companionship. When I am not able to be alone with trees, I dream of how one day I can disappear into the woods and simply become one of them.

Affinity for trees is undoubtedly embedded in the human psyche. The folklore, mythology, and symbology of trees are rooted in human cultures all over the world. For tens of thousands of years, people took refuge under trees, held council under trees, and depended upon trees for survival. People talked to trees, sang to them, danced under them. They tended, maintained, and harvested them for the gifts they provided, year after year, generations after generations. Trees were honored, revered, and worshipped; we lived *in kinship with* them. However, we have lost our intimate bond with trees, and both they and we suffer for it. While our modern, industrial world still depends upon certain species of trees, most trees have been demoted from sacred to the status of a cash crop harvested in rows like wheat [1], or a disposable urban landscape element planted for decoration, with a percentage allowed refuge within the borders of designated parks and wilderness areas. Otherwise, they appear to be there for the taking. Trees are undoubtedly one of the most vital parts of our interconnected global living system. There are currently an estimated 60,000–100,000 species of trees on the planet [2], the majority of which are in tropical forests. We don't know how many there are because, like with many other species, we simply can't count them all. We may never identify many of them before they are lost to deforestation. Carelessly, we allow most species to diminish in the name of human progress. Their regional deforestation has led to cultural collapse more than once in history. Today, trees are being lost not only due to over-harvesting and habitat loss, but also to the spread of invasive insect pests and diseases from imbalances caused by climate change.

Trees and forest products contribute an estimated $250 billion to the economies of the developing world alone [3]. But trees provide us, and the rest of the planet, with much more than just measurable products and ecological services; the psychological and spiritual gifts they provide are ineffable.

Why is it that we feel more relaxed, open, and at peace in the presence of trees than we do surrounded by a sea of concrete? Our sensorial bodies, our scientific minds, our feeling hearts, and our energetic spirits *know* the answer, and we hunger for it.

We modern industrialized humans live in a tamed world that has been mechanized, gadgetized, and sterilized; wildness has almost entirely been bulldozed, desacralized, and forgotten. Yet there are remnants of wild Nature all around us. Even in the most desolate, paved-over urban area, resident critters scavenge what they can; tenacious "weeds" bust up sidewalks; and ill-fated landscape trees with roots entombed in concrete support tiny microcosms of life invisible to us. But without continual human intervention, the built world submits to Nature's swift and willful reforestation of its ecological self [4]. Simply watch an abandoned urban lot re-vegetate over time: an act of ecologic integrity. In most climates, such a mini-ecosystem would eventually become a forest.

Ironically, because of humans' exploitation and, in some places, obliteration of forest ecosystems, trees now rely on us for their survival. In the case of some species, due to a lack of natural regeneration, humans are the only hope for tree propagation and restoration. But what about the restoration of our own ecological selves and the reforestation of our souls? As mammals, we are a part of Nature, even though we may feel *apart* from it. As modern, industrialized mammals, we have slowly dissociated ourselves psychologically from the kinship of all life [5]. But unconsciously and consciously, we are restless for balance, reunion, and healing, as are all the living systems of the planet that are currently in decline. We humans have a deep *need* for personal reconnection to the wild, not only because everything, including us, depends on it biologically, but because we depend on it psychologically and spiritually as well. We need the wild much more than we need the automated machines, petrochemicals, and economic systems we have manufactured to run our lives. We need to free the wild from the bondage of slavery and abuse we have imposed on it for thousands of years, and set it free in ourselves as well.

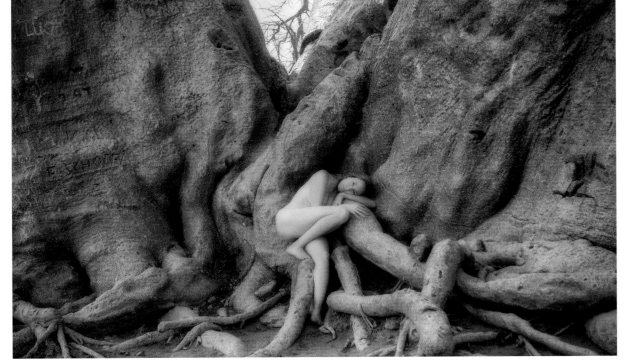

◄ Seven Sisters Sitting (2008)

African Baobab (Adansonia digitata), *Botswana. See also p. 19.*

Previous p. 6:
Sequoia Meditation V (2001)

A Giant Sequoia, see p. 65, in Merced Grove, Yosemite National Park, California, USA.

Therefore, in the context of this book, *the wild*, *wildness*, and *wild Nature* refer not to a Nature devoid of all human culture, but rather a Nature absent of an industrialized civilization that values exploitation, domination, and control over "the other" instead of coexistence within the family of things. Nature is wildness; wildness is Nature; Nature is the life force. *The wild*, in this sense, refers to lands, waters, and beings who have their own freedom, will[6], and integrity—those that have not been subjugated by humans. This wildness is the same mysterious life-force energy and matter that is continually creating life on Earth and infinite solar systems; we humans can neither fully scientifically define nor tame it all. In fact, we must return to a relationship of alliance—a conscious dialogue with Nature of asking questions and appreciating the mystery. At the same time, we must not romanticize the wild; for this same wildness holds the power of creation as well as destruction. The Earth has its own set of "rules"; we must come into accordance with them, for the Earth with all of its abundance and power demands a participation of reciprocity.

As allies with the wild, there are a myriad of actions we must engage in order to preserve and restore the integrity of the

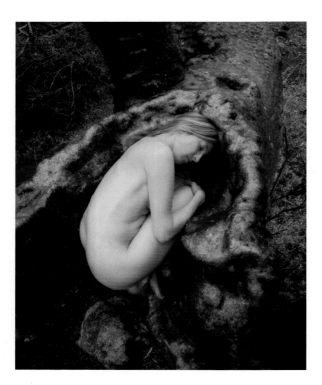

Pafuri Baobab Nest (2008)

African Baobab (Adansonia digitata), *Kruger National Park, South Africa. See also p. 19.*

planet: ecological restoration, environmental education, political advocacy and activism, policy reform, reducing our voracious appetite for natural resources, and powering down our energy consumption and our industrial growth society's false promise of infinite economic growth[7]. We must do this all with humility, integrity, ingenuity, creativity, soulfulness, and even daring. Along with these, we also must cultivate our personal relationship of love, reverence, and awe with the wild.

To reunite and belong once again to wild Nature is to choose to leave behind the artificial and managed world we humans have blindly constructed, consisting of automated buttons, touch screens, personal identification numbers, manipulated images, and decimal points. It means finding refuge from the built environment's light and noise pollution, chemical offgassing, and electromagnetic buzzing. By re-immersing ourselves in wild Nature, we desert the domesticated mind and return gracefully to the soul's vocabulary. Here we find sanctuary, freedom from the deadness of human construction, and relief from the insanity of our technology-addicted lives and the control we are constantly attempting to maintain. Our modern brains do not understand it cognitively, but our souls recognize the language of the Earth: the natural patterns, cycles, seasons, growth and decay, rhythms, systems—the relationships, the deep intertwinings. Our modern cultural consciousness is starved for and fascinated by the mysterious, complex beauty of the thriving, sometimes seemingly chaotic creative life force that is beyond our control. We hunger for a deep, rich engagement and interaction with that which is ancient to our mind, heart, body, and spirit.

Our mammalian bodies also remember and recognize the untamed world of wild Nature as our primal home. We yearn to experience all of our senses safely and feel fully comfortable in our bodies, interacting with the body of the landscape. We have cravings to walk barefoot, touch fur and feathers, climb trees and mountains, pick and gather, get wet in rain or bodies of water, dig in the moist earth or sand, explore cavernous holes, create beauty from found natural objects with our hands, even to take adventurous risks. These patterns are deeply ingrained in us. There are textures in Nature that are engaging and pleasurable to touch, colors and patterns beautiful to the eyes, fragrances intoxicating to the nose, edibles delicious to taste, sounds harmonious to our ears, and physical comforts in which we feel safe and held. These sensations may come in the delicate softness of pussy willow buds held between our fingertips; being blinded by a fruit tree exploding in spring's pink, blooming fullness; the intoxicating smell of sweet orange blossoms on the air; the taste and texture of a freshly

picked fig; the rustling sound of dried leaves gently chased by the autumn wind; the physical sturdiness of a trunk or flexibility of a limb while our bodies come into tandem with a tree while climbing it. Interacting with wild Nature also engages our psychic-intuitive-spiritual senses, warming our hearts and bringing inspiration, awe, and connection.

Such unexpected stimulation can affect us so deeply that we are able, for a few moments, to abandon our minds, release our isolation, and embrace the pleasures of shared aliveness as devoted companions, as lovers with the rest of the animate world. This is a kind of intimacy with wild Nature, a romance of the senses that can at times be physically sensual, passionate, and even erotic.[8] When we allow ourselves to be seduced by the sensuality of wild Nature, these raw experiences can nurture and enliven the body and reawaken the soul. This experience of surrender is akin to falling into the arms of our beloved. There we can recalibrate our attention to a state of enthralled reawakening and engage in secret conversation with the living Earth—until we once again feel we belong at home. This is a sacred and dynamic conversation with Nature as mutual subject, not object. It is the true act of intimacy as an intentional collaborative act of admiration, adoration, respect, humility, vulnerability, caretaking, and reciprocity. To be alive on the planet is to surrender to being her partner. To be in service of the Earth is to be enraptured by her.

Imagine if we truly open our hearts to Nature in all its wildness and fall in love over and over again with the sacredness of this life force? I invite us modern humans to become re-enchanted with Nature, to experience intimacy and soulful engagement with trees and our more-than-human companions on this Earth. I invite us to live life embodied in our animal forms, and in our animal psyches, in affectionate alliance with the animate land, soil, rocks, water, plants, fungi, lichen, animals, and invisible, microscopic world—the *Kin-dom* of Life.[9]

I have found my secret love in the plant kin-dom of trees. I encourage you to find your secret love in Nature as well—to find your wild within. As inspiration, I am offering you an anthology of my own private encounters with trees around the world, captured in intimate self-portrait photographs and photographs of others. Each photograph is accompanied by a short story of my encounter with the tree.

The tree species are organized by their common name, and are accompanied by their natural history to fulfill your curiosity, including other known names for the tree, distribution of the species, ecosystem type, maximum recorded age and size, distinctive characteristics, animal community, traditional and modern uses (including uses for medicine, food, fiber, tools and objects, art and ceremony, shelter), and threats to the species' survival. The better we understand our wild neighbors—friends, ancient elders, relations—the better we can appreciate, value, and be of service to these trees.

The tree species bookend five essays that each tell a different story about the human-tree relationship. "Lovers" tells of my own sensual and transmutational encounter with a tree; "Tree Affinity" explains the biology and psychology of why we are attracted to trees; "Touch Trees" examines a holistic view of arboriculture—the science of taking care of trees; and "The Goddess and the Green Man" tells the story of my own spiritual encounters with trees as these female and male Nature archetypes. The last chapter, "Rewilding," offers some accessible ways in which you can connect with Nature to rewild yourself.

Ecopsychology, in essence, is the discipline of understanding our modern disconnection from the Earth. The conscious and unconscious knowing that our home, planet Earth, is being destroyed, is causing massive psychological grief, anxiety, denial, trauma, dysfunction, as well as physical illness. Ecotherapy is the therapeutic practice: Nature is healing because we *are* Nature, and essentially, ecotherapy is coming Home.

I hope this collection of images and essays will inspire you to connect with trees, to fall in love with the wild, and to explore your own intimate encounters with Nature. As a photographer, I place myself in the landscape to show that Nature, in its inherent wildness, is where we belong—sometimes naked, sometimes vulnerable, in humility, with our shoes off and the wind blowing against our skin, ears open, listening to our lover, with all our heart and soul.

[1] Andreas Feininger, *Trees* (New York: Rizzoli International Publications, 1991) p.10.

[2] Colin Tudge, *The Tree: A Natural History of What Trees Are, How They Live, and Why They Matter* (New York: Three Rivers Press, 2005) p.16.

[3] Economic Contributions of Forests: Background Paper 1. United Nations Forum on Forests, Tenth Session, prepared March 2013.

[4] Alan Weisman, *The World Without Us* (New York: St. Martin's Press, Thomas Dunne Books, 2007).

[5] Jerome Bernstein, *Living in the Borderland: The Evolution of Consciousness and the Challenge of Healing of Trauma* (New York: Routledge, 2005).

[6] Dave Foreman, "Five Feathers for the Cannot Club" in *The Rediscovery of the Wild* (Cambridge: MIT Press, 2013).

[7] Richard Heinberg, *Power Down: Options and Actions for a Post Carbon World* (Gabriola Island, New Society Publishers, 2004).

[8] The concept of eros, while it may initially take the form of passionate desire and love, it is more truly a desire for "psychic relatedness"—for interconnection and interaction with other sentient beings. Ultimately, it is the desire for wholeness.

[9] *From Kingdom to Kindom: Acting as if We Have Relatives*. Lecture recording from Bioneers Conference of Brock Dolman, Brian Swimme and Paul Stamets, 2011.

A Note about the Natural History

Other Names: In this book, common names of tree species are intentionally capitalized, while indigenous names are capitalized only if they were documented as such. Note that many trees around the world have the same common name, but are in no way related species or even in the same genus. This is because European botanists or colonists who named and recorded trees on behalf of Western science named them after the properties of their timber, or species that looked like, or sometimes smelled like (albeit sometimes vaguely), species from their home countries. Following the vernacular names in parenthesis are either the tribes who used them, e.g. (Pomo), or regions, e.g. (Nigeria) or languages, e.g. (Sanskrit). This is why using the globally recognized Latin names is useful in certain situations. It is important, however, to also acknowledge the trees by the names they were called by local peoples for hundreds or even thousands of years. For some species there are likely more names than can be listed here.

Distinctive Characteristics: As this is not a tree-identification book, specific information such as shape or size of the overall tree, leaf, seed, fruit, and bark is not listed unless they are unusual or distinguishing to a general audience.

Distribution: Trees are indicated by what country they are native, and not locations where this species has become naturalized, or planted. Each photo caption indicates the location of where the tree was photographed, which in some cases, is not its native country. Elevation is also noted.

Ecosystem: This includes some of the other trees, shrubs and plants that share a community with the tree. All common names of tree species are capitalized intentionally in this section and throughout this book, while shrubs and plants are not.

Maximum Age: I did my best to be accurate and realistic with age data, without being overly conservative. The age of large trees is often overestimated around the world. Everyone would like a tree to be 10,000 years old! Some ages, because of limited or contradicting data, are listed as "estimated."

Maximum Height and Girth: These are based on the measurements of currently living individuals, and in some cases, historical trees no longer standing. Measuring technologies have varied throughout history, and are far more accurate today. The numbers presented here are based on the most accurate data that currently exists. Most circumferences were measured at diameter at breast height (DBH)—which is the standard for tree measurement in arboriculture and forestry—and not at the base, which, when a tree is buttressed, may be impressively larger. However, DBH is defined differently depending on where you are. Currently, in continental Europe, the UK, and Canada, DBH is measured at 4.26 ft. (1.3 m) above ground. In the US, Australia, New Zealand, Burma, India, Malaysia, and South Africa, it is measured at 4.5 ft. (1.37 m). Historically, diameters may have been measured at yet another height. It should be noted that circumferences may include buttresses, irregularities and concavities (including fire scars) as if the tree was solid. (Coast Redwoods and Giant Sequoias are also being measured at DTB—diameter at the top of a buttress where a tree becomes columnar.) Measuring and verifying

the tallest and largest trees in the world continues to be a hot point of contention; meanwhile, the trees continue to grow!

Animal Community: Many more animal species than are listed may live in, around or depend upon a tree. Animals (as well as trees) in this book are referred to as 'who' not 'it'.

Traditional Uses: This category refers to a living legacy from the indigenous peoples who utilized the trees prior to colonization or industrialization. The use of past tense verbs does not imply that these traditions are no longer engaged with. The amount of ethnobotanical information documented for species varies widely. Many traditional uses of these trees have probably been lost or undocumented. Where this information was available, the names of the tribes who used the trees in these ways are listed in parentheses, e.g. (Pomo) or regions (Nigeria). It is likely that more tribes than are listed used the tree in some way. While some species will have information in the Art and Ceremony section regarding how the trees were valued spiritually and religiously, the Natural History for each species here is primarily ecological, as there are many fine books that describe the sacred and mythological aspects of trees.

Modern Uses: "Modern" refers to the age of industrialization and after. This does not imply that traditional uses are no longer continued.

Threats and Conservation: While all species are at risk to some degree by pests, pathogens, climate change and habitat loss, what is listed here are imminent threats causing a species to be listed as 'Threatened' or 'Endangered'.

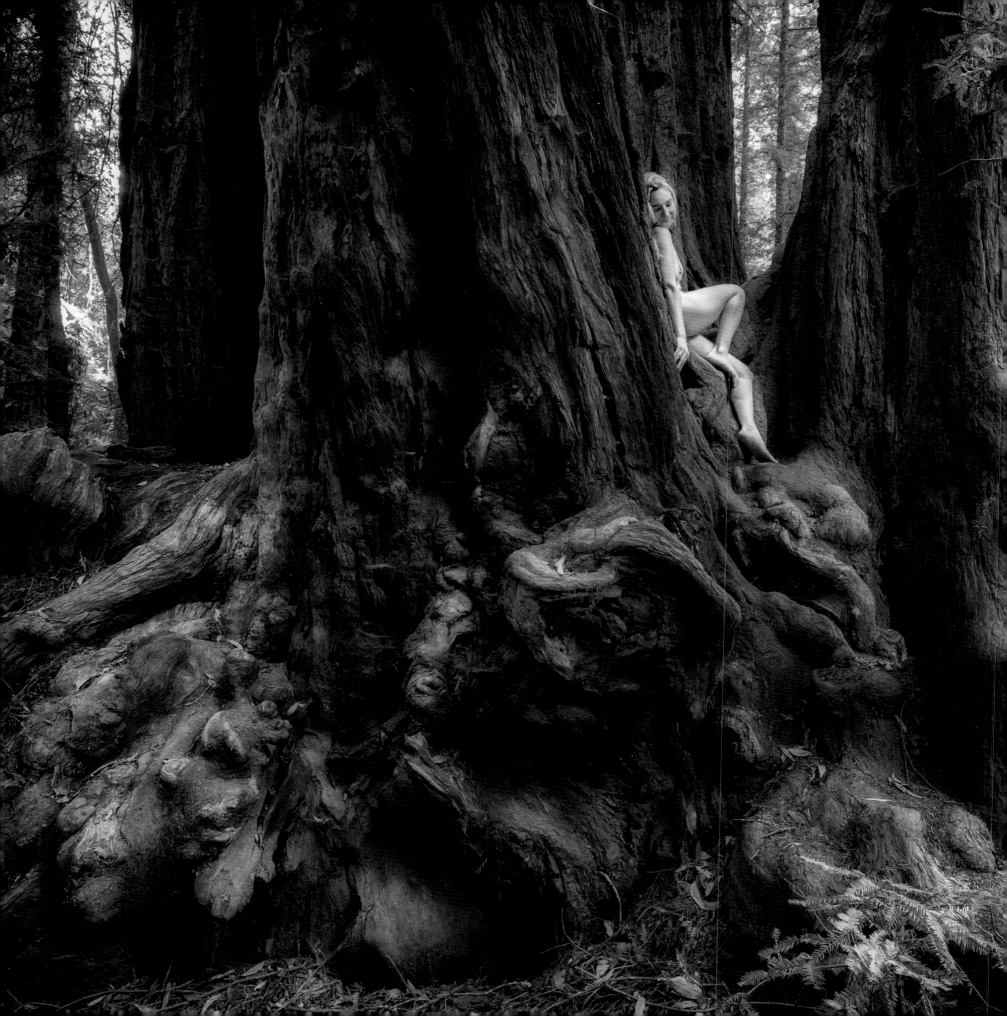

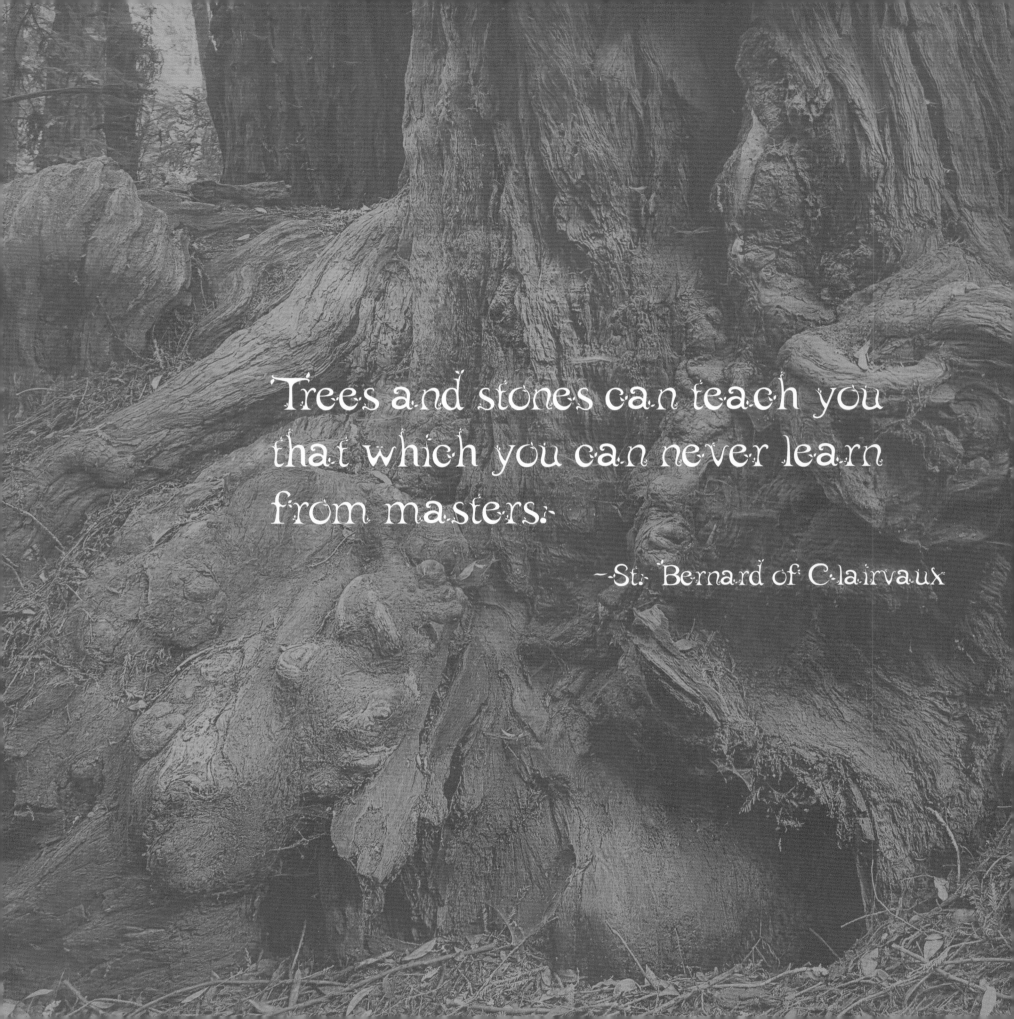

Trees and stones can teach you
that which you can never learn
from masters.

~St. Bernard of Clairvaux

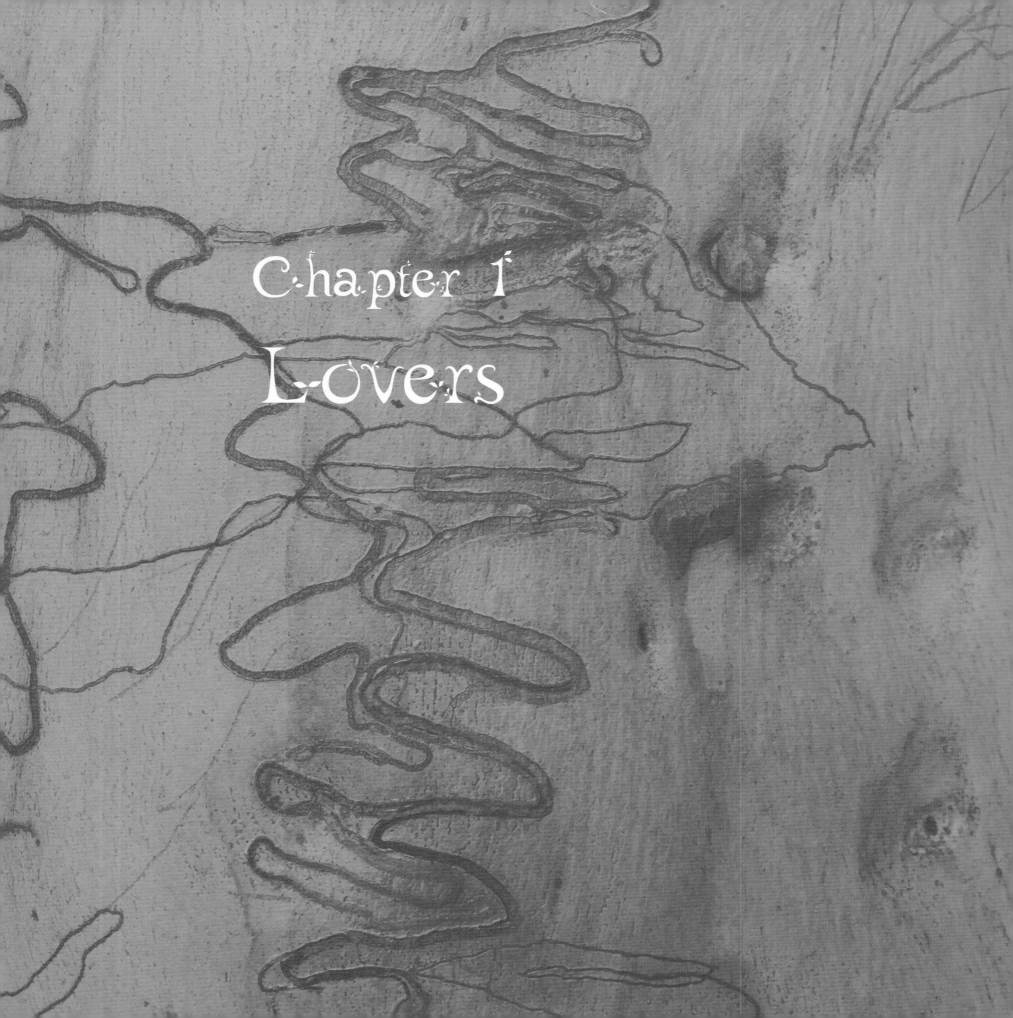

Chapter 1
Lovers

As a photographer of trees, I make pilgrimages around the world to meet individual trees I have researched, mapped, and sought out in a kind of botanical treasure hunt. As I move geographically and physically closer to each of these trees, my body becomes electrified with anticipation. I feel that I am about to meet a cherished old friend or a lover I have not yet met in the flesh. Sometimes these experiences are like blind dates, with only a photograph from a book to go by, and other times they are unexpected encounters, a surprise romance along the journey. But always, they are magical and meaningful connections, ineffable joys of the soul that can never be planned or repeated. What is this attraction and connection? How is it that I bond with certain trees as unique individuals with presence and personality? What are these emotive sensations that go directly and deeply into my heart?

At times, when I am walking in the woods, I become so immersed in my surroundings that I want to disappear completely from the human-constructed reality and *become* Nature. I experience an undeniable, almost cellular, yearning to merge with the forest, not merely to perceive its aesthetic beauty from a distance but to be nested *within* it, like a soul gestating in the safety of a womb or resting in the embrace of an imaginary beloved. No human has the capacity to make me feel so elated, so at peace, so at home. I long to transmute greenly into the landscape, melt into sun-baked desert rock and sand, evaporate into coastal fog and be carried back into the vastness of Mother Ocean—to abandon my human form, reorganize my cells, and synthesize seamlessly into each ecosystem I encounter, detaching myself from the linear mind that dictates separation of self from other.

Intently observing the natural world surrounding me, I envy the complex web of mycelium that spreads beneath decomposing logs, embracing all that is born within the rich, living soil. I wish to be likewise interconnected, held, and fed—to be part of the nourishing life force, which is both tender and wild. I dream of finding refuge in the forest, flittering from tree branch to tree branch, sheltered

within the leaves, ascending the canopy to touch those places outside of human reach. In the company of trees, I feel I have returned to the tranquil sanctuary of my soul's native habitat. There is no more doing, only *being*, in sync with Creation's cyclical rhythms: the tides, the moon, the soil, new growth and hibernation, the stars, sunrise and sunset.

When I first sight a treasured tree, I approach with deep curiosity and reverence as a fellow sentient being whom I recognize as fully animate and aware of my presence. I feel alive with purpose as I become magnetically engaged with this other being. My body becomes electrified, and my heart radiates with admiration as I set my mind to speak the language of intuition. Spirit to spirit, I introduce myself and explain my heartfelt intention to connect. Often, but not always, I sense responsiveness from the tree. Sometimes a tree seems sleepy and surprised, unaccustomed to human interaction after perhaps hundreds of years of solitude, but still gracious, welcoming me into its space. Sometimes a tree seems stoic and private, at other times gregarious and eager for contact. Often, a tree feels so magical and ethereal that I am simply humbled in its presence. Whether I am seeking guidance on a question or simply greeting a tree, I trust the more-than-human wisdom that is deeper than my rational thoughts. Usually this solicited or spontaneous advice is short, simple, and clear: it's about love.

When I sense a tree's receptivity, I move closer, circumnavigating the tree to take in its unique form and character and to see and feel where I may fit most naturally. This creative process is both aesthetic and intuitive; it involves a shift in consciousness. Taking a quick look around to assure our privacy, my heart pounding, I take a deep breath and remove each layer of my clothing, dropping it to the ground in a ritual to free my female animal body, emancipating it from my civilized, domesticated self. Although I may be hesitant at first to remove my shoes, my delicate feet quickly adapt to the cool earth and the complex texture of the world between my toes. Now I stand naked, vulnerable to the air and temperature, exposed for all the eyes of Nature to see. All my physical senses enliven to my surroundings: touching, smelling, seeing, hearing, and even tasting the earthiness on my lips. If the temperature is cool, tiny hairs awaken with the breeze, followed by goose bumps all over my body. If the temperature is hot, my skin begins to relax and melt into the surrounding air. I begin

to disappear from the human world as I touch the tree and step into another world. I am now using my body's cellular memory; there is something so familiar about this state of being—a heightened sense of awareness, focus, and sensorial pleasure. My intuitive senses continue to awaken. Will the tree welcome me as a new friend or a familiar lover? The camera stays off until the tree and I come to an agreement.

If I am welcomed, I do my best to gracefully climb and comfortably intertwine my body with the tree's body. Using bare hands and bare feet to grasp hold, my muscles work hard to move my body, like an animal other than human. I either fit, or I don't; I can't force it. To move in collaboration with a tree is to dance in effortless sync with a lover; there is a reciprocal exchange, a natural rhythm between the two balanced as one. Although it may appear that the tree is not moving, I feel actively held in the strong arms of my partner. For those moments, I feel like the tree's consort. Although the interface of supple skin with rugged bark may seem odd, my soft animal body eagerly responds sensually to our contrasting textures. As I grasp hold and shift my body's weight to find balance with the tree, I feel how we are vulnerable to each other. I attune to the subtle movements and patterns of cellular tissue, the shape and strength of another living body. My internal temperature rises as I feel more and more at home; I forget about modesty and am pleasured by the innate comfort of my nakedness in the open air. I yield, opening and expanding my awareness, sensing what is beneath the hard exterior layers of bark and inner cambium, the core of another breathing body. I feel that I could enter inside and find shelter within. Feeling the vibration of the tree's life force, I relax. I synchronize with the egoless pulse of Nature and reset my own rhythm to a more primal state.

I have climbed onto the sturdy shoulders of a woody giant, up into another world. I make my temporary nest with the wild company of the wind and the winged. Camouflaged among the leaves and branches, sharing space with the tiny six-leggeds who are traversing its crevices, I am aware that I have entered a microcosmic universe, the tree's own ecosystem. My spirit becomes activated, and I abandon the realm of the civilized: the compartmentalized, the managed, the manic, the mind. It is not an escape; it is a reunion. I feel that I am welcomed home—to a home that was home before I was me.

Previous p. 12:
Redwood Castle VI (2015)

Coast Redwood (Sequoia sempervirens). *Henry Cowell State Park, Santa Cruz, California, USA. See also p. 45.*

Why do I feel completely safe and held, high up in the tree I have foolishly climbed, clinging to it with bare feet, bare hands, and bare skin? There are moments when I doubt my actions. I look down and laugh nervously. I lift my hands and hold fast, considering anxiously that surely I must fall; this tree is so strong and grounded compared to my clumsy and fragile self. But each time I return to the arboreal world, I relearn through trust to move intuitively with swaying branches, as if magnetized. Whether I am ten or eighty feet off the ground, I somehow come to know with blind assurance that if I hold fast with care and reverence for this other living being, the tree will not *let* me fall. She will hold me like a marsupial mother carrying her baby joey.

I pray to the tree for support. I feel her strength and acknowledge it. In that moment, I release fear and surrender into safety. I am held. A wave of gratitude rushes over me, extinguishing my fears and alchemizing my adrenaline into peaceful trust. The respect I feel for the tree, I sense, is reciprocal. Our separation is only illusion. The tree reflects back to me the power that we share: *You are strong. We are strong together.*

Now engaged with the four elements, my bare feet freed from the prison of the paved-over world, I experience life from a tree's perspective. Being up in the canopy is not the same as life down on the ground. As I sit cradled among the branches in the summer's heat, I can smell the raw odors of mammals who have marked the tree with their scents. Their curious eyes, hidden from my view, are watching this strange new creature closely; they tell the others, I'm sure. The forest's inhabitants have kicked up dust from the dry earth below, and it coats the tree's limbs. Now it covers my fair, sterile skin as I grasp the tree's furrowed bark to anchor myself. Sap and sweat mix with forest dirt and specks of leaf litter that have pooled in the depressions where I perch. Looking deeper into a hollow cavity, I find a mass of tiny bones and fur—an owl pellet, perhaps. There is no turning back to the sanitized world now.

If I were to lie still here long enough, eventually epiphytic moss would take to my skin, a wild vine would wrap itself around the trunk of my torso, and lichen would set up camp in my hair. In time, my supple skin would harden into plates of bark. Our patterns of cracking would eventually become indistinguishable. I'd become part of the weather, too, absorbing the rays of warm sun for breakfast, drinking up the misty fog and exhaling it back in the cool, moist evenings.

My linear understanding of time and space would shift to a biological language. The dried leaves on the tips of branches would become music in the wind, singing tree to tree about the oncoming weather. I'd sync into the rhythm of seasonal growth and rest, going inward and slowing down. When the light returned, I'd sprout new leaves and extend my roots, disperse pollen and produce ornaments of swollen fruit during the long days of sunshine. By autumn, I'd be dripping with ripeness cascading down to the rich earth. A myriad of opportunistic organisms just above my roots would devour my mulch of crisp, broken leaves. A circus of insects would nest in my crevices and call me home, while raucous scrub jays and resolute woodpeckers would steal acorns brashly from my branching hands and hide them miles away to be later forgotten. With luck, a few out of my thousands of seeds would take root and grow, escaping the mouths of mammals or the extremes of too much or too little water or sun; under the perfect conditions, my offspring would regenerate into forest, continuing the cycle. I'd feel the fullness of the life force within me and all around me. I would truly become the Green Woman.[1]

My consciousness expands into this new synthesis of self and tree. In a flash, I comprehend more deeply the interdependence of the web of life, from the tiniest microorganisms beneath the earth all the way to the stars and back. A tree is not just a single magnificent being; it is one with ancient rocks and stones, slow-growing lichen, mycorrhizal networks, tenacious thorny plants and their rambling roots, and infinitesimal invertebrate critters. It is the big burly beasts and the warm air they exhale, the songbirds' clear songs as they call in the dawn's thanksgiving, and the elegant patterned dance of migrating pollinators in search of sweet nectar. A tree is moisture willfully transpiring from its canopy-forming, shape-shifting clouds, and warm sunlight traveling millions of miles. Trees are one with the teasing pull of the cycling moon, soft pooling rain, the winds and storms, the ocean tides crashing against cliffs and avalanches of rock debris carving out mountainsides. Each tree with their shallow networks of filament roots is intertwined with neighboring trees, connected with their relative trees miles and miles away, with the aged weathered ones, with the dead decaying gracefully on the forest duff, and the new generations of fallen seeds willing life into being, pushing up through fertile humus.

The grandness of one tree results from this connective exchange, from the diversity of its infinite relations acting in concert. Even the largest, most ancient of beings don't experience themselves as separate beings. They may appear to us as distinct, recognizable energy focused into forms and patterns, but in reality they are an orchestration of elements in constant transmutation and collaboration. The strength, beauty and vitality of the life force arise from the miraculous existence of each and every being (animate and inanimate) mutually engaging and holding one another up in cycles and systems that flow beyond the understandings of science. The more connections, the stronger a being's presence. *Without the other, there is no one.*

---

[1] See Chapter 4 of this book: The Goddess and The Green Man for explanation of the Green Woman.

Moreton Bay Figness (2015) ▸

Moreton Bay Fig (Ficus macrophylla), *New South Wales, Australia. See also p. 94.*

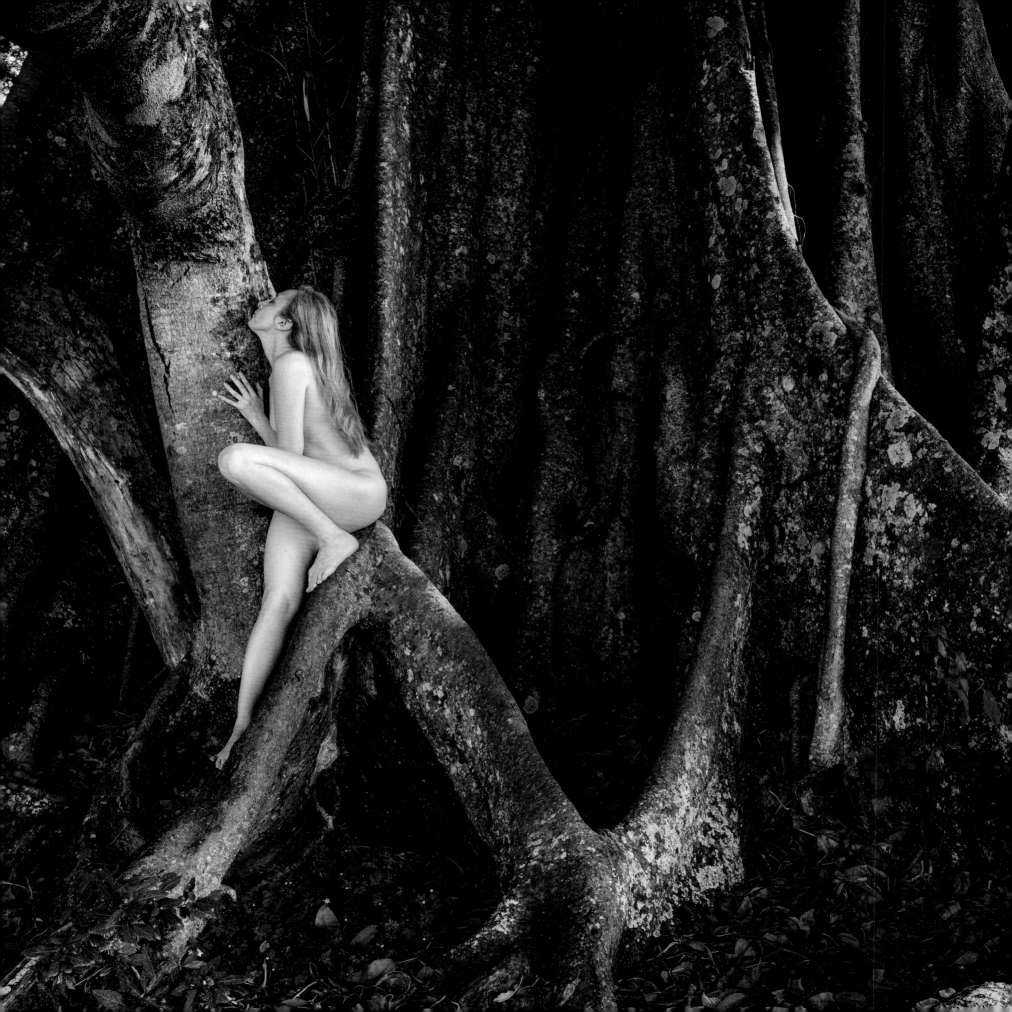

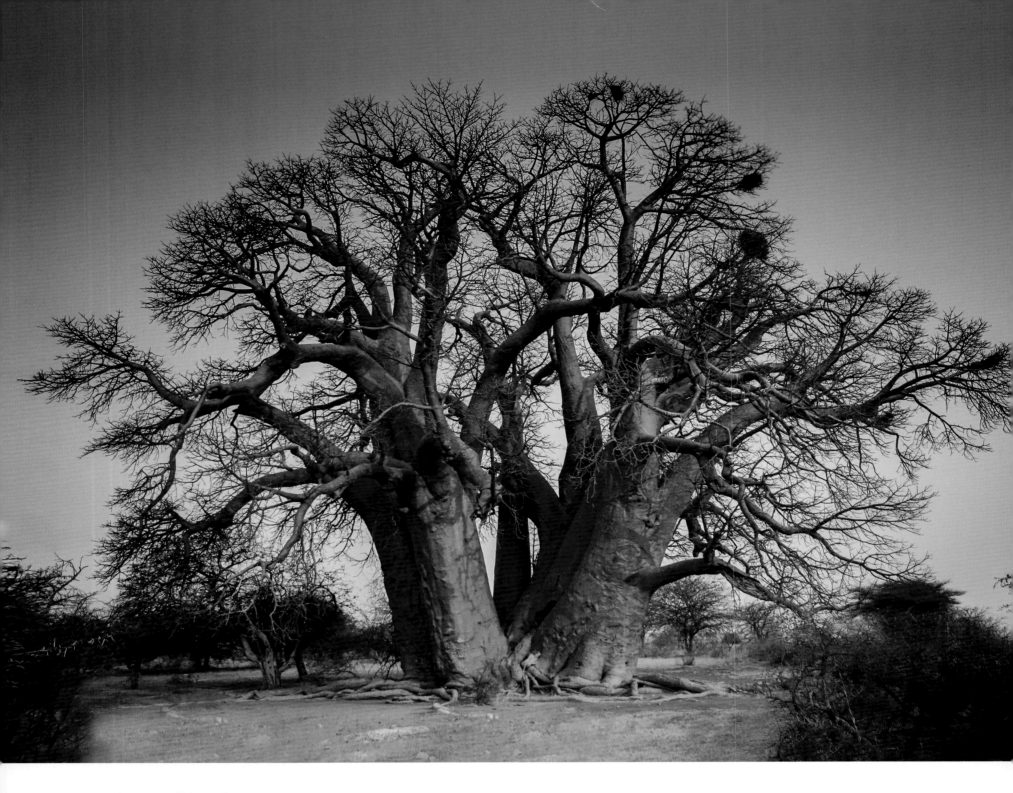

Seven Sisters Full (2008)

Known as "Chapman's Baobab," named after an elephant hunter who used the tree's hollowed nook as a mailbox in 1852, this 2,000+-year-old tree is known to locals as "Seven Sisters" because of her six giant trunks and a sprouting seventh. Her historical significance as a landmark for passing travelers is tattooed into the pink flesh around her entire circumference. She is covered in the graffiti of human initials, including those of anthropologist Dr. David Livingstone, who made their marks. Still, the walls of her beautifully wrinkled elephantine bark were smooth to the touch and a joy to rest against.

Along the base of her trunk, exposed roots fanned out like a spaghetti bowl of giant octopi tentacles. These above-ground root systems serve to catch water in the arid desert, while being wonderful playful appendages to entangle with.

The dry heat of the desert allowed me to slow down and comfortably melt into her skin as I spent the hours of sunrise and sunset naked, entangled within her roots and towering trunks. In a moment of awe and praise for the grandeur of this ancient individual, the medicine she offered was the eternal wisdom

that a tree is not just a tree in itself, but is also the desert, all the organisms of that ecosystem, the entire universe, and everything in it. I am the tree, as well, and it is all interconnected.

African Baobab

Adansonia digitata

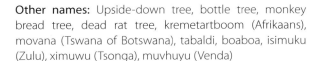

Other names: Upside-down tree, bottle tree, monkey bread tree, dead rat tree, kremetartboom (Afrikaans), movana (Tswana of Botswana), tabaldi, boaboa, isimuku (Zulu), ximuwu (Tsonga), muvhuyu (Venda)

Distinctive Characteristics: The deciduous Baobab dominates the landscape with its majestic presence, standing solitary out on the grasslands. Its branches are almost always concentrated at the top, giving the look of roots reaching into the air, or of a tree planted upside down, especially when bare. Once they reach maturity, their trunks have a distinctively chunky or bloated look, but when young, are thin and inconspicuous. As the trees age, their smooth bark becomes wrinkled and pock-marked, appearing like elephant skin. Measured by trunk circumference, African Baobabs are some of the biggest trees in the world. Baobab wood swells to store water, and mature trees can hold as much as 32,000 gal. (120,000 l) within their spongy tissue. After wounding, they can regenerate bark to some extent. Because they contain up to 75% water, Baobabs collapse into a heap of rotting fiber when they die, decomposing rapidly within a few months to a year. Fortunately, the soft, fibrous wood is useless as lumber.

Distribution: Native to 31 countries on continental Africa. Elevation: below sea level to 3,428 ft. (1,045 m).

Ecosystem: Baobabs live as solitary trees in arid, savanna woodlands and grasslands in Sub-Saharan regions. Occasionally associated with Tamarind, Mopane, and Acacia. Parasitic mistletoe can live in the canopy.

Maximum age: Estimated at 2,000 years, based on radiocarbon dating, and often exaggerated. Their indeterminate rings make accurate dating impossible; they appear deceptively older than they are because they hold water.

Maximum Height and Girth: Up to 60 ft. (18 m) in height, 118 ft. (36 m) in circumference.

Animal Community: African Baobabs provide habitat and food for many animals in different regions of the continent. Bees often nest in them. Birds supported include the red-crested turacos, great white pelicans, rosy pelicans, bateleur hawks, black-chested snake eagles, secretary birds, brown harriers, Verreaux's eagle owls, barn owls, hammerkops, Marabou's storks, red-winged starlings, gregarious swifts, kingfishers, rollers, barbets, parrots, lovebirds, hornbills, mosque swallows, weavers, bee-eaters, spinetails, honey-guides, and guinea fowls. Reptiles supported include pythons, boomslangs, vipers, mambas, cobras, monitor lizards, baobab geckos, and flapnecked chameleons. Mammals supported include monkeys, baboons, chimpanzees, bush babies, elands, nyalas, kudus, hyenas, porcupines, bush pigs, genets, honey badgers, squirrels and fruit bats. The leaves and twigs are a favorite food of elephants when other food is scarce; they use their tusks to rip off the moist, fibrous bark, and can eat a hole straight through a trunk.

Each white, showy flower blooms for one night, hoping to be pollinated by a fruit bat or bush baby in order to produce fruit.

Traditional Uses:

Medicine: The bark, roots, leaves, fruits, flowers, and seeds have been widely used to treat many conditions in, including hemorrhoids, hemorrhaging, colic, dermatitis, acne, burns, high blood pressure, poisoning, viruses, and allergic reaction. Specifically:

Bark: A concoction including the bark was used to treat swollen limbs (Namibia) and sickle-cell anaemia (Nigeria). The stem bark was made a heart tonic with diuretic properties (Nigeria). A bark decoction was gargled for toothaches (Nigeria). Bark resin was rubbed into the roof of the mouth (Ghana). Dried and powdered mashed bark was used to treat malaria. The ash from the burnt bark was made into soap (East and West Africa). The bark was also used to treat male sterility and as an aphrodisiac (Tanzania); to bring down high fever in infants (Tanzania, Kenya); to treat rickets (Malawi, Congo, East Africa); to reduce childbirth pain (Zimbabwe); and to improve night vision (Mali). The bark was made into clothing (West Africa), and waterproof hats that double as drinking vessels (Senegal and Ethiopia).

Leaves: Fresh or dried leaves mixed with oil have been used to treat skin diseases of the head (Somalia). Leaves have also been used to treat rheumatism, external bleeding, as an astringent, and as a salve to treat numbness of the limbs (Senegal); to treat diarrhea, tumors, ear and eye disorders, and inflammation, and as a purgative (West Africa); as a paste to treat parasites (Mali, Burkina Faso, Nigeria); as a blood cleanser and to treat backache and fatigue (Senegal, Mali).

Flowers: An infusion was made to treat digestive and gynecological disorders (Mali).

Fruit pulp: Has been used as an insect repellent (Nigeria), and to treat external bleeding (Senegal), constipation (Mali, Malawi), nausea (Gambia), smallpox (West Africa), nervous system complaints (Mali), hangovers (Malawi), coughs (Tanzania), measles, and fever.

Seeds: A paste of the crushed and roasted seeds was used to treat diseased teeth and gums (West Africa) and to treat diarrhea (Ghana). The oil was used to treat pain and eczema (Tanzania); a powdered form was used to treat hiccups (South Africa); and a compress was used to treat sores (Senegal) and fever.

Roots or bark of roots: Used for treating stomachache (Tanzania), urinary disease (Fulani), sore throat (Chewa), mental disorders (Senegal), gynecological disorders (Mali), and asthma (Togo), and to bathe babies for smooth skin.

Gum: Used for treating wounds (Mali, Senegal).

Baobab Flower (2008)

The flowers of the African Baobab are creamy white and showy, hanging down about 8 in. (20 cm) from a stalk, like a pompom with a ruffled showgirl skirt above it. From October through December in spring, they bloom for just twenty-four hours, in hopes of being pollinated by a bat. After blooming, the flowers drop to the ground; in the heat of the desert, they close up, dry, and preserve into a brilliant rusty-crimson color.

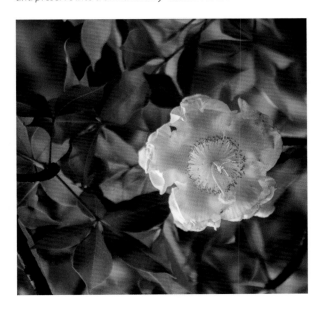

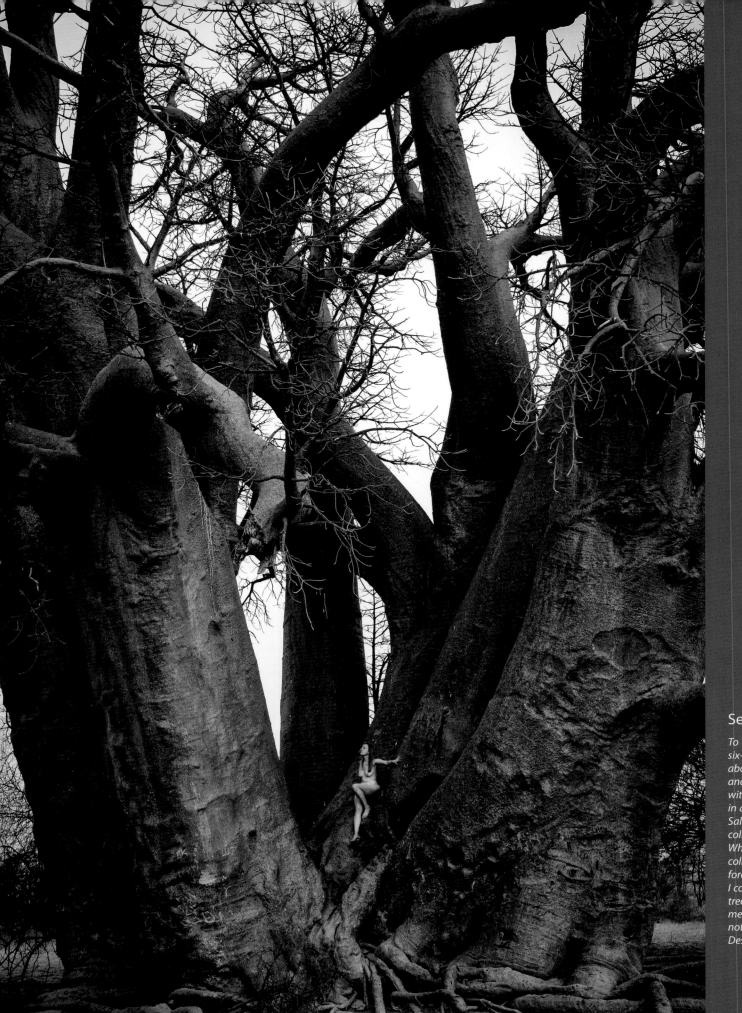

Seven Sisters Sepia (2008)

To get to this tree, in September, 2008, I flew in a little, six-seater propeller plane that hovered like a tiny bird above the white, crusty Makgadikgadi Salt Pan, once an ancient lake the size of Switzerland. This magnificent tree, with a base circumference of 85 ft. (25.90 m), held court in a dry savanna in the Kalahari Desert, on the Ntwetwe Salt Pan in Botswana, until January, 2016, when she collapsed from the center like a blossoming lotus flower. When Baobabs finally fail, after only a few months their collapsed, pulpy wood decomposes and disappears forever. In massive shock from this news, I grieved because I could not take other treegirls to this tree some day. This tree was my greatest TreeGirl mentor, and she proved to me that, indeed, her magnificent body and life force, was not hers to call her own, but belonged with the Kalahari Desert and beyond.

African Baobab

Food: The large gourd-like seedpods contain a dry, sour pulp (that dehydrates naturally inside the pod) high in vitamin C that was made into candy, porridge, or a lemony drink. The seeds were winnowed of their husk and eaten fresh, roasted, dried, fermented or sprouted; they were also pressed for oil. The seedling roots and young leaves were eaten fresh or dried. The flowers and nectar were eaten or made into a drink.

Fiber: The inner bark was harvested to make very strong rope and cordage for mats, hammocks, baskets, netting, fishing and game nets, bark cloth, strings for musical instruments, and hats.

Tools and Objects: The empty pods were used as containers, dishes, drinking cups, spoons, and paint pots. The bark gum was used as an adhesive in Tanzania, and the pollen mixed with water was used to make glue. Strips of bark were used to make beehives. The seeds and burnt ashes of the seeds were used to make soap and hair wash. The Senegalese made a board game out of the seeds.

Art and Ceremony: Baobabs are the focus of much taboo, folklore, and spiritual reverence in Africa, and are often sites of communal meeting or ritual. The blossoms were used as decorations during festivals. Bark fiber was used to make artificial hair for girls' rite-of-passage ceremonies. The large pods were carved and decorated, and made into musical instruments and masks. The roots were made into a dye (East Africa). The seeds were strung onto necklaces (West Africa).

Shelter: These naturally hollow trees have been used as water storage tanks, mailboxes, prisons, pubs, lavatories, storage, shelters, semi-permanent dwellings, watch towers, hunting perches, refuge from predators, shrines, burial sites, stores, bus stops, hideouts—and even trash receptacles.

Modern Uses: The oil is becoming popular for making natural beauty products. It is currently being marketed as a healthy new superfood, in powder form and in naturally sweetened fruit chews, because of its high vitamin C and fiber content. Tea bags of dried powdered leaves are sold as a health food in Europe. Fruit drinks are sold in Malawi and Gambia.

Threats and Conservation: Although their status is not officially listed as threatened, Baobabs are threatened by climate change, desertification, and, in some places, deforestation for cattle grazing, agricultural development, and mining. They are a protected species in South Africa.

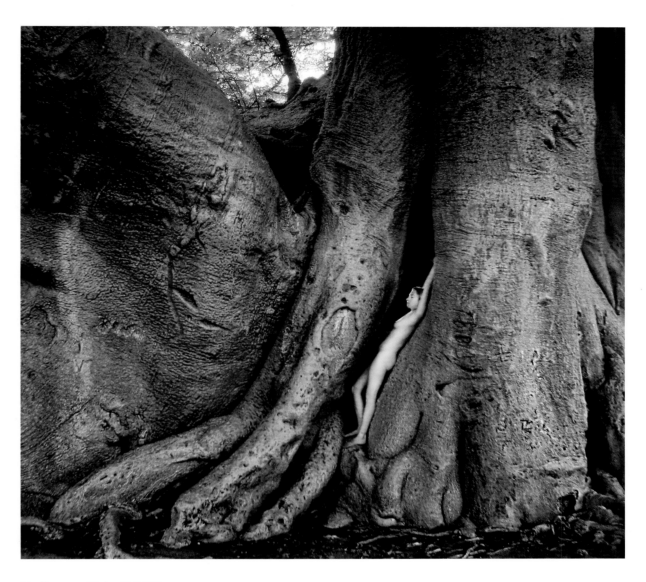

Holboom Side (2008)

This colossal, ancient African Baobab is known as Holboom, or "hollow tree" in the Afrikaans language. She is indeed hollow, completely cracked open from perhaps 2,000 years of aging and weathering. The tree's gray, wrinkly bark weighs heavily on the ground, like the foot of an enormous elephant. Stepping inside a naturally heart-shaped doorway, I landed on a floor of desert soil and dried crimson Baobab flowers that had showered down after their one night of blooming creamy white. I stood in awe of the world I had stepped into. The "walls" of the trunk's interior and her mammoth branches were worn smooth like weathered, polished bone, and inside I felt as if I were climbing around on a giant carcass. The tree is a castle ruin with an adventurous playground of windows with seats, hollowed-out nooks like tiny rooms, and steps to perches providing refuge from the hot sun and wild desert beasts.

Less well known, and rarely visited compared to many other Baobab giants in Africa, Holboom is located in the remote northeastern corner of Namibia, near Tsumkwe in the area known as Bushmanland, inhabited and caretaken by the San people. This tree may indeed be the Baobab with the largest girth in the world, at 119 ft. (36.12 m), measured by Thomas Pakenham. When I saw this tree in his book, The Remarkable Baobab, I knew I had to meet her myself. The journey to find her took my guides and me two days, with two safari vehicles caravanning to this remote location in an ancient Kalahari desert woodland of Baobabs—where no one else would take me. My gift was to have the tree all to myself, without any other visitors for two days and nights. Sleeping on top of one of the vehicles to protect myself from hyenas that might take me for a midnight meal, I lay under the canopy of a neighboring Baobab as people had done for thousands of years. The relationship between humans and Baobabs may be among the oldest continuous relationships, as the San people carry the oldest genetic traits of humans. Who knows how much longer she will last, or if I will ever see her again?

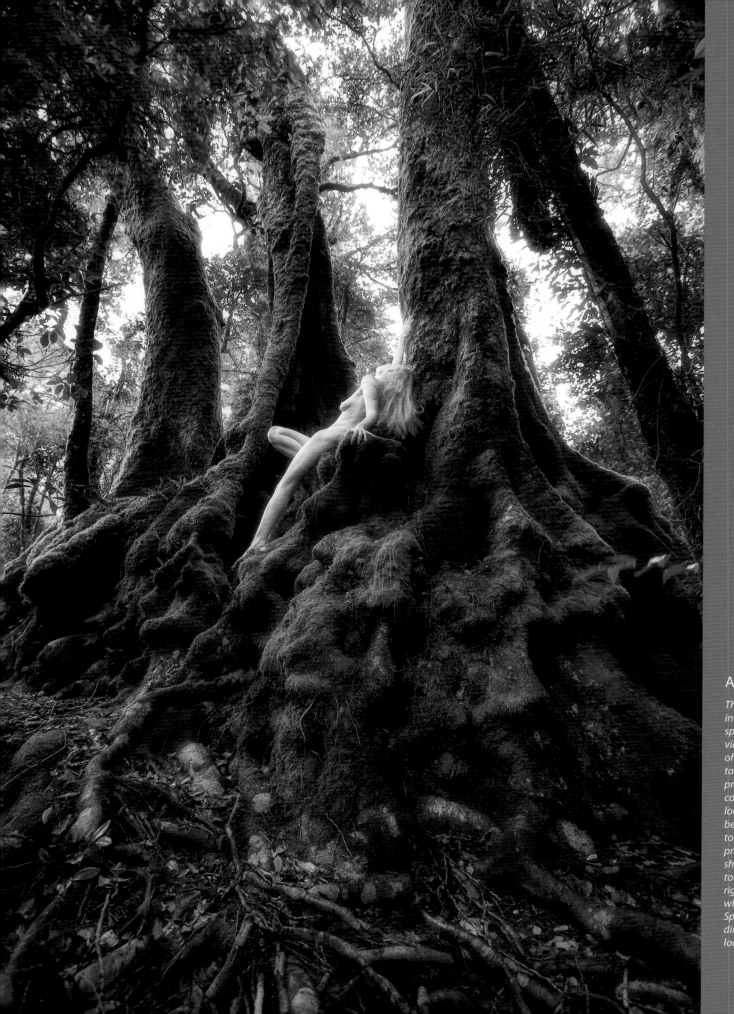

Antarctic Exaltation (2015)

This magnificent ancient Antarctic Beech tree, 42 ft. (13 m) in circumference, is made up of three separate trunks that sprout out of an ancient base—the result of regeneration via self-coppicing after the original trunk failed hundreds of years ago. The trunks, leaning crookedly upward, stand tall like three wise sisters. Gnarled and nooked, the tree provides habitat for many organisms. The craggy trunk, completely covered in thick moss, lichens, ferns, and fungi, looks and feels like a miniature fairyland. Not completely benign, this tree—like most of the rainforest—is home to some tiny, sneaky leeches, one of whom attacked my privates while I was curled up during my blissful photo shoot. Despite the risks and the chilly climate, it was a joy to explore this perhaps 2,000+-year-old triple goddess right before dusk, and again early the next morning, while hiding from passing trekkers. The tree is located in Springbrook National Park, New South Wales, Australia, directly off the busy "Best of All" track on the way to the lookout (named after the spectacular view, not the tree).

Antarctic Beech

Lophozonia moorei

Other Names: Negrohead Beech, Mountain Beech, *Nothofagus moorei*

Distinctive Characteristics: This gnarly, evergreen (or partially deciduous) tree is prone to continuous regeneration by vegetative growth, or self-coppicing, when major parts of the tree break off at their base. It is very common to have many trunks emerging out of the same crown like a fairy ring. The tree's genus is 200 million years old from the era when with the supercontinent of Gondwana—Antarctica, Australia and South America—were one land-mass (hence the name Antarctic Beech). Residing in a moist habitat, it is often covered in ferns, fungi, mosses, and epiphytes. It is the only host for the rare orchid *Dendrobium falcrostrum*.

Distribution: Native to Australia, in limited patches, along the border of southeastern Queensland and northern tablelands of northeastern New South Wales, from the headwaters of the Manning River north to the McPherson Range. Elevation: 1,640–5,085 ft. (500–1,550 m).

Ecosystem: High altitude, cool temperate rainforest.

Maximum Age: Estimated at over 2,000 years.

Maximum Height and Girth: 164 ft. (50 m) in height; 42 ft. (13 m) in circumference. Note: Since these beeches have the capacity to regenerate themselves, producing a fairy ring, this maximum circumference may indicate multiple trunks fused together.

Traditional Uses: There were most likely traditional uses of this tree.

Modern Uses: The hard, durable pink wood has high commercial value for making cabinets, pianos, and turned-wood objects.

Threats and Conservation: This species is in danger due to habitat loss and climate change, although not officially listed as threatened.

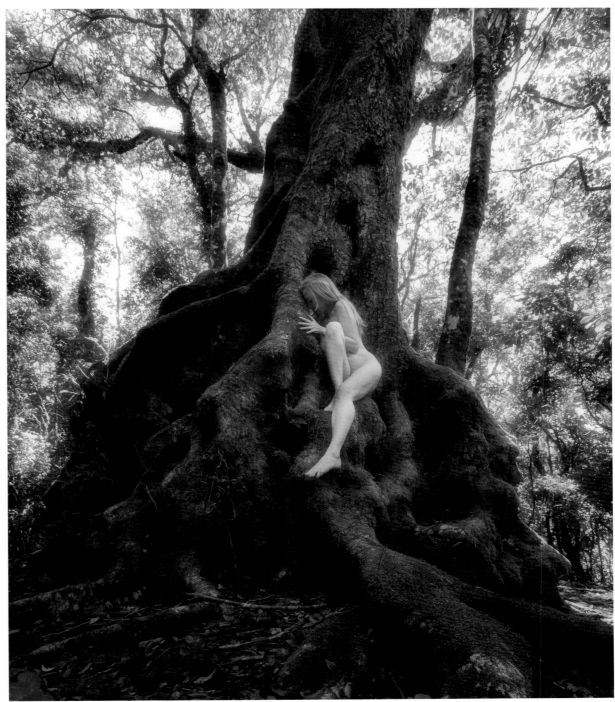

Antarctic Beech Bonding (2015) ▸

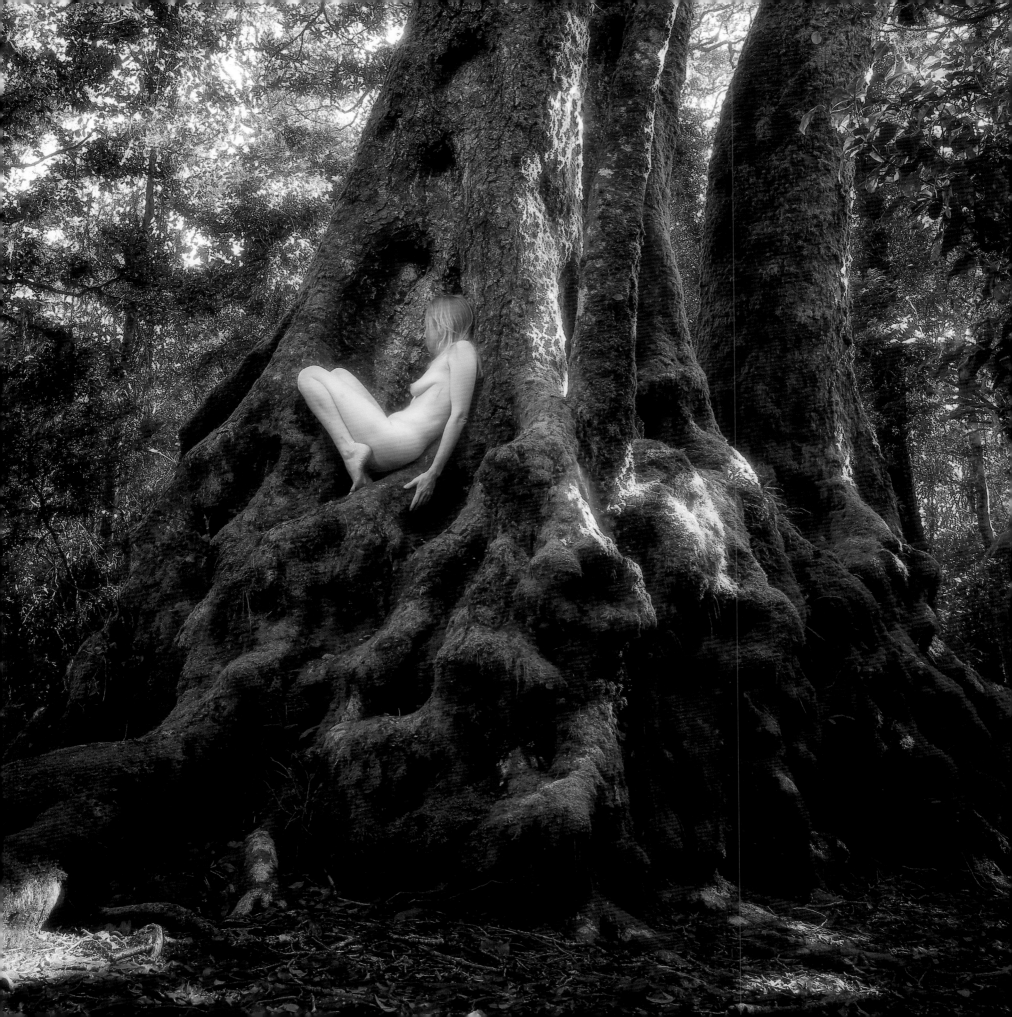

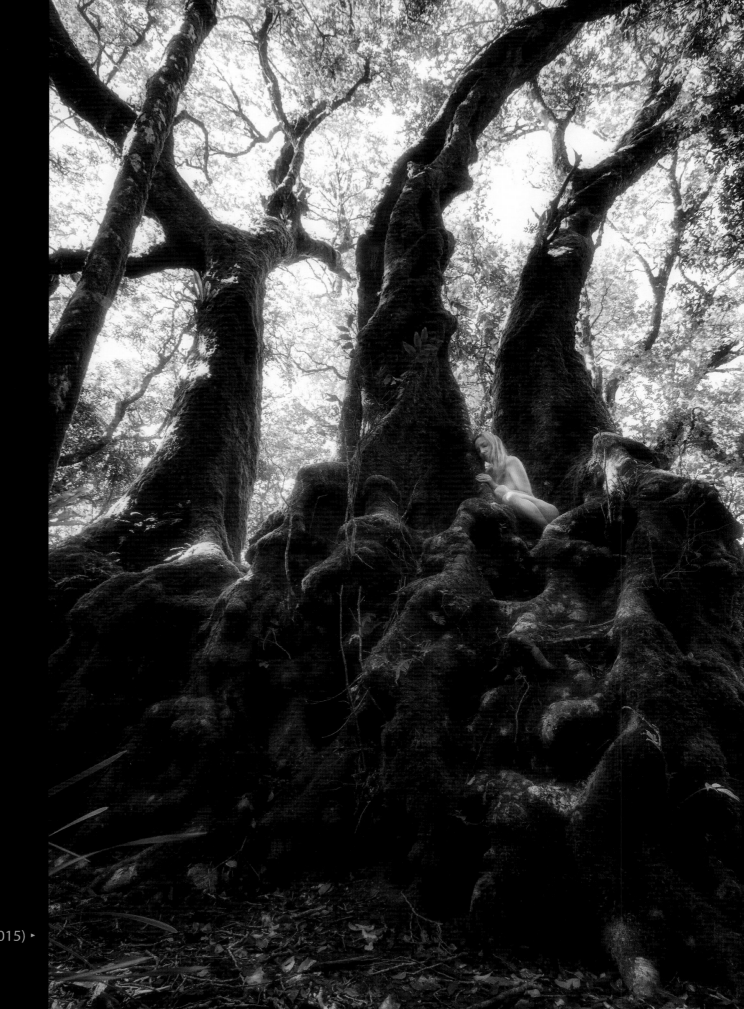

◂ Beech Niche Nook (2015)

Antarctic Beech Ethereal (2015) ▸

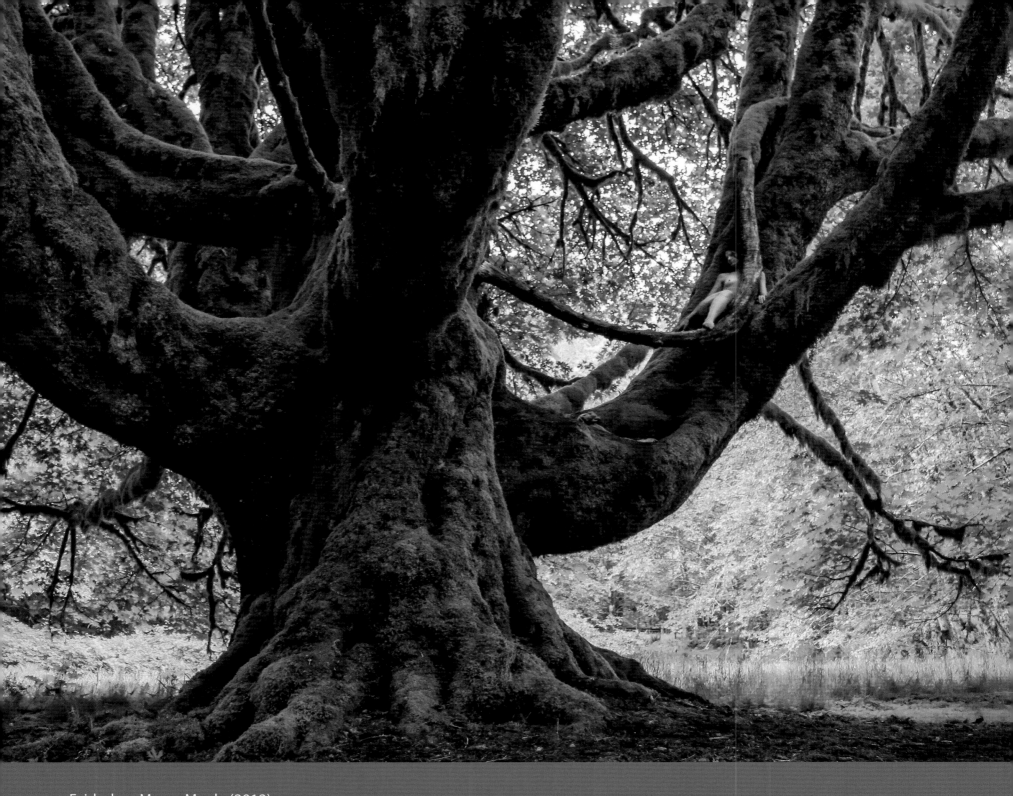

Fairholme Mossy Maple (2012)

The lush, furry coat of moss on this Big Leaf Maple is not uncommon in the rainforest of the Olympic Peninsula, where life grows literally upon other life, prevailing even in the smallest of crevices. Moss, an epiphytic plant that needs shade and dampness to thrive, can absorb up to twenty times its weight in water, and is indicative of a healthy, robust forest.

While not the size champion of its species, this particular tree nevertheless stands out from the crowd in its stature. The moss-covered trunk and arms are inviting to climb, seducing passersby

to engage its thick, soft carpet with their bare hands, bare feet, and bare everything. One could easily spend hours napping cozily upon its strong, dramatic arms. This welcoming, sensual tree lives peacefully near the boat ramp and parking lot of the Fairholme campground on Lake Crescent, just within the boundary of Olympic National Park, Washington State, USA.

Big Leaf Maple

Acer macrophyllum

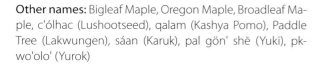

Other names: Bigleaf Maple, Oregon Maple, Broadleaf Maple, c'ólhac (Lushootseed), qalam (Kashya Pomo), Paddle Tree (Lakwungen), sáan (Karuk), pal gön' shē (Yuki), pkwo'olo' (Yurok)

Distinctive Characteristics: This is the largest tree of all the maple species. Its deciduous leaves are also the largest of any maple, ranging from 4 to 14 in. (10 to over 36 cm) across, with five deep lobes and large pointed teeth. The seeds are held in double samaras with wings 1½ to 2 in. (4 to 5 cm) long. In wetter climates, they are often covered completely in thick moss and lichens.

Distribution: Native to southernmost Alaska to southern California, mostly near the Pacific coast, as well as inland, in the foothills of the Sierra Nevada range. An isolated population may occur in central Idaho. Elevation: sea level to 6,000 ft. (1800 m).

Ecosystem: Found within riparian forests, mixed conifer or oak woodlands. Grows with Pacific Madrone, Douglas Fir, Grand Fir, Ponderosa Pine, Port Orford Cedar, Redwood, Sitka Spruce, Western Hemlock, White Fir, Western Redcedar, Black Cottonwood, California Bay Laurel, Coast Live Oak, Oregon White Oak, Red Alder, White Alder, and Willow.

Maximum Age: Approximately 200 years.

Maximum Height and Girth: 115 ft. (35 m) in height; 25 ft. (7.6 m) in circumference.

Animal Community: The leaves are important browse for deer and elk, and the seeds are eaten by squirrels, chipmunks, and some songbirds.

Traditional Uses:

Medicine: An infusion from the bark was used to treat tuberculosis (Yuki, Klallam), and part of the tree was used to treat sore throats and internal conditions (Saanich).

Food: The sprouted seeds were boiled and eaten (Costanoan, Nlaka'pamux). The inner bark was eaten (Coast Salish). The sap was boiled to make syrup, and shoots eaten raw (Thompson). The leaves were used as a cooking flavoring with deer, seal, or porpoise meat (Cowichan, Saanich).

Fiber: The inner bark was used to make baskets (Concow). The flexible branches were used as a coarse twine warp and weft in making baskets and as coiling thread for sewing (Maidu). Bark was used to make cordage (Cowlitz). Women's skirts were made out of the bark fibers (Concow, Karuk, Tolowa).

Tools and Objects: The leaves, bark, and wood were used for a number of cooking purposes. A disposable basket-like carrier was made for acorn dough by lining inner bark with leaves (Maidu). Leaves were used to line hot rocks for cooking acorn bread in an earthen oven (Wintu). The wood was used for smoking salmon (Swinomish, Snohomish, Chehalis, Quinault), while the leaves were made into mats placed in baskets to cook salmon and other food in earth ovens, and also to cover dried, winter-stored salmon (Karuk, Skagit, Snohomish, Lummi). Wood was fashioned into bowls (Nitinaht, Swinomish) and spoons (Karok, Swinomish). Small pieces of wood were also made into a dice-type gambling game (Pomo, Kashya Pomo) and into cradle boards (Lummi, Swinomish). Wood was used to make paddles and spindle whorls.

Art and Ceremony: Wood was used to make decorative carvings (Haida, Tsimshian, Tlingit); toys, games, and rattles (Hesquiat); and masks and rattles (Nitinaht, Southern Kwakiutl).

Shelter: The limbs were used for house construction (Cahuilla).

Transportation: The wood was made into canoe paddles (Clallam, Skagit, Snohomish, Nitinaht).

Modern Uses: Big Leaf Maple is not highly valued as lumber material; it is often intentionally knocked over and left in place when stands of redwood and Douglas Fir are commercially harvested. The wood is prized by craftsmen to make fine furniture, flooring, musical instruments, carved bowls, and veneer.

Threats and Conservation: Big Leaf Maple is a host for the fungus-like microorganism *Phytophthora ramorum*, which leads to sudden oak death. Trees may be adversely affected by the pathogen but are usually not killed by it. Otherwise, not threatened.

Mossy Maple Moss Woman (2012)

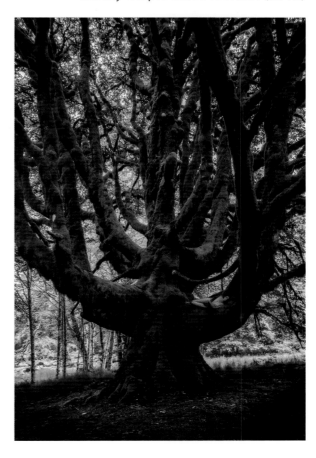

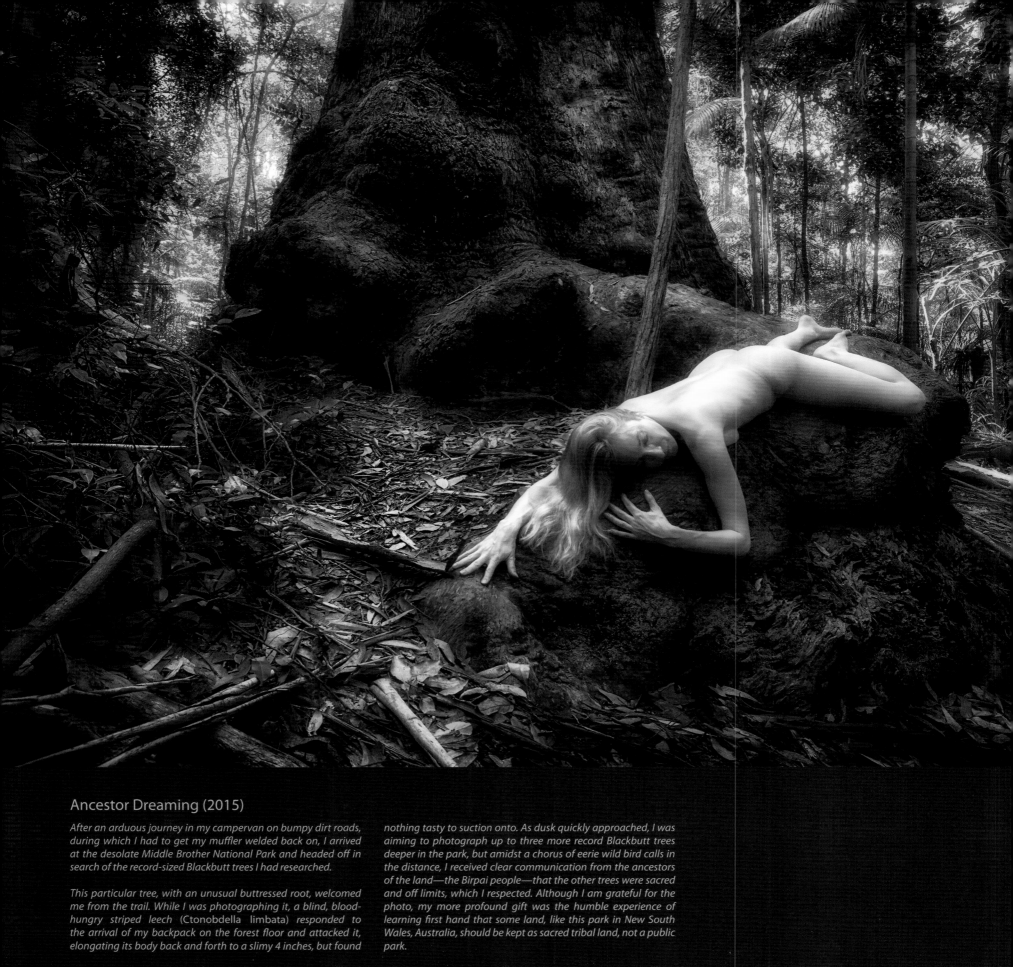

Ancestor Dreaming (2015)

After an arduous journey in my campervan on bumpy dirt roads, during which I had to get my muffler welded back on, I arrived at the desolate Middle Brother National Park and headed off in search of the record-sized Blackbutt trees I had researched.

This particular tree, with an unusual buttressed root, welcomed me from the trail. While I was photographing it, a blind, blood-hungry striped leech (Ctonobdella limbata) responded to the arrival of my backpack on the forest floor and attacked it, elongating its body back and forth to a slimy 4 inches, but found

nothing tasty to suction onto. As dusk quickly approached, I was aiming to photograph up to three more record Blackbutt trees deeper in the park, but amidst a chorus of eerie wild bird calls in the distance, I received clear communication from the ancestors of the land—the Birpai people—that the other trees were sacred and off limits, which I respected. Although I am grateful for the photo, my more profound gift was the humble experience of learning first hand that some land, like this park in New South Wales, Australia, should be kept as sacred tribal land, not a public park.

Blackbutt

Eucalyptus pilularis

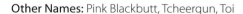

Other Names: Pink Blackbutt, Tcheergun, Toi

Distinctive Characteristics: The common name Blackbutt supposedly refers to the charred black base of the tree after bushfires. This evergreen tree has hard, leathery foliage adapted to prevent moisture loss. The lower part of the trunk is covered in dark, fibrous, spongy bark that can peel away in strips, while the bark higher up and on the branches is smooth and cream-colored, occasionally scribbled by insect larvae.

Distribution: Native to southeastern Australia, on the far south coast of New South Wales up to southeastern Queensland. Elevation: sea level to 2,500 ft. (800 m).

Ecosystem: Found in grassy coastal forests and wet sclerophyll ecosystems, which are characterized by open stands of tall eucalypts, and an important element of several endangered communities, including Blue Gum high forest, Illawarra lowlands, Bangalay sand forest, and grassy woodland, with other tree species including Tallowwood, White Mahogany, Red Mahogany, Grey Ironbark, Coast Grey Box, Brush Box, Turpentine, Spotted Gum, Sydney Peppermint, and Sydney Red Gum.

Maximum Age: Over 200 years.

Maximum Height and Girth: 230 ft. (70 m) in height; 49 ft. (15 m) in circumference.

Animal Community: This species of *Eucalyptus* is one of the trees that koalas depend on for food. As they tend to hollow at a young age, the become an important habitat for animals. The tree's nectar and pollen are highly sought after by fruit bats, blossom bats, honeyeaters, and lorikeets. The seeds are sought after by gang-gang cockatoos.

Traditional Uses: There were most likely many other traditional uses of this tree, in addition to the following known ones.

Medicine: The gum resin, known as kino, is an astringent.

Shelter: Some hollow cavities of burnt out trees were used by aboriginal women to give birth in.

Modern Uses: Economically, Blackbutt is an important hardwood, logged from natural forests or plantations. It is used for flooring, poles, railroad ties, building construction, furniture, decking, fencing, landscaping, wood chips and paper pulp. The pollen of the tree is utilized by bees in commercial honey production.

Threats and Conservation: Not threatened.

Blackbutt Leaves and Flower Buds

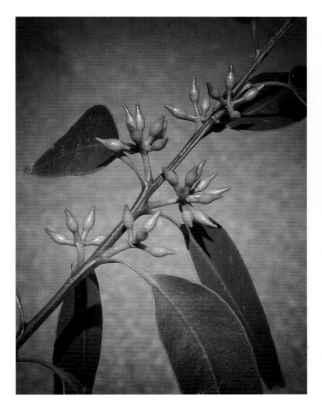

Blackbutt Trunk

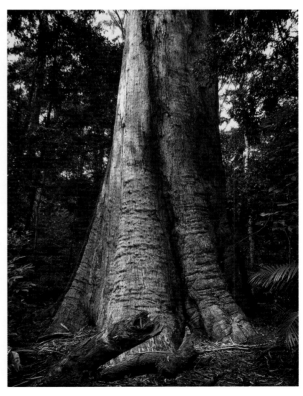

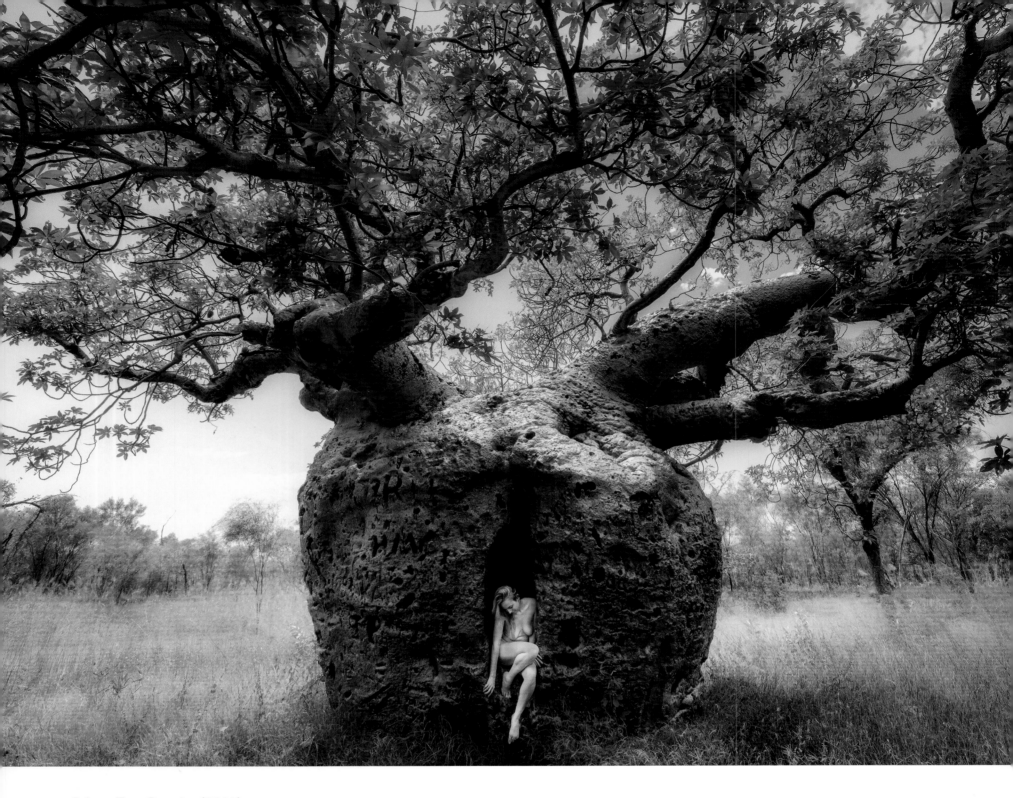

Prison Tree Respite (2015)

The famous Boab Prison Tree (The Australians call them Boabs.), a few miles outside Derby in Western Australia, is one of the most famous Baobabs in the world because of its odd shape. It is its own species, different than its cousins in mainland Africa and on the island of Madagascar. Historically, the tree was unlikely to have been a prison for aboriginals at all, as the name and associated tales suggest. As I approached the interpretive sign, I was greeted by a small green tree frog, perhaps an ancestral spirit of the land, who seemed to be the gatekeeper to this special place. The tree was measured at 45.43 ft. (13.85 m) in circumference and 28 ft.

(8.53 m) tall, and reputed to be 1,500 years old, but is probably much younger. (All species of Baobabs, with their characteristic bulbous shape, seem deceptively older than they are due to their water retention.) Still, the tree is fantastic: Looking like some sort of giant tuberous vegetable, covered in pock marks and old graffiti carvings, it has a hole that seems to be a perfect portal to another world. The inside of the tree is completely hollow and womb-like, with a little crack of a window at the top. When I visited the tree in late March of 2015, the temperature was probably over 100°F (38°C)—not unusual for the Kimberly region—and I

felt very comfortable being naked in the heat. The only risk was being seen. The only other tree-watcher there was a lone, elder gentleman in his campervan, who told me of another Boab, "suitable for a treegirl." How did he know my identity? I suspect that he was actually an angel or dryad in disguise, appearing only for me. Alas, I ran out of time to make the epic drive to the other record-breaking Boab, so I guess I will have to return someday.

Boab

Adansonia gregorii

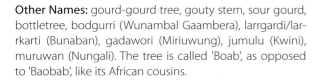

Other Names: gourd-gourd tree, gouty stem, sour gourd, bottletree, bodgurri (Wunambal Gaambera), larrgardi/lar-rkarti (Bunaban), gadawori (Miriuwung), jumulu (Kwini), muruwan (Nungali). The tree is called 'Boab', as opposed to 'Baobab', like its African cousins.

Distinctive Characteristics: The iconic, deciduous Boab is easily identified by its bottle-shaped trunk, and as they age, a barrel shape. Branches are concentrated at the top. Boabs store water in the fibrous tissue of their trunk. The smooth grey bark is easily scarred and pock-marked.

Distribution: Native to northwestern Australia, in the Kimberley region of Western Australia, from Broome eastward to the Fitzmaurice River region in the Northern Territory. Elevation: sea level to 1,000 ft. (300 m).

Ecosystem: Usually grows alone in open woodlands and grasslands, but also found in monsoon forests, thriving along creeks and drainage channels. Sometimes found with Bloodwood.

Maximum Age: Over 500 years. Like all species of Baobab, its indeterminate rings and hollowing make accurately dating the tree impossible. They appear deceptively older than they are because they hold water. This species does not attain the age of its African cousins, *Adansonia digitata*.

Maximum Height and Girth: 50 ft. (15 m) in height; 55 ft. (17 m) in circumference. This species does not attain the size of its African cousins, *Adansonia digitata*.

Animal Community: Rock wallabies and kangaroos feed on the fruit and disperse the seeds. Scaly-tailed possums feed on the leaves, flowers, and fruits. Flying foxes and fruit bats, who roost in the branches, eat the flowers and pods. Several widespread species of birds have been observed either eating, resting, or nesting in Boabs, including crows, black kites, magpie-larks, grey-crowned warblers, kestrels, fork-tailed kites, zebra finches, white-faced herons, black-breasted buzzards, honey-eaters, corellas, owls, and owlet-nightjars. Reptiles such as goannas, geckos, skinks, and green tree frogs may also use the tree as habitat.

Traditional Uses:

Medicine: The bark was used as an antiperiodic and anti-pyretic. The flowers were used as a fever reducer. The pulp was used as an anti-acid, to treat nausea, and given to mothers of newborns. The green leaves were laid on the fire as a mosquito repellent. Sickly children's bellies would be rubbed up against the trunk for healing.

Food: The gourd-like fruit pods contain an edible, lemony dry pulp that can be chewed alone, mixed with water and sweetener as a drink, mixed with edible gum from other plants, or may be cooked into custard or a kind of bread. The edible seeds were eaten raw, ground up, or roasted on hot coals. Young plants have a large edible tuberish tap-root. The leaves were eaten. Mucilage exuding from damaged wood was eaten or made into a drink, as well as the flowers, either fresh or fermented.

Fiber: The bark was made into cordage, cylindrical bee-hives, and carrying cradles.

Tools and Objects: The gourd-like seedpods were hollowed out and used as containers. Stored rainwater from the hollows was tapped for emergencies. The gum and pollen were used to make glue. Scoops for gathering and eating honey out of hives (mops) were made out of masses of the fiberous roots.

Art and Ceremony: The soft, thick bark of the trunks has been carved with zoomorphic, anthropomorphic, geometric, and symbolic imagery (dendroglyphs), and painted with mineral paints. The gourd-like seedpods may have been used as musical rattles for ceremony. The pods were decoratively carved, although this use may not be traditional. String made from bark was tied together to make crosses for dance ceremonies. The staminal tube from the flower was used as a paintbrush to apply body paint.

Shelter: Two famous "prison" Boab trees are said to have been used to keep aboriginals in confinement for short periods of time. They are also said to have served as a temporary dwelling, store and post office.

Threats and Conservation: Not threatened.

Decorated Boab Pod

This Boab seed pod is carved with images of the creator spirit, Wandjina, historically painted as pictographs on cave walls and sacred to the Mowanjum people.

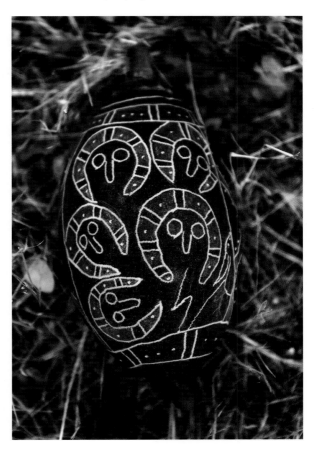

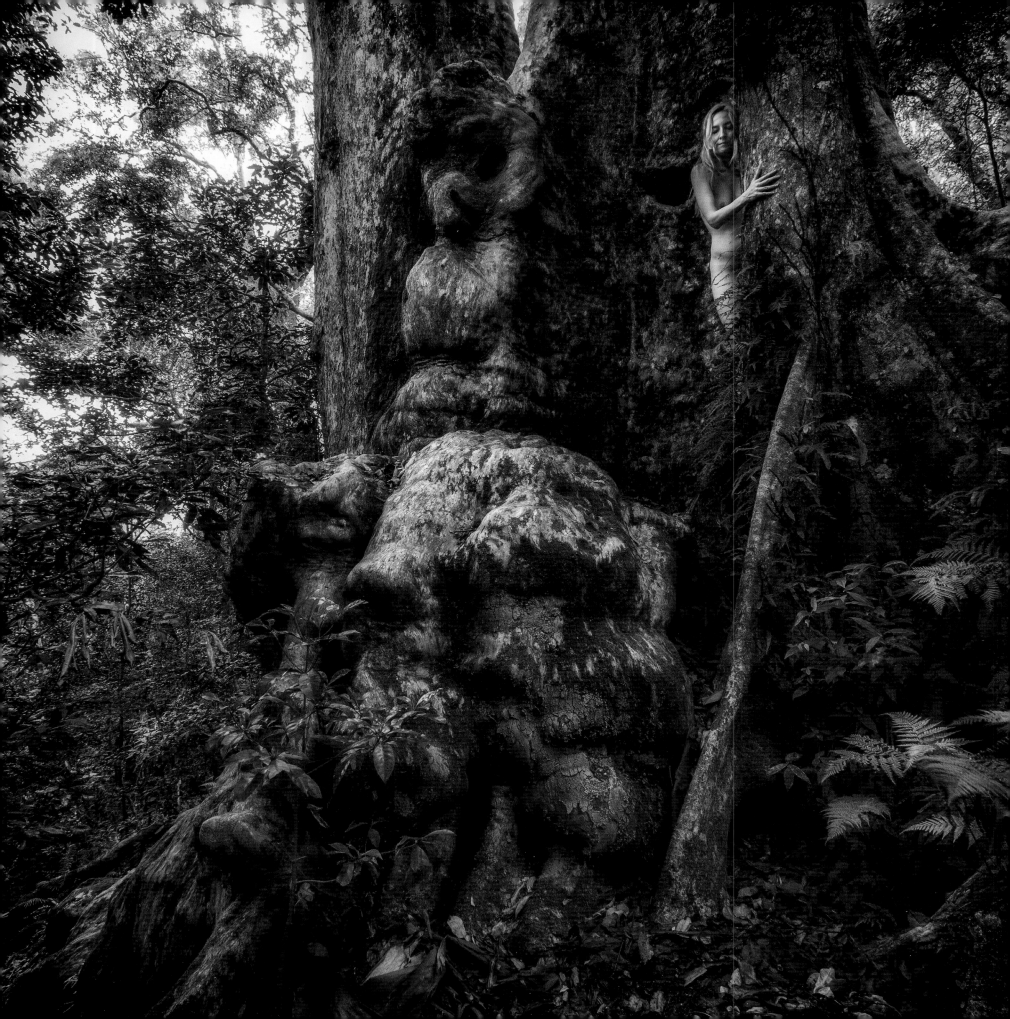

Brush Box
Lophostemon confertus

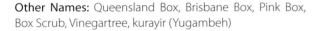

Other Names: Queensland Box, Brisbane Box, Pink Box, Box Scrub, Vinegartree, kurayir (Yugambeh)

Distinctive Characteristics: This evergreen tree is resilient against pests, smog, drought, and poor drainage. It is popular as a street tree, although it does not reach its full potential in size in this environment.

Distribution: Native to Australia, from northeast coastal New South Wales northward to northeast coastal Queensland, from Newcastle to Fraser Island, and in isolated stands to the Windsor Tablelands. Elevation: 500–2,800 ft. (150–850 m).

Ecosystem: Dry or wet sclerophyll forest (characterized by open stands of tall eucalypts and hard, leathery, evergreen foliage adapted to prevent moisture loss); in valleys, ridges, mountains, and coastal headlands; in ecotones between rainforest and tall, open eucalypt forest. Associated with Flooded Gum, Tallowwood, and Turpentine.

Maximum Age: About 1,500 years.

Maximum Height and Girth: Up to 177 ft. (54 m) in height, and over 40 ft. (12 m) in circumference.

Animal Community: This tree is a food plant for the larval stages of the common red-eye, rare red-eye, and eastern flat butterflies.

Traditional Uses: There were most likely many traditional uses of this tree.

Modern Uses: Popular as timber, flooring, and wharf decking.

Threats and Conservation: Not threatened.

Brush Box Sanctuary II (2015)

The adventure to reach this Brush Box included a 6-mile (10 km) trek on a very poorly marked sequence of trails, which included crossing a wide, rocky rushing creek. The long day was worth the work, however, as I was greeted by a large blue crustacean on its way from a nearby stream. After it put up its giant claws in frightened defense, it realized that I was not going to eat it with my camera, and continued heading for a napping spot within the base of the giant tree. I continued with my purpose of communion with the elephantine Box tree, climbing and nesting on its gigantic, blobby burls. This hard-to-find Brush Box is in the Green Mountains section of Lamington National Park, New South Wales, Australia.

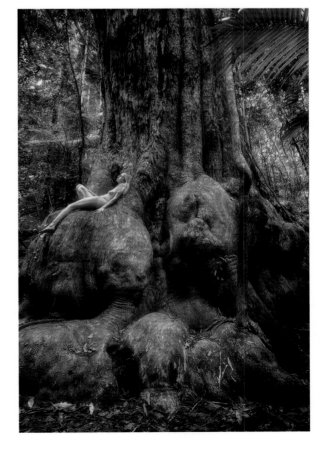

◂ Brush Box Hideaway II (2015)

This ancient Brush Box tree, estimated at 1,500 years old, is a world unto itself: two major trunks, covered in faces and eyes, ferns and vines; giant buttressed roots; bulbous burls; and snake-like roots entangled with large rocks. It is the perfect tree for any treegirl: Each side of the tree has places to climb and rest. I did not want to leave, but I had spent the entire day in the Park, and it was getting dark. The spirit, or spirits, of the tree were very kind to let me climb all over it, and I would happily return for a reunion. This Brush Box tree is an easy twenty-minute walk from the main trail heads in the Green Mountains section of Lamington National Park, New South Wales, Australia.

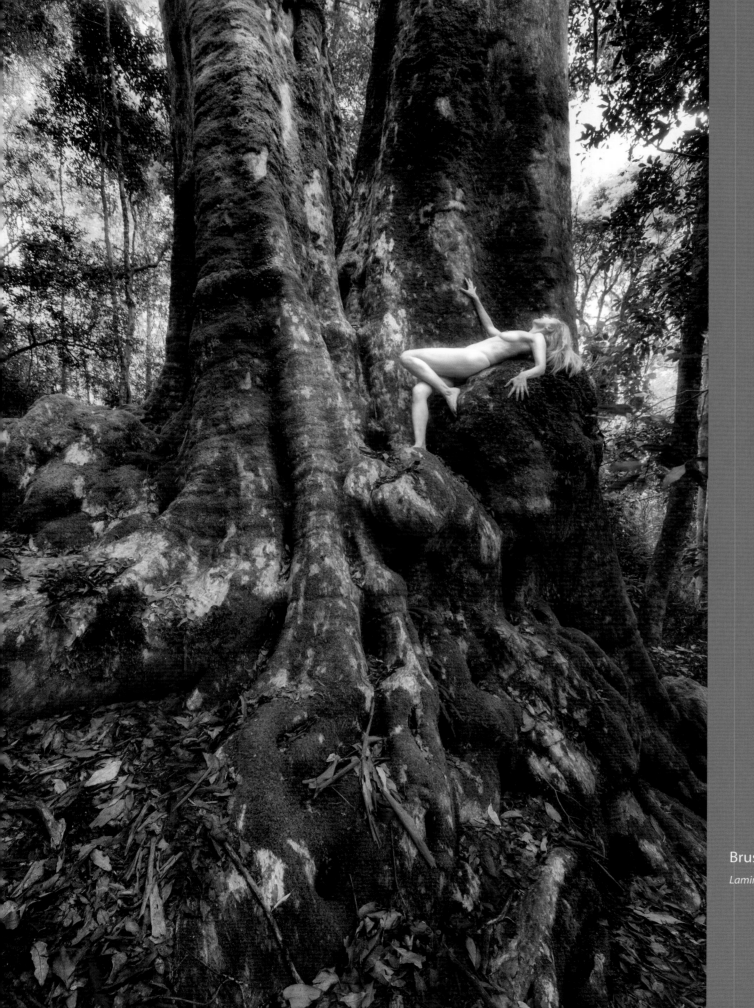

Brush Box Bliss (2015)

Lamington National Park, New South Wales, Australia.

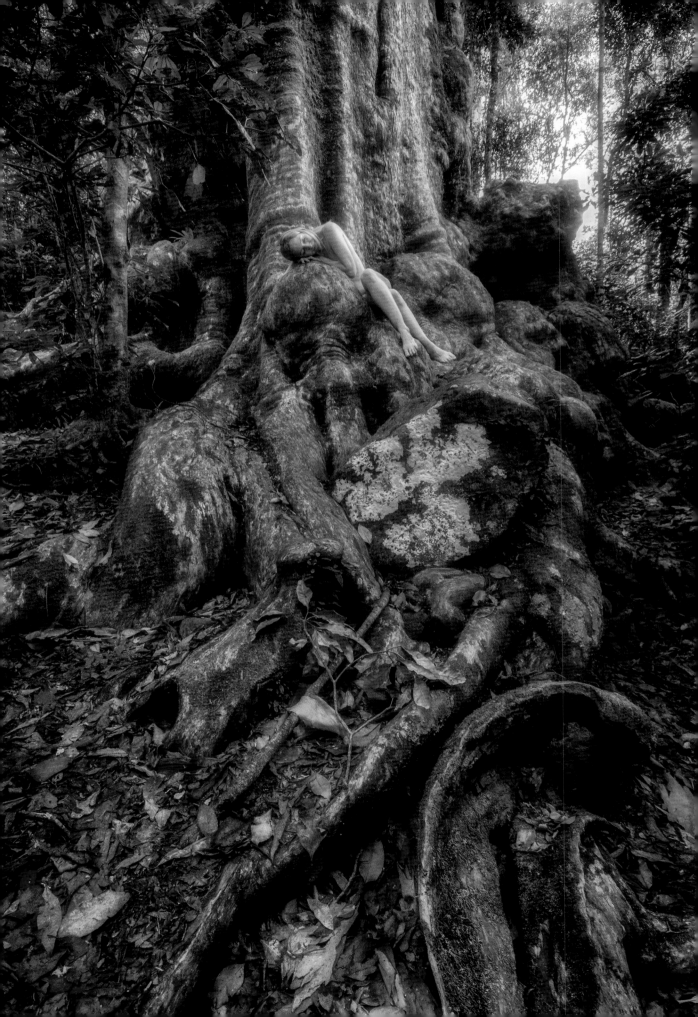

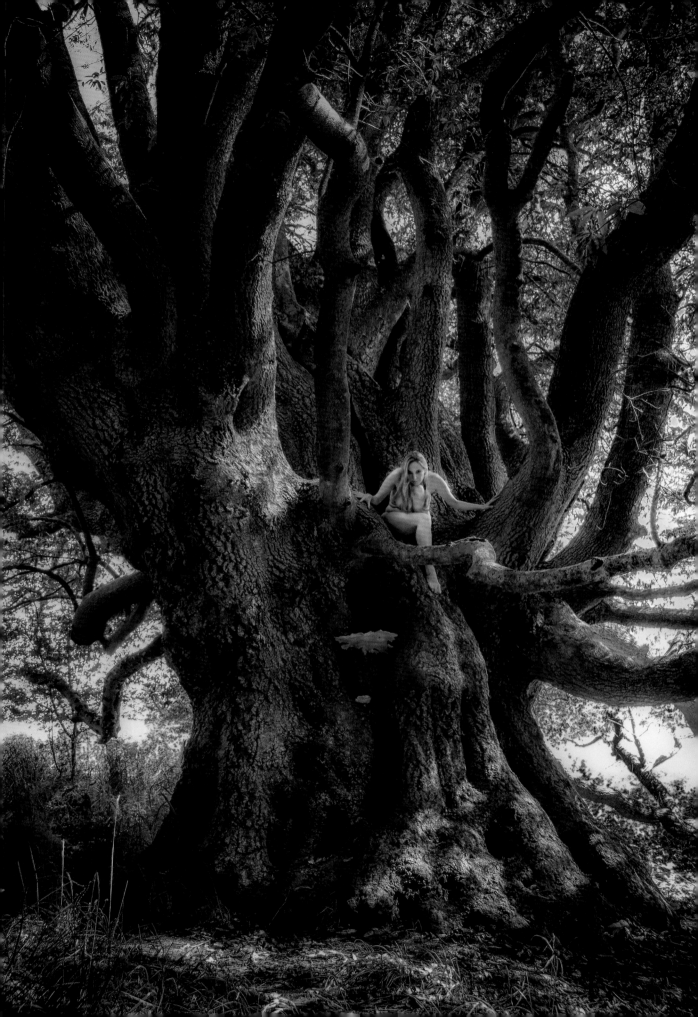

California Bay Laurel

Umbellularia californica

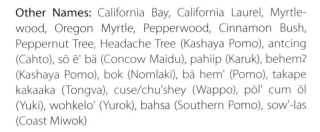

Other Names: California Bay, California Laurel, Myrtle-wood, Oregon Myrtle, Pepperwood, Cinnamon Bush, Peppernut Tree, Headache Tree (Kashaya Pomo), antcing (Cahto), sō ē' bä (Concow Maidu), pahiip (Karuk), behem? (Kashaya Pomo), bok (Nomlaki), bā hem' (Pomo), takape kakaaka (Tongva), cuse/chu'shey (Wappo), pōl' cum ōl (Yuki), wohkelo' (Yurok), bahsa (Southern Pomo), sow'-las (Coast Miwok)

Distinctive Characteristics: This evergreen, shade-tolerant tree has a single or multiple trunks with an open, dome-shaped crown. The shiny, dark-green leaves are narrow, long pointed ovals with smooth edges; leaves can reach 4 in. (10 cm) long and 1.2 in. (3 cm) wide. Small yellow-ish-green flowers are held in an "umbel," a number of short flower stalks, equal in length and spreading from a common point, somewhat like umbrella ribs. The tree's fruit, the bay nut, is a round to olive-shaped green berry about 1 in. (2.5 cm) long; it matures to a purple color with a cap that resembles a golf tee. Under the thin, leathery skin is a bit of green flesh coating a hard, thin-shelled edible pit, in whole resembling a miniature avocado (the trees belong to the same Lauraceae Family).

The tree is similar to its Mediterranean cousin, the culinary Bay Laurel (*Laurus nobilis*), which is smaller in size, with generally narrower leaves containing sweeter oils. All parts of the California Bay Laurel, especially the leaves, contain a distinctively aromatic camphor-like volatile oil that has cooling, irritating, germicidal, and insecticidal qualities. The fragrance is much more aromatic than that of its Mediterranean relative, and it can easily cause headaches that last for days, and can send over-zealously inhaling hikers to the emergency room.

Distribution: Native to the southern tip of Oregon, down the coastal forests throughout the entire length of California, and inland in the western foothills of the Sierra Nevada range. Elevation: sea level to 5,000 ft. (1,500 m).

Ecosystem: Found in redwood forests, mixed evergreen forests, oak woodlands, and chaparral in coastal areas with Douglas Fir, Redwood, Madrone, Valley Oak, Black Cotton-wood, California Sycamore, Big Leaf Maple, Boxelder, Interior Live Oak, Coast Live Oak, Canyon Live Oak, Blue Oak, California Scrub Oak, California Buckeye, and manzanita. Also found in the foothills of the Sierra Nevada range with Grand Fir, White Fir, Coulter Pine, Sugar Pine, and Western White Pine.

Maximum Age: Approximately 500 years.

Maximum Height and Girth: 108 ft. (33 m) in height; 31 ft. (9.4 m) in circumference.

Animal Community: California Bay Laurel provides nesting and cover for game and songbirds, and hiding for deer, wild pig, black bear, and various small mammals. It is browsed by black-tailed deer and livestock. Squirrels and Stellar's jays eat the fruit.

Traditional Uses:

Medicinal: Crushed fresh leaves were inhaled as pain relief for headaches and nasal congestion, though the volatile oils in the leaves may also cause headaches (Cahuilla, Coast Miwok). A poultice was also applied to the head for headaches (Miwok, Yuki, Mendocino Indian). Fresh leaves were placed in water and boiled to make aromatic steam to treat colds and sinus infections (Karok). The light-green tips of new growth were used as a poultice to treat toothache (Lake Miwok). A leaf poultice was used for shingles. A tea was used for sore throats and colds. Leaf oil was used to treat earaches and sores and to prevent allergies in the spring; it was also used to relieve colitis and ulcers. Women used an infusion of the plant for pain after childbirth (Karok). A decoction of the plant was used as a wash for head lice (Mendocino Indian). An infusion of leaves was used as a bath (Mendocino Indian) and a poultice was applied for rheumatism (Pomo, Kashaya Pomo). The leaves were taken as a decoction or poultice for stomachaches (Mendocino Indian, Coast Miwok). Kashaya Pomo doctors would sometimes hit a patient with little branches while singing as a treatment for pain, headache, or colds. A decoction of the leaves was used for menstrual cramps (Kashaya Pomo). A poultice made from flowers was used to reduce swelling. The burning leaf smoke and vapor was used to treat many diseases and to fumigate the house after sicknesses. Leaves were made into an infusion for cramps from diarrhea, food poisoning, or gastroenteritis. A diluted tincture or strong tea can be used as an antimicrobial or antifungal on skin; and a bath may be taken with the leaves for arthritis and joint pain. A repellant tea was made of the root bark, and smoke from burning leaves was used to keep insects out of acorn granaries and houses. Feather-work and baskets were stored with leaves to repel insects. Used also as a flea repellent (Costanoan, Kashaya Pomo, Mendocino Indians).

Roasted bay nuts

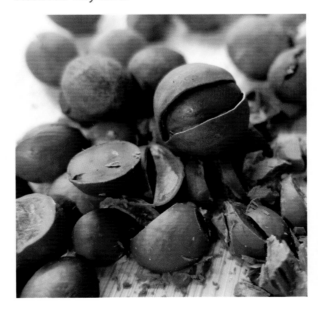

Bay leaves (2016)

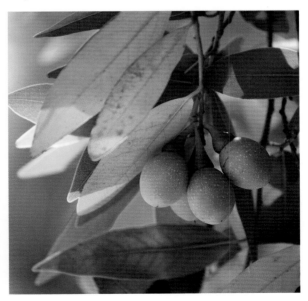

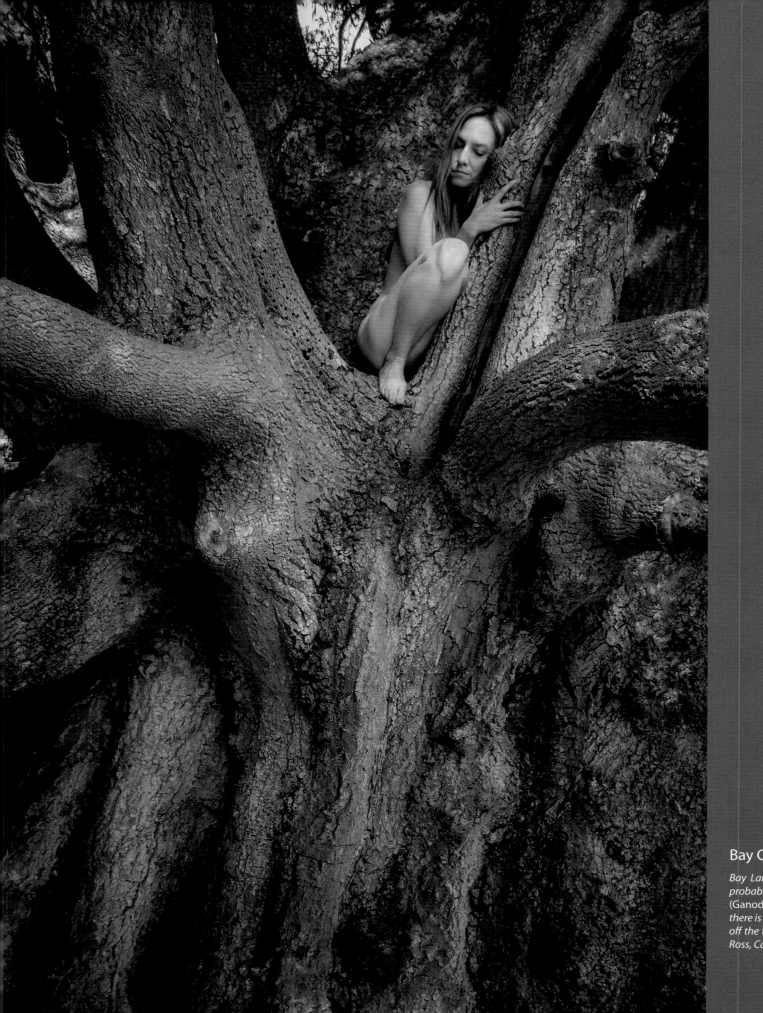

Bay Castle Perch (2012)

Bay Laurels age to about 500 years, and this tree was probably close to that. The large artist conk fungus (Ganoderma applanatum) at the base indicates that there is plenty of rot inside. This little-known tree is hidden off the trail near the orchard in Fort Ross State Park, Fort Ross, California, USA.

California Bay Laurel

Food: Both the fruity flesh under the skin and the nut itself are edible. The fruit is palatable raw for only a brief time when ripe; if too ripe, the flesh quickly becomes bruised, like an overripe avocado, and the volatile aromatic oils are so strong that the fruit is inedible. The shelled nuts, which look like the pit of an avocado, are roasted (to remove pungency) in hot ashes and eaten whole, or pounded and sun-dried to make flat cakes that can be eaten right away or stored for winter's use. Roasted nuts or cakes are eaten with greens, buckeye meal, acorn meal, mush, or seaweed. They were also ground into a powder and roasted to make a beverage with the taste of unsweetened coffee or burnt cocoa. While the leaf can be used in cooking, it is spicier and stronger than the Mediterranean seasoning and used in smaller quantity.

Tools and Objects: The wood was used to make bows (Western Mono).

Art and Ceremony: The plant was used in many ceremonies. Leaves and branchlets were used for ceremonial purification, and branches were fashioned into drumsticks. Children threw leaves into fires to hear them crack like firecrackers (Karok). The foliage was placed on a fire during the Brush Dance to drive away evil spirits (Karok). Leaves were rubbed on the body before hunting, to hide human odor (Kashya Pomo). Small, leafy branches were hung in houses to ward off harm or were burned to dispel bad luck in the home (Kashaya Pomo, Yurok). The smoke was waved over people as they left the home. The wood was used to make split-wood clapper instruments for dance circles (Costanoan).

Modern Uses: "Myrtlewood," as it is marketed in Oregon, is sought after by woodworkers around the world. It is considered an excellent tonewood (used to construct the back and sides of acoustic guitars and violins). The beautiful wood is also used for cabinets, furniture, paneling, and veneer. Burls are used for making turned bowls, spoons, and other small tourist items.

Threats and Conservation: California Bay Laurel is one of the main foliar hosts of the pathogen *Phytophthora ramorum*, which causes sudden oak death, and it is also affected by a less severe disease known as ramorum dieback. Although the tree is not threatened, its population may decline due to removals to prevent the continued spread of the disease.

Bay Castle Companion (2012) ▸

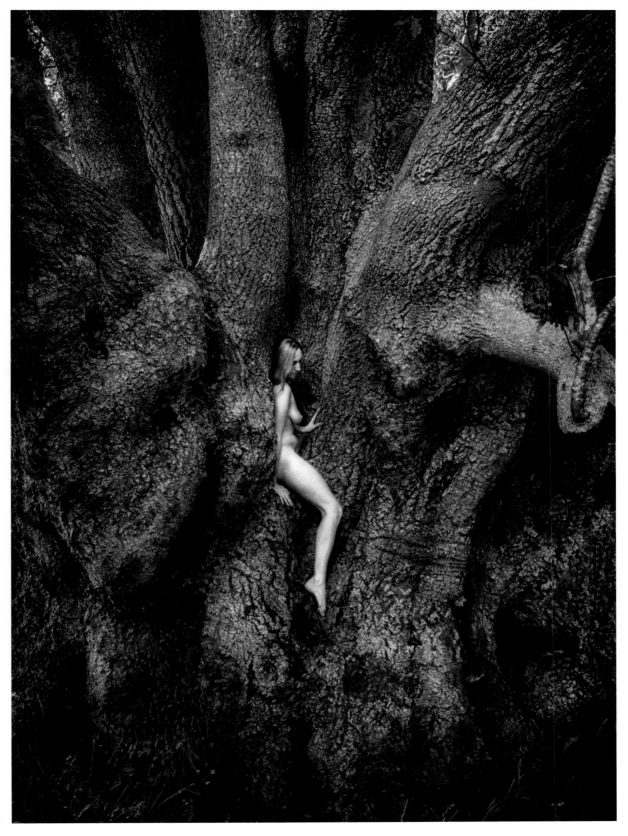

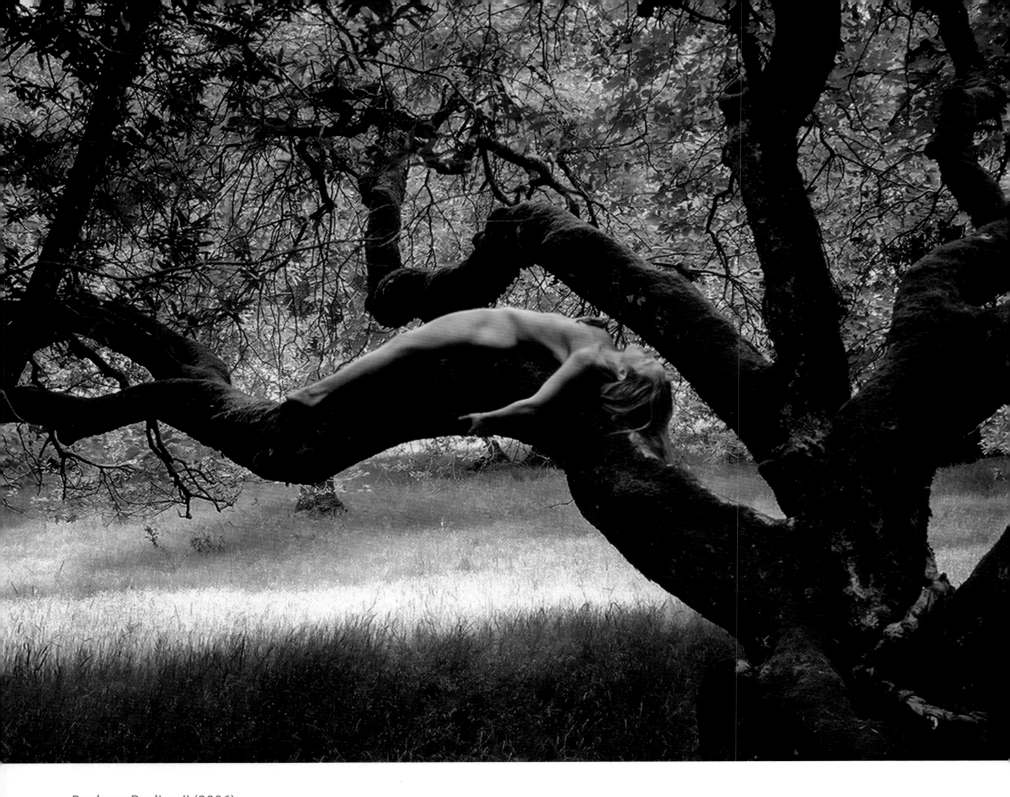

Buckeye Recline II (2006)

This Buckeye lives near a stream in Olompali State Historic Park, Novato, California, USA. Native California Indians resided on this land for about 1,500 years, gathering the Buckeye nuts to make into food and fish poison. At one point it was an important trade center for tribes, and may have been the largest Coast Miwok village in the area.

In the 1960s it was briefly a hippie commune, and a retreat for famous rock stars of the era. It is doubtful that they knew what to

do with the nuts, but they undoubtedly enjoyed the wild beauty of the land.

California Buckeye trees are distinctively multi-trunked. This one's moss-covered, twisted arms arch to make the perfect resting spot for a limber treegirl.

California Buckeye

Aesculus californica

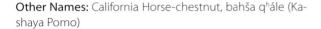

Other Names: California Horse-chestnut, bahša qʰále (Kashaya Pomo)

Distinctive Characteristics: The California Buckeye is identifiable by its beautiful 6–10-inch-long erect clusters of pinkish-white flowers in Spring, distinctive palmately compounded, deciduous leaves with 5 to 7 leaflets, and large, poisonous seeds known as buckeye nuts, that are shiny outside of their husk. These deciduous trees are notable for their early leaf drop in summer; they are often mistaken for being sick or even dead. They can be easily identified in late summer, fall and winter with their thick, light colored bare branches with hanging light-colored ornament-looking seed pods.

Distribution: Native to western and northern California, reaching into Southern Oregon and the foothills of the Sierra Nevada range. Elevation: sea level to 5,600 ft. (1700 m).

Ecosystem: Often grows on slopes alone or with other trees in mixed woodland, oak grassland, or mixed evergreen forests, intermingled with Valley Oak, Oregon Oak, Coast Live Oak, California Bay Laurel, Blue Oak; and at higher elevations with Black Oak, Gray Pine, Ponderosa Pine, and Interior Live Oak.

Maximum Age: Approximately 300 years.

Maximum Height and Girth: 46 ft. (14 m) in height; 14 ft. (4.2 m) in circumference.

Animal Community: The pollen and nectar of the Buckeye flowers are vitally important to native bees and butterflies, but toxic and deadly to the European honeybee.

Traditional Uses:
Medicinal: Smashed nuts were used for a hemorrhoid remedy (Kawaiisu, Costanoan). A decoction of bark was used for toothaches (Costanoan, Mendocino Indian).

Food: Although not as popular as acorns, Buckeye nuts were traditionally prepared for eating by first boiling or roasting them, and then by leaching out their bitter tannins with water. The potato-like meal, called dihsa, was said to be good with meat, seafood, and baked kelp (Kashaya Pomo, Luiseno, Ohlone, Costanoan, Kawaiisu, Mendocino Indians, Miwok, Tubatulabal, Yana, Yuki).

Tools and Objects: The raw nuts, which contain a toxin, were also ground and tossed into water to stun fish for easier catching (Costanoan, Pomo, Kashaya Pomo). The wood was used to make bows and drills for fire-making (Kashya Pomo). The wood was made into bowls (Kawaiisu, Mendocino Indians).

Threats and Conservation: Not threatened.

Buckeye Flowers (2016)

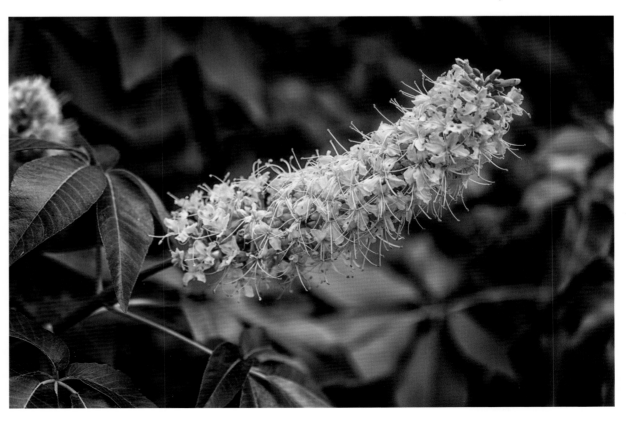

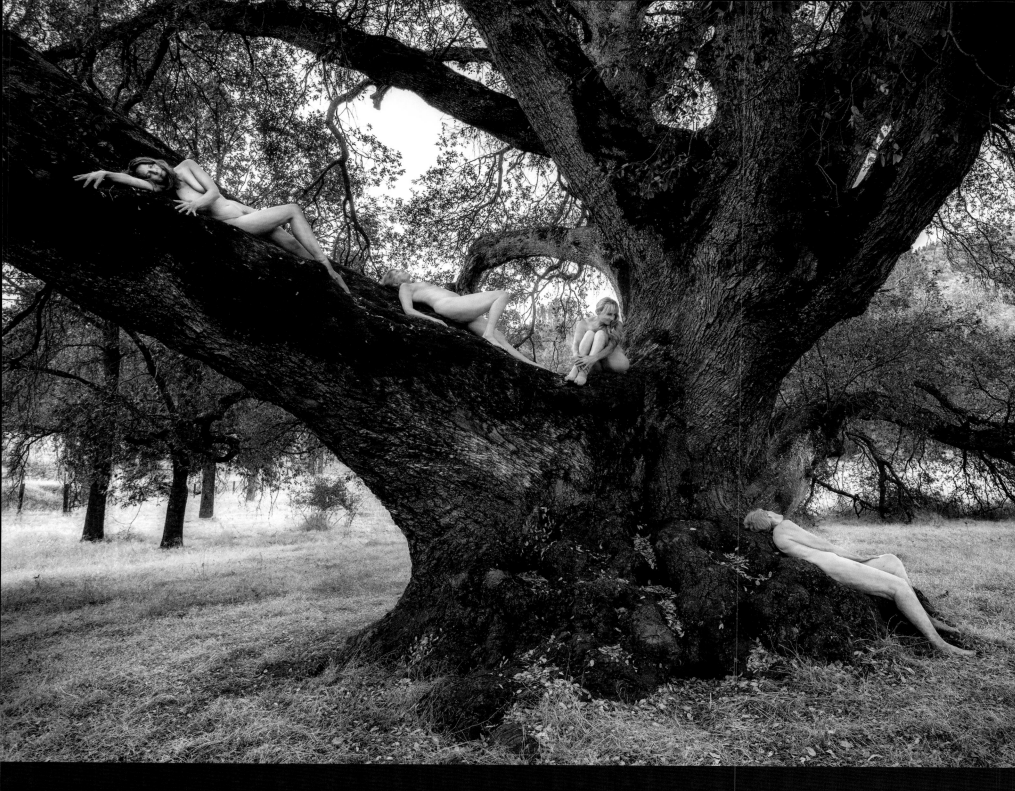

Canyon Live Oak Maidens I (2015)

This giant Canyon Live Oak was adored by tree-lovers for many, many years, until too much affection led the landowners to

naked women peacefully lounging in a tree? It was the first time TreeGirl ever got busted for "trespassing" (even though the tree

Canyon Live Oak

Quercus chrysolepis

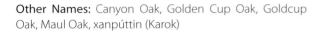

Other Names: Canyon Oak, Golden Cup Oak, Goldcup Oak, Maul Oak, xanpúttin (Karok)

Distinctive Characteristics: This evergreen oak can be distinguished from other Oaks in its region by the fine, yellow, fuzzy hairs on the underside of the leaves, which may be toothed or smooth on the same tree, and the golden velvety inside of the thick, shallow, warty, bowl-shaped acorn caps.

Distribution: Native to southern Oregon and California, along the northern coast ranges, in southern California, (including the Channel Islands) and the Sierra Nevada range, with small, scattered patches in Mexico, Nevada, and Arizona. It is the most widely dispersed of all California Oaks. Elevation: between 1,640–5,000 ft. (500–1,500 m) in southwestern Oregon; 330–4,500 ft. (100–1400 m) in northern California; and up to approximately 9000 ft. (2,700 m) in southern California.

Ecosystem: Canyon Live Oak occurs in pure stands, as the dominant tree species or as part of variable plant communities such as mixed evergreen forest, riparian hardwood forest, and oak woodlands, on steep canyon slopes and bottoms, ridge tops, riparian areas, or flat areas with Interior Live Oaks, Oregon White Oak, Blue Oak, Pacific Madrone, California Bay Laurel, Big Cone Douglas Fir, Douglas Fir, Ponderosa Pine, Jeffry Pine, Sugar Pine, Coulter Pine, California Buckeye, manzanita, and Huckleberry Oak.

Maximum Age: 300 years.

Maximum Height and Girth: 110 ft. (33 m) in height; 31 ft. (9.4 m) in circumference.

Animal Community: Canyon Live Oak provides habitat for perching, nesting, foraging sites, shade, and cover for numerous species of birds, mammals, reptiles, and amphibians. Acorns are consumed by a variety of mammal species, such as California ground squirrels, dusky-footed wood rats, western grey squirrels, western harvest mice, red tree voles, deer mice, white-footed mice, pinyon mice, California pocket mice, broad-footed mice, cactus mice, black-tailed jackrabbits, beaver, brush rabbits, red-backed voles, Sonoma chipmunks, California voles, porcupines, pocket gophers, black-tailed deer, and black bear, as well as birds such as western scrub jays, Steller's jays, band-tailed pigeons, wild turkeys, mountain quail, and crows. Other bird species inhabit and forage in Canyon Live Oak woodlands, such as California spotted owl, Nuttall's woodpeckers, acorn woodpeckers, ravens, white-breasted nuthatches, oak titmice, black-headed grosbeaks, and Baltimore orioles. Many reptiles also use the tree for habitat, including San Bernardino Mountain kingsnakes, Gilbert's skinks, western fence lizards, and southern alligator lizards, as well as amphibians such as San Gabriel Mountain slender salamanders, yellow-blotched salamanders, large-blotched salamanders, black-bellied slender salamanders, California slender salamanders, and foothill yellow-legged frogs. While the tree is utilized by countless invertebrate species, it is a specific host for insects such as the moth *Neocrania bifasciata*, and butterflies such as the California sisters, golden hairstreaks, Propertius duskywings, mournful dusky-wings, and gold-hunter's hairstreaks.

Traditional Uses:

Medicine: A poultice of the acorn flour or ashes from the bark was used to treat burns (Karuk).

Food: Many tribes ate the acorns as a food staple.

Modern Uses: Early settlers used the wood for fuel and for making farm implements, mauls, boats, carriages, wagons and furniture. It has also been used more recently for wall paneling.

Threats and Conservation: This Oak is susceptible to the disease sudden oak death (caused by *Phytophthora ramorum*), but its status is not threatened.

Canyon Live Oak Maidens II (2015)

Canyon Live Oaks are common in the foothills of the Sierra Nevada range. However, a tree of this size is extremely rare. They are estimated to live about 300 years, although this one appears much older. We did not have the tools to take an accurate measurement of age or girth, but her circumference was notably record-breaking. This regal Oak lived a long, secluded life in Mariposa, California, USA. Tragically, a few months after this photo was taken, she split apart and fell to the ground in a storm.

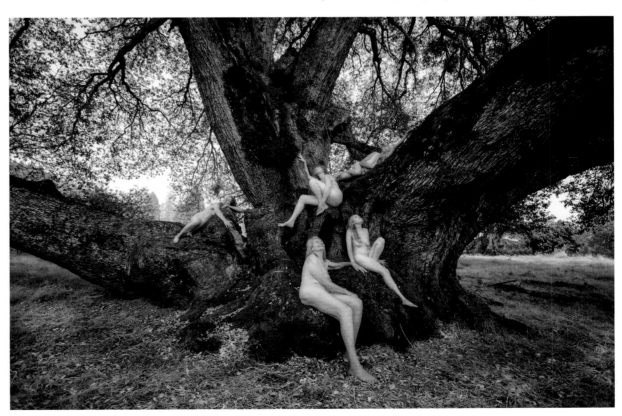

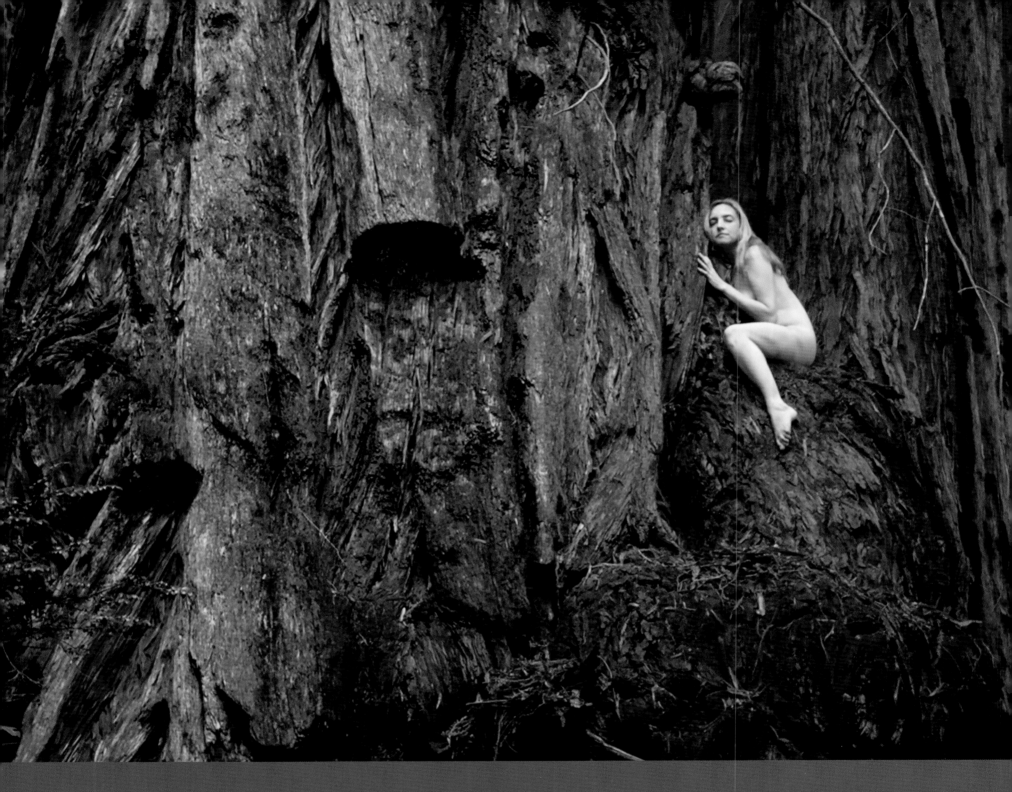

Grandmother Redwood II (2012)

Redwood trees are common here in northern California, although we don't have as many big, ancient ones as there are farther up the California coast. This tree was a bit daunting to climb, but once I was up there, it seemed like a perfect perch. I didn't want to come down, but it was getting dark. The bark was so soft and the massiveness of the tree felt so comforting, I just wanted to live there in that forest. This tree, of which I don't have the measurements, seems to be two trees grown together at the base. It lives on private property in Occidental, California, USA.

Coast Redwood

Sequoia sempervirens

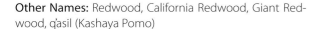

Other Names: Redwood, California Redwood, Giant Redwood, q̓asil (Kashaya Pomo)

Distinctive Characteristics: No flora species in California is more iconic than the Coast Redwood. These trees are legendary for their majestic beauty, towering height, habitat biodiversity, rot-resistant wood, and epic role in environmental activism. Coast Redwood is currently the tallest species of tree on Earth, while its cousin, the Giant Sequoia (*Sequoiadendron giganteum*), which lives exclusively in the Sierra Nevada range of California, is generally larger-girthed and is known for being the world's largest tree by wood volume. There have been recent discoveries of individuals equal in girth to the largest living Giant Sequoias. The species name *sempervirens* means "evergreen" or "everlasting," and is also known for its longevity.

Mature Coast Redwoods are distinguishable from other coniferous trees by their tall, columnar trunks, horizontal drooping branches, and narrowly conical crowns. Trees grow in giant stands that cut through fog and sunlight, appearing like ethereal cathedrals. The soft, fibrous, rusty-red to gray bark is up to 1 ft. (0.3 m) thick, with a furrowed texture that is often painted lightly with moss and lichens. This magnificent fire- and insect-resistant bark, along with foliage that doesn't start till high above the ground (up to 100 ft./30 m in mature trees), contributes to the trees' longevity. These trees keep growing until near death, with their rate of wood production increasing as the tree ages (because they have more leaves). Another notable characteristic of the Coast Redwood is its shallow root system. Even the roots of the most ancient individuals extend about only 6 ft. (1.8 m) deep, while their wide-spreading lateral roots interconnect with those of other redwoods for mutual support.

Coast Redwoods can reproduce from seeds (assisted by, but not dependent upon, fire) or from dormant epicormic sprouts that form live burl growths at a tree's base. The cones are small, ¾–1½ in. (1.9–3.8 m) long. The adventitious sprouts, which sometimes grow on living trees, are stimulated to grow into full trees when the base of the trunk is cut, burned, or naturally fallen, regenerating a "fairy ring" of trees sprouting along the circumference of the dead parent tree. Because these young trees are technically genetic clones, a circle of trees can continue to expand out for thousands of years, and be considered part of the same tree.

These water loving trees survive the dry season by collecting about half of their water needs from fog, letting it fall as "fog drip" to the forest floor, adding 12 inches (300 mm) or more of precipitation during the dry season. Like all trees, they also transpire moisture back into the air, helping to create water vapor that in turn creates clouds. In this way redwoods play a large part in regulating the climate of their coastal areas, and keeping the temperature cool. One can imagine a cooler climate in northern California over a hundred years ago before the majority of the old growth was logged. Redwood forests also store massive amounts of carbon.

Distribution: Native to a narrow range of about 450 miles (724 km) from the southwestern tip of Oregon to California's Monterey Bay, and restricted within 35 miles (56 km) of the coast. Elevation: sea level to 3,000 ft. (920 m).

Ecosystem: Found in moist coastal and riparian areas along the fog belt (but not directly on the seashore) within flood zones, on steep slopes, and at the top of ridges. Coast Redwoods are found in pure groves with well-developed understory shrubs and plants, as well as being the dominant species in mixed coniferous forests, which, depending on region, can include Western Hemlock, Douglas Fir, Big Leaf Maple, Vine Maple, Sugar Pine, California Bay Laurel, Pacific Madrone, Tanoak, Oregon White Oak, Coast Live Oak, California Black Oak, Pacific Yew, California Nutmeg, Red Alder, Western Dogwood, Torrey Pine, Grand Fir, Western Redcedar, Port Orford Cedar, hazelnut and manzanita.

Maximum Age: Over 2,500 years.

Maximum Height and Girth: The tallest living Coast Redwood, and tallest tree in the world is currently the Hyperion tree, at 379.1 ft. (115.72 m) in height. New discoveries of trees claim a circumference of 91 ft (27.7 m), matching the girth of the current largest Giant Sequoias.

Animal Community: Coast Redwood forests provide habitat for a large variety of creatures, including black bears, mountain lions, bobcats, gray foxes, eagles, Pacific tree frogs, and the rare Pacific giant salamander. Endangered fish such as salmon depend on these ecosystems for their survival in the wild. The Northern spotted owl and marbled murrelet are federally protected threatened and California-endangered bird species whose existence depends upon old-growth habitat. Not only live trees, but also downed trees and standing snags are vital for habitat and regeneration of the forest ecosystem. Individual old-growth redwood canopies are entire ecosystems in themselves, as discovered by botanists and research scientists Stephen Sillett and Marie Antoine. They ascended trees in the ancient Grove of Titans in Northern California and found a unique ecosystem hundreds of feet off the ground that included huckleberry thickets, extensive epiphytic plants including mosses and lichens, thick soil, and animals such as voles and salamanders who had never touched the ground below.

Traditional Uses:

Medicine: The young foliage was warmed and then applied as a poultice as a treatment for earaches, and the gummy sap was taken from the bottom of the tree, soaked in water, and drunk as a health tonic (Kashaya Pomo, Pomo). A tea from the needles can be used as an expectorant, and as disinfectant for urinary tract infections. A poultice of the bark was used as a treatment for venereal diseases. A tincture of the inner bark was taken to purify the blood.

Fiber: Sprouts from burls were used for basketry material (Costanoan).

Redwood Needles and Cones (2016)

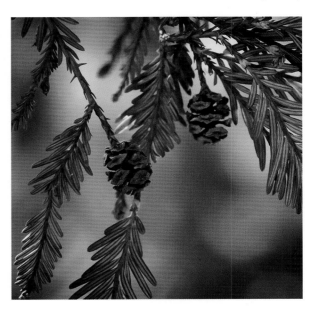

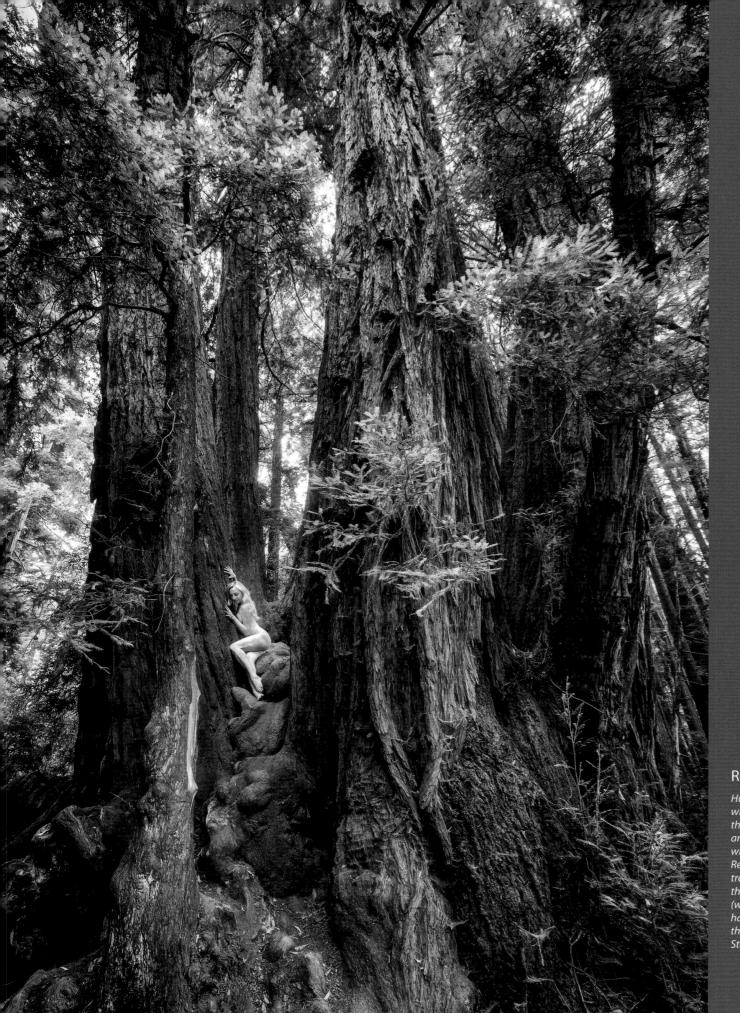

Redwood Castle II (2015)

Hundreds of years ago, a different tree grew here, and when it died or was cut down, the epicormic sprouts that grew out of the base of the tree, formed a fairy ring around the stump. Eventually, the fairy ring grew into what appears to be one tree with multiple trunks. Since Redwoods can resprout indefinitely, it is possible that this tree kept regenerating itself for thousands of years. With the help of some local tree-loving friends, the Regen family (who lovingly know these woods like the back of their hands), we spent a wonderful, adventurous day playing in the trees. This castle lives right on the trail in Henry Cowell State Park, Santa Cruz, California, USA.

Coast Redwood

Tools and Objects: Pieces of bark were used to play dolls.

Shelter: Slabs of the thick and insulating redwood bark were used to make simple structures (Miwok, Maidu, Kashya Pomo, Ohlone, Yurok, Hupa, Tolowa, Wiyot, Whilkut, Nongatl, Mattole, Sinkyone).

Transportation: Canoes were carved out of downed redwood (Yurok, Mendocino Indians).

Modern Uses: Coast Redwood is one of the most valuable timber species harvested commercially in California and southern Oregon, and is exported worldwide. It is prized for its beautifully rich-colored, lightweight, straight, rot-resistant wood. It has been extremely popular for lumber for houses, decking, and shingles. Old-growth trees are prized even more for their tight-grained durability. Redwood burls are often illegally cut from live trees (which damages the trees), and also salvaged from logged stumps, to make tabletops, veneers, turned bowls, and tourist kitsch. Bark is used for garden mulch.

Threats and Conservation: Although not classified threatened or endangered by the U.S. Endangered Species List, they are considered endangered according to the IUCN Redlist. Threats include climate change, air pollution and logging. The clear-cutting and logging of Coast Redwood forests endangers entire watersheds, increasing erosion and landslides and releasing silt into fragile streams and rivers, many of which are habitat for Coho salmon, Chinook salmon, and Steelhead trout. These magnificent fish, once widely abundant and a major food source for California indigenous tribes, are now federally threatened species due to over-fishing and the devastating effects of logging. Only 5 percent of the original redwood forests remain, and only 2 percent of the original old growth is left, of which only half is protected. While some ancient groves are protected in national and state parks, many acres are awaiting selective logging or clear-cutting on private land owned by individuals or by corporations such as Weyerhouser, Green Diamond Resource Company, Sierra Pacific Industries, and Humboldt Redwood Company. Even trees on our public lands are not safe. The U.S. Forest Service is one of the largest clear-cutters and loggers in the U.S., with the annihilation to our prized ecosystems paid for by our tax dollars. Activists have been working nonstop since the 1960s within forests and for organizations to protect redwood forests. Many have devoted their lives, and some given their lives, to defending them. Their work—*our* work—is nowhere near finished.

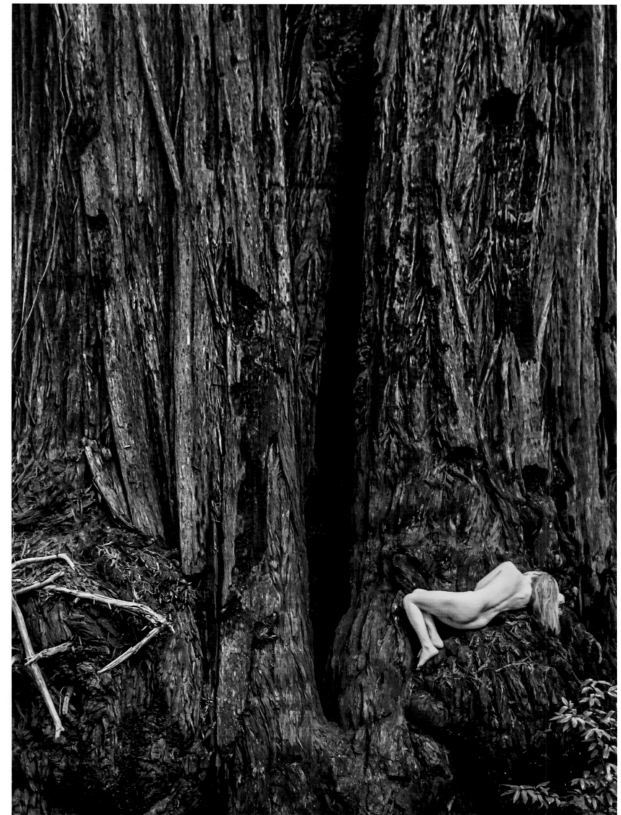

Grandmother Redwood Fae (2012) ▸

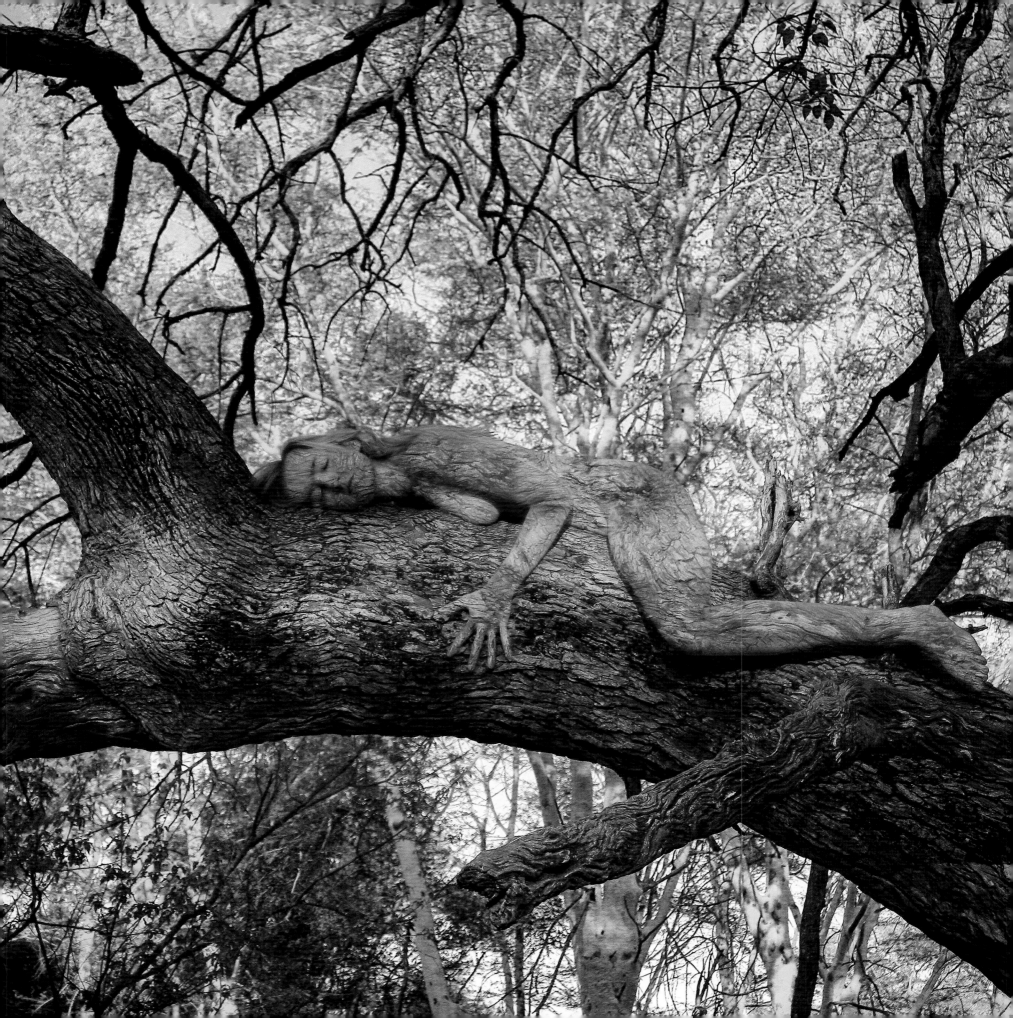

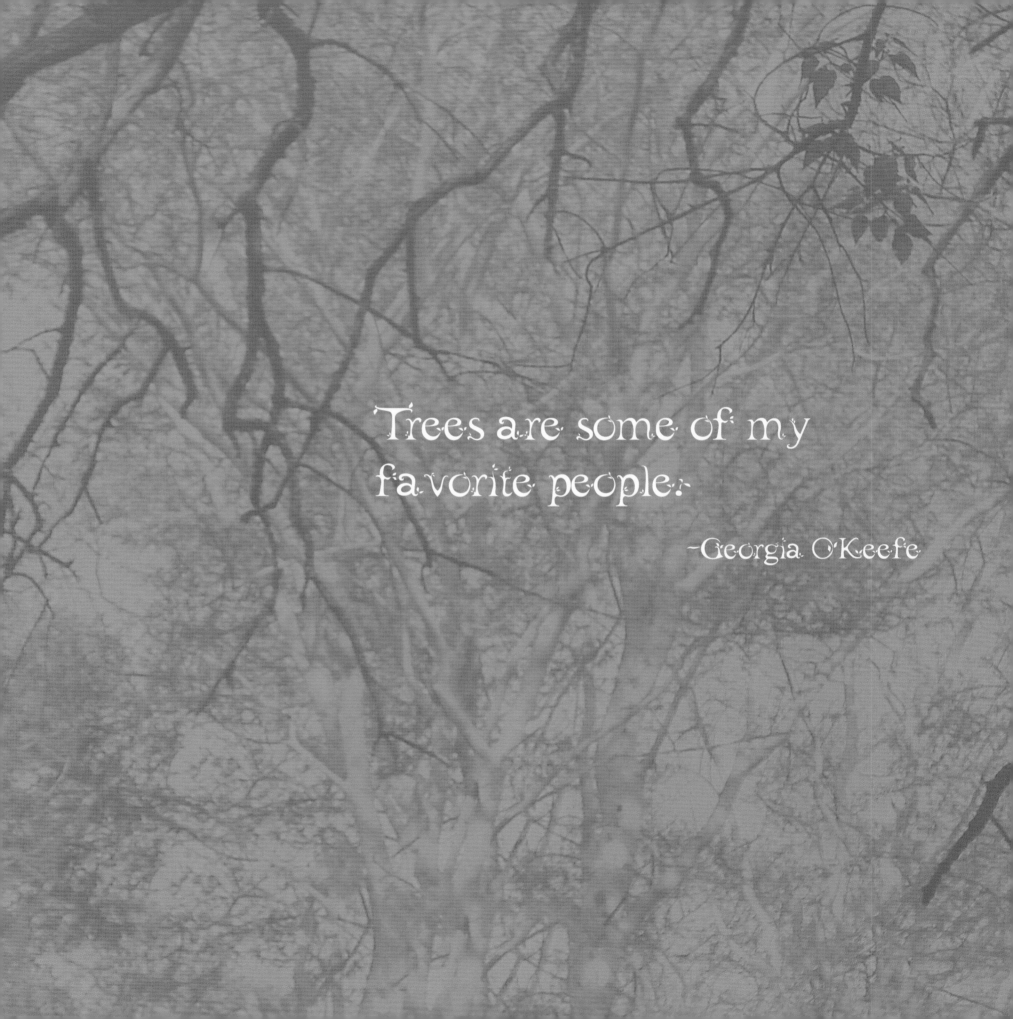

Trees are some of my
favorite people.

-Georgia O'Keefe

Chapter 2.
Tree Affinity

We enjoy being lost—or perhaps found—in wildness and the grandness of Nature. Genetically, we remember Nature as home, and we yearn for the company of our wild neighbors. We have an affinity for other living things, and especially for trees. Aesthetics are often an initial part of this attraction, but the pull is about much more than what suits the eye. We feel affinity viscerally—in our heart, psyche, and body.

Each individual tree, like each individual person, radiates its own presence, though we may not quite be able to define what that is. There are trees that command attention by their location, size, form, texture, color, fragrance, or size. Some are undeniably impressive and irresistible to any passerby. A 2,000-year-old Giant Sequoia (*Sequoiadendron giganteum*), resident only to the Sierra Nevada in California, is one such awe-inspiring, mammoth tree. As the record-breaking tree in terms of volume (as well as being among the top species in terms of girth, height, and age[1]), the Sequoia's grandness has the power to stop us in our tracks, commanding respect and inspiring instant humility. We cannot help but ponder the nature of our own lives, and of life itself. As if transported back to the Jurassic Age of the dinosaurs, we feel we could be crushed at any moment, inconsequentially, if the tree's trunk suddenly moved. And yet these trees also radiate the silent beauty and peaceful presence of a divine being, so powerful that at first contact we are lost for a few seconds, until we finally regain our breath and normal state of consciousness…but do we want to?

What is the instinct that makes a person want to crawl inside the hollow of a tree giant, ascend onto a sturdy low branch like mounting a horse, or climb straight up a trunk to the top like a squirrel? What impels us to investigate a tree's intricate bark, build a tree house, or simply put our arms around a tree and hug it? To hide, to take refuge, to embrace with abandon, to lean one's back against a sturdy trunk, to physically ascend the mystical ladder of the axis

Leadwood Leopardess (2008)

This Leadwood tree (Combretum imberbe), with glorious chartreuse Fever Trees behind it, has scratchy bark that breaks up into blocks looking like a crocodile skin. It is apparently a favorite hangout tree for leopards. Although I could not see them, I was very aware of the smell and energy of the wild creatures intently watching me. This tree is near Pafuri in Kruger National Park, South Africa.

mundi from the earthly plane to the upper realms, branch by branch, or energetically ground into the rootedness of the world beneath the soil by sitting at a tree's base—these are natural responses for both young and old when encountering a great tree. What exactly is this "natural" instinct?

Perhaps it arises from the intimate relationships humans have had with trees for hundreds of thousands of years. In the days when people knew their neighbors, trees were part of the community, with their own stories to tell and gifts to give. Trees defined a place and, like other organisms, depended on their ecological community for survival. Many modern folk have inherited an instinct for intimately interacting with Nature—numbed-down from electronic stimuli compared to our pre-industrial ancestors, but still an innate, practical, and even spiritual drive to connect with the "other." I have heard that rock climbers and mountaineers experience this type of resonance, yearning to scale complex geologic formations. Surfers feel a similar pull to ride big waves, and kayakers long to be carried by untamed rivers. You can't keep them away; they must connect. They long for freedom and, at the same time, for a kind of merging. This primal, kinetic attraction to Nature is similar to the experience of love: being drawn so intensely to another, engaging passionately in tactile energetic movement or resting peacefully in a gentle stillness, abandoning oneself to something grander. Nature-lovers express their yearning for a particular place they have experienced or want to visit, or simply their idyllic love for "the mountains," "the ocean," "the desert," or "the forest." It is a genuine attraction, felt in the core of one's being. This natural instinct to be closer to Nature is an ineffable experience, personal and often sacred, primal and genetically universal.

The act of climbing, whether trees or mountains, has been identified as a "core human interaction pattern," or "nature language," in Peter Kahn and Patricia Hasbach's book *Ecopsychology: Science, Totems and the Technologic Species*. The authors affirm that human patterns of interacting with elements in Nature—being near or on moving water, gathering around a fire, sleeping under the stars, tracking and hunting animals—are as old as the human species themselves and are ingrained in our psyches and bodies[2]. Indeed, these ingrained relationships cannot be buried or evolved away from no matter how technologically dependent we become. With the exception of some arctic and desert regions, *Homo sapiens* have lived in direct relationship with trees for over two hundred thousand years. So

it's fair to say that the tree-human connection is genetically and psychologically ingrained in us.

While most industrialized people have lost their daily familiarity with trees, those of us who relate directly with trees as arborists, botanists, tree workers, restoration ecologists, foresters, orchard farmers, or land stewards engage in extremely intimate work. We may assist trees as they are born from seeds, plant seedlings, and help propagate them by transplanting or grafting. In order to ensure their survival, we cut, prune, remove limbs and branches, nourish them with water and nutrients, and treat for pests and disease. Sometimes we fail, intentionally or unintentionally, to keep an individual alive. We are responsible for giving and taking the life of single trees—or, sometimes, entire groves or forest ecosystems. Sometimes we perform surgery with artistry and consciousness, sometimes carelessly, with butchery. We act as midwives, paramedics, surgeons, parents, friends, admirers, advocates, lovers, and sometimes as killers. It doesn't get much more intimate than that.

Intimacy usually refers to an authentic and reciprocal emotional and energetic bond with another human being, often including affectionate physical contact. This bond could be experienced with a lover, a partner, a child, a dear friend, or even a new acquaintance with whom we've had a deep encounter. However, we also experience intimacy when we engage in a close, caring, respectful relationship with a companion animal, our garden, a beloved deity or spiritual system, our home, or a place we visit often. The longer we've known a person, being, or place, the more intimacy we tend to feel, based on history and experience. But while the amount of time and intensity of close interaction can be important factors, they aren't the only ways for intimacy to grow. Intimacy may or may not include love. Family members, soldiers, and even strangers who survive traumatic or life-threatening situations together can develop intimacy based on the intensity of their shared experience. People who undergo an intense ritual or a once-in-a-lifetime rite of passage together may quickly develop intimacy.

However it develops, an intimate relationship often involves a deep sense of comfort, based on trust, authenticity, empathy, compassion, vulnerability, transparency, security, and belonging. Intimacy resides not only in the heart and mind, but is embedded in a whole-body somatic experience. The communication exchanged and senses engaged include not only the verbal, physical, and emotional, but the spiritual or psychic-intuitive as well.

There are ineffable spiritual encounters with Nature that embrace a kind of elation and harmony. Mystic mountain man John Muir recounted this eloquently:

"My fire was in all its glory about midnight, and, having made a bark shed to shelter me from the rain and partially dry my clothing, I had nothing to do but look and listen and join the trees in their hymns and prayers."

The reverence, magnetism, and deeply experienced intimacy of Muir's lifelong amity with Nature, known as well by mystics and animistic cultures throughout the ages, has almost been lost in our modern world. To hear the song of the wild, to listen closely enough to hear the rapture of the life force within the silence or the storm, to know its language as your own secret, this is a great ecstasy.

E.O. Wilson popularized Erich Fromm's term biophilia, the "love of living things," which refers to our innate affinity with Nature.[3] He proposed that biophilia is an instinctual drive—that the deep affiliations (or *philias*) humans have with the rest of life are rooted in our biology. Human beings, he suggests, subconsciously seek out connections with other living systems because those relationships help us, and everything else, survive. Biophilia is commonly seen in people's love and care for companion animals, their fascination with wild animals, and their care for houseplants, landscaping, or gardening. Similarly, it explains why ordinary people sometimes risk their lives to save others—human or animal, and even trees and natural places. He proposes that the emotions of empathy, compassion, and love are actually biological and genetic.

Biophobias or ecophobias, in contrast, are the fears that people have of certain organisms, such as poisonous spiders, snakes, and plants, or even bodies of water or certain types of terrain, such as cliffs. These instincts were most likely passed on genetically for survival and evolution. While some of these instincts are still useful, many such fears prohibit people from valuing specific organisms or Nature as a whole. In industrialized societies, children are unconsciously and consciously taught to fear most organisms that seem to threaten their territory or their bodies, and therefore that they have the right to kill anything foreign that gets in their way.

For pre-industrial societies, as well as the few indigenous cultures that have survived, there was and is no distinction between Nature and human. The place called home was considered a community of beings who live together in the same area (even though some members were considered unwanted). A tree was a neighbor you knew by name—not as "John" or "Jane" tree, but as a being that has a presence and possesses familiar characteristics that are known in relationship to others.

As each generation becomes more and more urbanized, there are fewer and fewer opportunities to cultivate a bond with Nature as our ancestors did. Ten years after it was coined, the term "nature-deficit disorder" has nearly become a household word. It and similar diagnoses are becoming common in the periodicals of mental health. In our modern global culture of false connectedness (e.g., social networking via the Internet), we lack our own ecological identity and sense of place—a feeling of intimately knowing the nature and ecology of the place where we live and belong. This is largely due to Western civilization's increasing dependence on and addiction to technology. Indeed, much of the childhood and adult state known as 'wonder' has been sacrificed to the Google® gods. One could say that these inanimate monsters of artificial intelligence have sucked the life force out of millions of innocent victims. Even for budding young naturalists, it is all too easy to quickly snuff out the curiosity of a birdcall with a smartphone App, and move on to the next target. Done. The world becomes a game to conquer instead of a relationship.

As a result, we often find ourselves imprisoned in mental self-absorption or escaping to the no-place fantasy world of cyberspace and the age of misinformation housed in electronic gadgetry. Humans are bonding with machinery and avatars instead of with living organisms. Research indicates that our increasing exposure to computers and the Internet is changing how our brains do and don't function.[4] Studies find that teens are lacking the development of interpersonal social skills, including empathy, due to dependence on texting as the dominant form of communication.[5]

This generalized biophobia and ecophobia is, in reality, a culturally prescribed alienation from anything that is not human-made or within human control. To an extent, despite our species' clever innovations and inventions, we've forgotten that we are not actually in control of the rest of the universe, no matter how many machines we produce. How many of us spend hours of our day "being in control" of our electronic devices—pushing buttons to make them do what we want them to do—and then feel angry and depressed when the device doesn't do what we've *told* it to do? But in the technological age, we have become numb, impatient, and uncomfortable with the subtler levels of experience—seeing the true darkness of night, or

being at ease in it, for example. The more time we spend inside buildings staring at screens, the less we accustom our senses to interacting with the rest of the living world as our ancestors did.

We've become eco-illiterate, lacking comprehension of how the rich biodiversity of life, as well as the climate, the fertility of the soil, and the interdependent life-support systems—the water, carbon and nitrogen cycles—function together to sustain all of life. How many of us can point in the direction of north wherever we are or indicate the pattern of the sun or moon across the sky as it changes in each season? How many of us could locate an edible wild plant in our neighborhood or find our way around without a GPS? Being ecoliterate also means knowing that everything we humans have created has come from the Earth and that, in actuality, *all* of our resources (read: other beings) are finite, no matter how 'renewable' they supposedly are. Our mass ignorance says that we have a right to all the smartphones we want, without having to care for any humans or other organisms being sacrificed to mine the required rare earth elements and mineral ores, or the toxic waste being created by the manufacturing process. In denial, we also believe that we will always have electronic technology available to us, as if there were an infinite amount of raw materials to extract, petrochemicals and water to manufacture them, and fossil fuels to transport them globally into our hands.

This estrangement from the more-than-human world stems from generations of an increasingly warped value system and a lack of direct personal experience with wild Nature. Domination—the kind of domination that has made civilizations possible—arises from fear and alienation. These phobias and disconnects propel us through the taming, extraction, and extermination of other life forms as we construct our built environment, resulting in global ecocide. Sadly, the disregard for this loss is also due to what Peter Kahn termed "environmental generational amnesia," which occurs when members of each generation construct their conception of what is environmentally normal based on the "natural" world they encountered in childhood. So, with each ensuing generation, the amount of environmental degradation increases, but each generation tends to take that degraded condition as the normal experience.[6] It is the same as being raised from birth in a dysfunctional or abusive household (society)—one may never know what healthy really is. Indeed, one could say that biophobia and ecophobia are becoming the rampant psychosis—the madness of humanity.

So how do we retrieve our affinity for the more-than-human world? How do we unplug and wake up from the cultural techno-trance? What can we do to reset and restore our minds and bodies? Fortunately, there is a psychological science to becoming sane. Here is one strategy: in the 1970s, psychologists Stephen and Rachel Kaplan developed what they called Attention Restoration Theory [7], which is still the foundation for many of the studies demonstrating that simply viewing nature out a window—or even just seeing pictures of Nature—is healing to the cognitive mind. (Of course, spending time *in* Nature is the most healing.) The premise behind the theory is that demands placed on the brain by processing information (such as repetitive computer use) leads to what its called attentional fatigue, and often stress. Too many things in modern society demand our attention, with increased pressure for multi-tasking. Overstimulation, including media distractions of meaningless 'infotoxins' [8] also breeds fatigue in an immediate way by taxing the brain and reducing the quality and quantity of sleep. [9] In fact, attentional fatigue affects many aspects of our functioning; it decreases overall effectiveness, our ability to solve problems, inhibition, adaptability, tolerance, patience, reflection, and our ability to see the big picture. It also alters perception in a way that increases distractibility, irritability, anxiety, chance of error (sometimes life-threatening), and impulsivity [10]. Sadly, these states of mind are becoming all too common, to the point that we forget how detrimental they are to ourselves, our families, and the planet. This also means we have less brainpower and time available for everything else that actually depends on our attention, including the increasing effects of the ecological crisis. Instead, we see our mental and physical fatigue as necessary evils in maintaining our fast-paced societal and cultural roles. Besides, we have caffeine!

Attention Restoration Theory holds that psychological recovery is found in time spent in Nature. [11] The four components of a restorative environment are [12]:

1. *Being away*: freedom from routine mental activities that require directed attention. Idyllic places such as bodies of water, mountains, or forests can provide a "refreshing change in scenery."

2. *Fascination*: "soft fascinations," such as sunsets, moving clouds, tree branches blowing in the breeze, ocean waves, a starry sky, birds, or falling snow. Attending to these patterns is considered interesting, effortless, and aesthetically pleasurable.

3. *Extent*: an environment that is rich enough to engage the mind—to constitute a whole other world, providing enough experience (and diversity) to command attention.

4. *Compatibility*: roles compatible and suitable to people's intentions in specific environments. These include ingrained human patterns such as the predator role of hunting and fishing, the locomotion role of hiking and boating, the "domestication of the wild" role in

◂ Brush Box Repose (2015)

*Brush Box (*Lophostemon confertus*), Lamington National Park, New South Wales, Australia. See p. 33.*

gardening and animal husbandry, the observer role in birdwatching, or the survivalist roles in fire-building and constructing shelter. In other words, a person does not go shopping in the middle of a wilderness area (although some national parks provide that activity—however, a shopper is, at core, a gatherer, and gathering is a core skill in survival.)

Attention Restoration Theory affirms that experiencing our connection with Nature is a human *need*. It is not just a way to improve one's mood; it is a vital ingredient in healthy human functioning.[13] This may be because we humans have spent 99% of our evolutionary history outside of buildings, in constant interaction with the weather, the ground, water, plants, and moving creatures. Sadly, most urbanized humans spend 99% of their time indoors, disconnected from all of these.

The inherent feature of living systems is the interplay between organisms and their environment. The word "interplay" is important here; it means interaction, relationship, and exchange. Life is not only attracted to (*philia*) other life; life is dependent upon engaging with other life, even if an organism is a mile underwater or deep within the walls of our guts. Many of us forget that we are not autonomous creatures. We cannot function for long within a glass bubble on Mars, in solitary confinement, or underground without exposure to daylight. And who would want to? Yet some days that we spend indoors are not so far from that kind of isolation. Like infants in orphanages who die from lack of interaction, I suspect that future humans would easily perish without intimacy with wild Nature. Our biology depends on it; our souls hunger for it.

We are so interconnected with the web of life that we feel *the presence of* other living beings, whether we can "see" them or not. They are probably aware of our presence as well. We can also feel the presence of topography, such as mountains, flat plains, cliffs, or bodies of water and our place within or near them. Being present within Nature does much more than just reset our attention to what we view out of the safety goggles of our regulated awareness. The perception and interaction within that presence is the crucible of cultivating felt intimacy.

It's important to add that not all of Nature is good for us all of the time. There are some places, organisms, situations—and even a few species of trees—that pose a potential hazard to our safety (as opposed to healing), and we should engage with them with due respect, caution, or, in some cases, total avoidance.

More than thirty years after the Kaplans' research, growing numbers of studies show that not only are experiences in Nature healing,[14, 15] but greater biodiversity produces greater psychological benefits.[16] Many forest ecosystems (or even individual trees) are especially high in biodiversity—rich with visual, olfactory, auditory, and tactile opportunities for both sensory stimulation and restful pleasure.

Engaging with trees is one of our most powerful and accessible ways of coming home to our extended family of the more-than-human world. Spending time with forests, wooded areas, or individual trees can help us restore our attention, calm our nervous systems, and shift our phobias to "philias". Whether suburban street trees or a whole wild forest, almost all of us can find trees to connect with (with the exception of some ecosystems).[17] When you encounter a tree, take the time to look deeply at the intricacy of its structure, patterns, colors, and textures, and sense the life it supports. Feel its presence and its life-force. Shift your consciousness to the trees' perspective and watch as a story unfolds—as multiple stories unfold. Look all around you and notice how that tree is connected to virtually everything else around it, above it, below it, and beyond it. Listen. Wonder. What conversations are happening here? What symphonies are being sung? *Be part of that story.*

These affinities and philias can restore our health, our sanity, and our sense of belonging within the wildness of the world that exists amidst our maddening daily routine. Follow your instincts, your curiosities, and your fascinations. Wake up from the trance state; choose to turn the power switches to Off. Step out of the house, the office, the car, and even your shoes. Be open to new relationships. Be open to falling in love.

[1] The Giant Sequoia holds the record for being the largest single tree on Earth in terms of volume, but the current record holders for individuals in terms of height is the Coast Redwood, in width is the Montezuma Cypress, and in age is the Bristlecone Pine.

[2] "A Nature Language" from Ecopsychology: Kahn Jr., Peter and Patricia H. Hasbach, Ed. *Ecopsychology: Science, Totems and the Technologic Species* (Cambridge: The MIT Press, 2012).

[3] E.O. Wilson, *Biophilia* (Boston, Harvard University Press, Third Ed., 1984) and Ed. Stephen R. Kellert and E.O. Wilson, *Biophilia Hypothesis* (Washington D.C.: Island Press, 1995).

[4] Nicolas Carr, *The Shallows: What the Internet Is Doing to Our Brains* (New York: W.W. Norton & Co., 2010).

[5] "Heavy Internet users score low on emotional intelligence, which is a measure of how one uses emotions in solving problems. Emotional intelligence is predicated on the ability to use verbal and nonverbal cues to monitor the emotional state of others. ... (can) remove the conscious perception that others may be in need." Selhub, Eva M.; Logan, Alan C. (2013-06-25). *Your Brain on Nature: The Science of Nature's Influence on Your Health, Happiness and Vitality* (p. 37). HarperCollins Canada. Kindle Edition.

[6] Kahn, Peter H., Jr.; Hasbach, Patricia H. (2012-08-13). *Ecopsychology: Science, Totems and the Technological Species* (p. 319). MIT Press. Kindle Edition.

[7] Stephen Kaplan, "The Restorative Benefits of Nature: Toward an Integrative Framework." *Journal of Environmental Psychology* (1995).

[8] "All of the modern-day infotoxins (the information of dubious quality) lure us in under the guise of interest, but because they are not important and do not hold fascination, they tend to become an energy sink." Selhub, Eva M.; Logan, Alan C. (2013-06-25). *Your Brain on Nature: The Science of Nature's Influence on Your Health, Happiness and Vitality* (p. 58). HarperCollins Canada. Kindle Edition.

[9] Selhub, Eva M., Logan, Alan C. (2013-06-25). *Your Brain on Nature: The Science of Nature's Influence on Your Health, Happiness and Vitality* (p. 39). HarperCollins Canada. Kindle Edition.

[10] Stephen Kaplan, "The Restorative Benefits of Nature: Toward and Integrative Framework." *The Journal of Environmental Psychology*, 1995, p. 171–2.

[11] Stephen Kaplan, "The Restorative Environment: Nature and Human Experience" in *The Role of Horticulture in Human Well-Being and Social Development*. Portland, OR: Timber Press (1992) Diane Relf, Ed.

[12] Stephen Kaplan, "The Restorative Benefits of Nature: Toward and Integrative Framework." *The Journal of Environmental Psychology*, 1995, p. 172–4.

[13] Stephen Kaplan, "The Restorative Environment: Nature and Human Experience" in *The Role of Horticulture in Human Well-Being and Social Development*. Portland, OR: Timber Press (1992) Diane Relf, Ed.

[14] Tsunetsugu, Yuko, Bum-Jin Park and Yoshimfumi Miyazaki, "Trends in research related to 'Shinrin-yoku' (taking in the forest atmosphere or forest bathing) in Japan". *Environmental Health Preventative Medicine*, 2010, 15:27–37.

[15] Townsend M. and Weerasuriya R., *Beyond Blue to Green: The benefits of contact with nature for mental health and well-being.* Beyond Blue Limited: Melbourne, Australia (2010).

[16] Fuller, Richard A, Katherine N. Irvine, Patrick Devine-Wright, Philip H. Warren, and Kevin J Gaston, "Psychological benefits of greenspace increase with biodiversity." *Biological Letters*, August 22, 2007, vol. 3 no. 4, p. 390–394.

[17] Even some lucky prisoners in the Department of Correction in Washington State system are part a horticultural therapy program. *http://sustainabilityinprisons.org/* (accessed October 14, 2014).

Sequoia Meditation III (2001)

This pair of unidentified, intertwined trees is on a Giant Sequoia ▶
trail in Calaveras Big Trees State Park, California, USA.

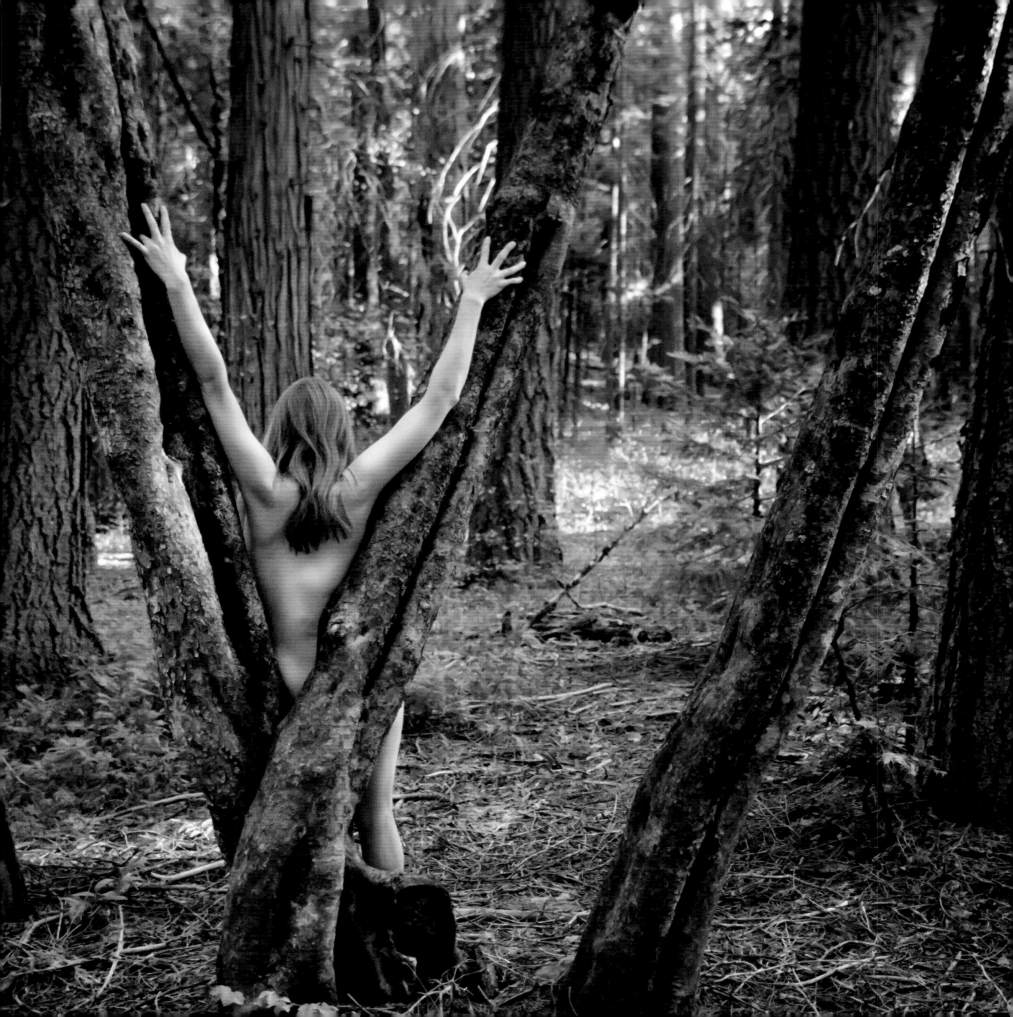

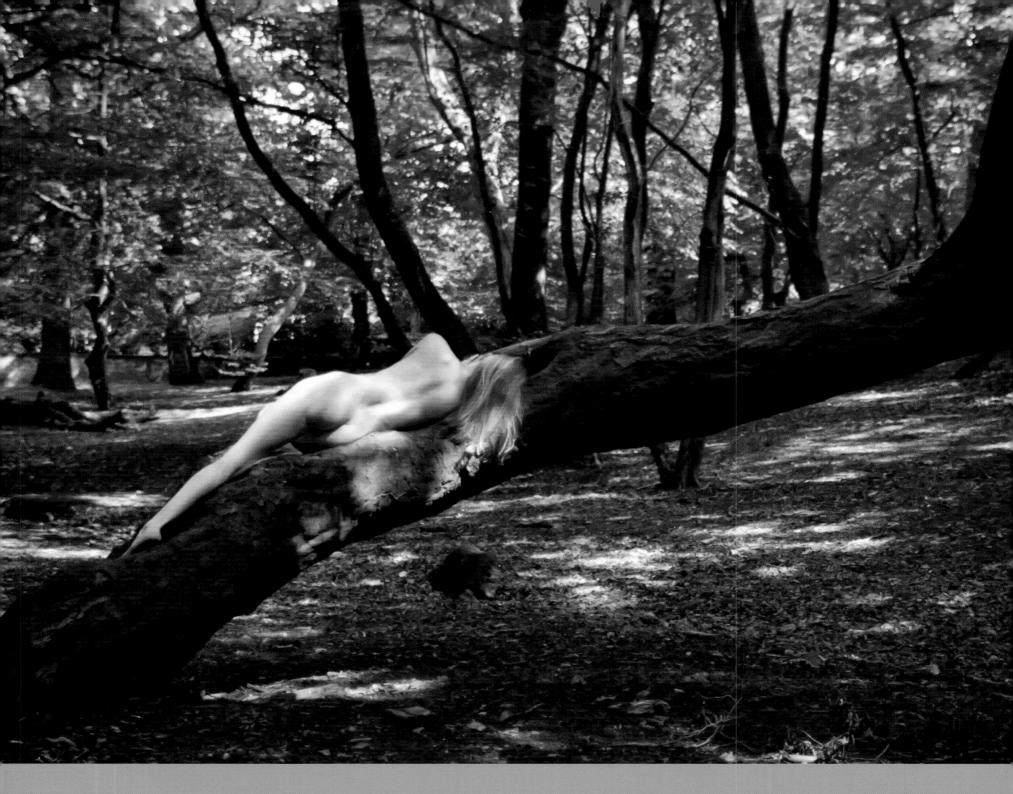

Back on London Log (2002)

As is the case of many trees I am attracted to, I knew exactly where I wanted to be for this shot. This wild-growing, reclining European Beech was so inviting, I could not resist it, even though it was the middle of the day in a city park. I had to be stealth. This photo was taken back in the day of my film camera, before I had a remote control. I had a mere ten second delay to get into place and took just one shot. Although it appears as a lengthy lounge, it was only a mere moment's rest in the quiescent Kings Park, London, England.

European Beech

Fagus sylvatica

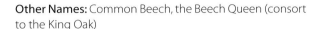

Other Names: Common Beech, the Beech Queen (consort to the King Oak)

Distinctive Characteristics: This common deciduous tree has a distinctive smooth gray bark that easily scars to a darker color. As it increases in age, the European Beech can form a gnarled, flared base and a shallow root system, often with large, exposed, snake-like roots spreading out in all directions, giving the tree much character. With great age, trunks can hollow and split at the base into what appears to be distinct trees.

Distribution: Native and common to Southern England and most of mainland Europe into Turkey. Elevation: sea level to 5,900 ft. (1,800 m).

Ecosystem: Prefers moist, shady woodlands. Usually found in pure stands, but also with English Oak, Sessile Oak, Sweet Chestnut, Large-Leaved Lime, Common Alder, Sycamore, Scots Pine, Norway Spruce and White Fir. Shades out wildflowers and understory plants.

Maximum Age: Approximately 300 years.

Maximum Height and Girth: 160 ft. (49 m) in height; 31 ft. (9.4 m) in circumference.

Animal Community: Beechnuts are an important food source for a variety of rodents and birds.

Traditional Uses:
Medicine: Folk medicine includes: use as an antacid, an antiseptic, an expectorant, as well as to reduce fever, and as a treatment for toothaches and certain skin diseases.

Food: After leaching out the tannins and roasting the nuts, their flavor improves and they can be ground to make a kind of flour. Bitter beechnuts are slightly toxic if eaten in large quantities due to their tannins and low concentrations of trimethylamine. The oil was pressed for cooking.

Tools and Objects: In European folklore and folk medicine, the beech has quite a history. The Anglo-Saxon word for *book* has the same origin as the word beech: "Bok" or "Buche"; it is said that the bark was cut into thin slices to write upon to become the very first books in Europe. In 19th-century England the nuts were pressed for oil in lamps.

Art and Ceremony: Beech branches were believed to make good divining and dowsing rods and, according to Druidic lore, pieces of beech were also worn as a good luck charm.

Modern Uses: Today, the wood is used as a building material for furniture, flooring, and handles for implements. In France the nuts are still sometimes roasted to make a coffee substitute.

Threats and Conservation: Not threatened.

Avebury Beech Green Woman (2009)

This group of European Beech trees, with their exposed, cascading, entangled, snake-like roots, tower over the outer henge at the megalithic Avebury Stone Circle like a magical band of druids. This famous Neolithic monument, which is the largest stone circle in Europe, is a UNESCO World Heritage Site, a sacred Pagan site, and a geomantic power spot. Indeed, the entire area of Wiltshire is full of power spots, including Stonehenge, and annual crop circles appear there. These mystical trees themselves seem like characters out of a Tolkien novel. I had to be quick in shooting, as there was a steady stream of unsuspecting tourists on the path. The tree is on the outer circle of the henge, where Green St. meets the corner of the southeast quadrant in the village of Avebury, southern England.

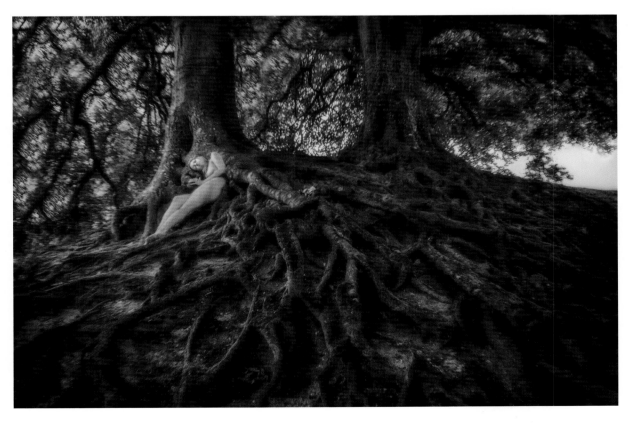

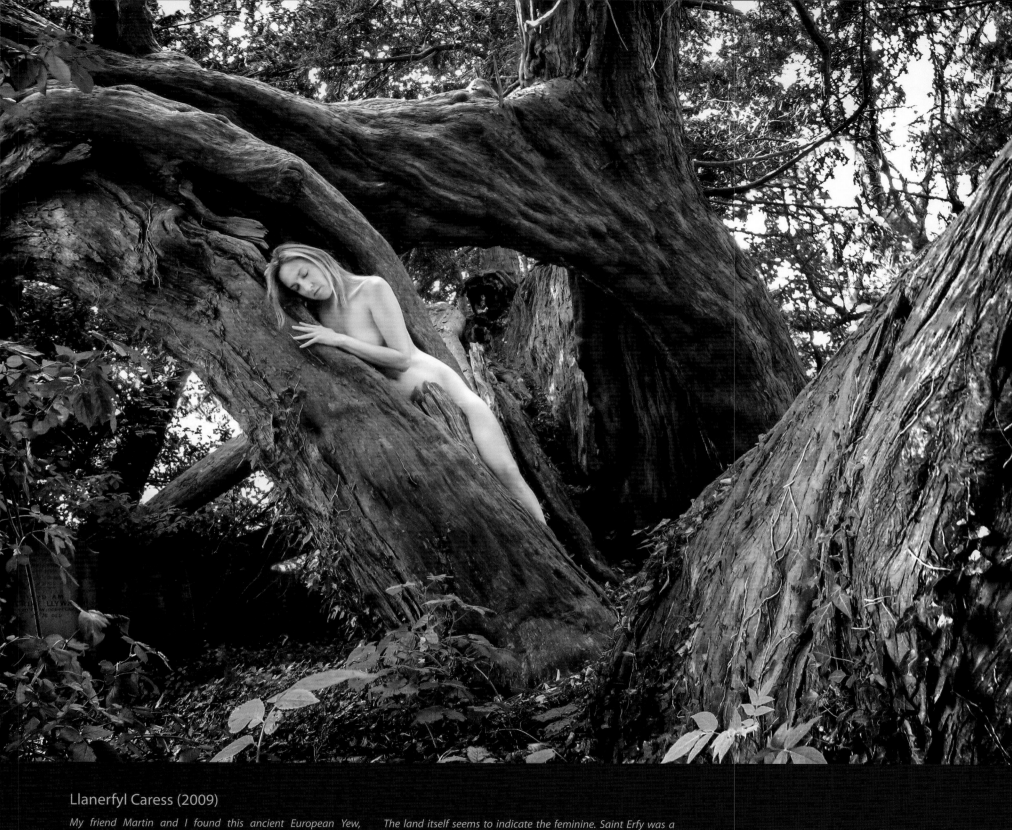

Llanerfyl Caress (2009)

My friend Martin and I found this ancient European Yew, appearing neglected and forgotten in the tiny, quiet churchyard of St. Erfyl, in Llanerfyl, Powyrs, Wales. This ancient tree has split into four fragments appearing as separate trees. Based on its form, it could be 3000 years old. The largest fragment, 18 ft. (2 m) in circumference, is known as the Patriarch Yew. However, only one remnant of the trunk is male, while the other three are female, so perhaps it should be renamed the Matriarch Yew? (Changing genders is not uncommon in very old Yew trees.)

The land itself seems to indicate the feminine. Saint Erfy was a virgin who founded an earlier church in the 3rd century at a holy well (also feminine, and reputed to be under the tree). This was one of my favorite trees to merge with because of its twisted, rippling form and perfect casket-like hollow to be reborn in.

European Yew

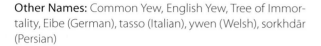

Taxus baccata

Other Names: Common Yew, English Yew, Tree of Immortality, Eibe (German), tasso (Italian), ywen (Welsh), sorkhdār (Persian)

Distinctive Characteristics: Yews are known for their incredible regenerative ability at any stage of life, earning the name "the Tree of Life." They form epicormic sprouts, budding from anywhere just beneath the bark on the trunk. As a Yew ages, it is naturally prone to hollowing out, creating a cavern as large as 10 ft. (3 m) across, sometimes eventually becoming a shell of its former self or a ring of fragments which may, centuries later, have become encased in secondary wood and thus may appear as separate trunks. The wood of these slow-growing trees is hard and durable, and these empty shells actually have greater tensile strength than a younger, solid trunk. Yews can send down tendrils from the center of the inside of a hollow, rooting into the ground and creating one or more entirely new trunks within the old one, eventually replacing it, after thousands of years.

These shade-tolerant evergreens are unlike most other conifers; in place of cones, they bear a unique single seed surrounded by a fleshy covering called an *aril*. These small, red, berry-like fruits covering the seeds appear only on female trees (although some trees have been known to change sex), and are the only nontoxic part of the Yew. All other tree parts are highly poisonous to humans and most animals, due to alkaloids called taxoids.

Once widely abundant in Europe, the remaining ancient Yews that were spared the axe in the 13th to 16th century survived primarily in formal gardens and in the sanctuaries of church graveyards, monasteries, and abbeys, many of which are thought to be former pagan sites.

Distribution: The adaptable genus of *Taxus* has a wide climatic and geographic range through much of the northern hemisphere. *Taxus baccata* is native to western, central and southern Europe including the British Isles, to northwest Africa, northwest Iran and Turkey. Elevation: up to 8,200 ft. (2,500 m).

Ecosystem: In temperate climates, the Yew may grow as a solitary tree or in clusters, or in mixed coniferous or broadleaf forests as an understory tree or shrub, with Beech, Oak, Ash, Maples, Firs, Spruces, Hornbeams, Lindens, and Elms, with understory companions such as Holly, Hawthorn and Myrtle. It is also associated with juniper, hazel, fern, bracken, wild strawberry, ground ivy, blackberry, common nettle, woodruff, and cyclamen.

Maximum Age: Debatable; possibly over 3,000 years. The European Yew is the oldest continually living tree species in Europe (15 million years), while the adaptable and resilient genus *Taxus* dates back 140 million years. The European Yew also has the longest lifespan of any tree in Europe, and one of the longest in the entire world. However, dating ancient hollow trees by counting rings is impossible.

Maximum Height and Girth: 92 ft. (28 m) in height; 40 ft. (12 m) in circumference.

Animal Community: While a wide variety of birds and mammals eat the red flesh of the aril, only deer, rabbits, and hares are able to eat the leaves without harm. Mammals such as grey squirrels, door mice, badgers, foxes, martens, wild boars, and even brown bears eat the arils. Birds such as starlings, thrushes, blackbirds, robins, jays, and pheasants eat the aril and disperse the seeds, while green finches, bullfinches, marsh tits, great tits, and nuthatches eat the seeds. The species feasting on Yew trees in turn attract larger mammalian and avian predators, thus supporting the food chain. Bees create hives and owls create nests in the hollows. The yew gall midge, specific to the European Yew, creates tiny artichoke-like galls on the tips of branches, but is no threat to the tree.

Traditional Uses:

Medicine: Despite its deadly toxicity, the European Yew was used cross-culturally as a potent folk medicine for thousands of years. In ancient Rome, yew poison was used as an antidote for snakebite. Hildegard von Bingen (1148–79) used the smoke to treat colds. It was also used as an insecticide and to treat skin ulcers, insect bites, rabies, worms, and parasites.

Tools and Objects: The oldest wooden artifact yet found in the world (produced in the Hoxnian interglacial period, 200,000–300,000 years ago) is a hunting weapon made from yew wood. Artifacts of Yew wood from pre-historic to pre-industrial times include bowls, cups, plates, ladles, spindles, needles, looms, awls, nails, wedges, tool handles, and spoons. Yew was one of the woods used by the Egyptians for making sarcophagi and carvings. Yew wood was the wood of choice for Stone Age hunting bows (the oldest dating to the sixth millennium BC), and also one of the most important trees in the history of European warfare, as its wood was used exclusively for making longbows, the principle weapon employed in Europe for about 400 years. The part of Yew wood where the heartwood meets the sapwood is both sturdy and flexible, and so was favored as bow material. When all the Yews in England had been cut down, the Yew trade in mainland Europe became both big business and big politics. This led to the near-destruction of the trees throughout the rest of Europe. By 1595, Queen Elizabeth had decreed that the longbow be replaced by guns—not for more efficient weaponry, but due to the lack of suitable Yew staves. In the 17th and 19th centuries, the rare wood was prized for making fine furniture, veneers, inlays, and grandfather clocks.

Yew Arils

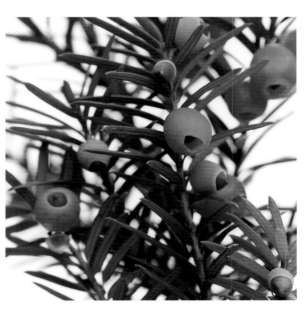

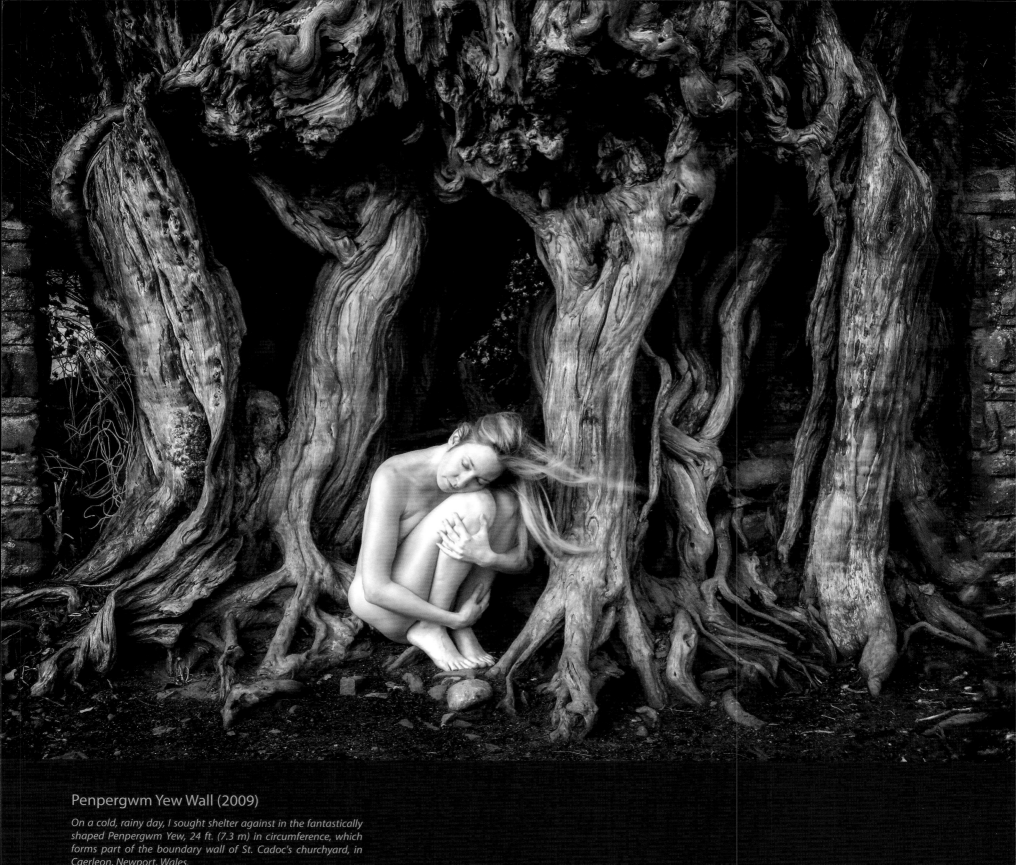

Penpergwm Yew Wall (2009)

On a cold, rainy day, I sought shelter against in the fantastically shaped Penpergwm Yew, 24 ft. (7.3 m) in circumference, which forms part of the boundary wall of St. Cadoc's churchyard, in Caerleon, Newport, Wales.

European Yew

Art and Ceremony: The Yew is found throughout in archeological history, mythology, and folklore wherever it grew. Yew needles have been found artistically rendered on artifacts cross-culturally throughout the ancient world. The oldest wooden musical instruments found are a set of Yew pipes dated between 2102 and 2085 BC. Yew wood was prized to make lutes. Sticks were used for divination (Celtic), magic wands, and sprigs for dowsing. Yews are important to Druids past and present. It is the 20th tree in the Celtic Tree Ogham alphabet. The Yew may have been the original evergreen Christmas tree at Germanic Yuletide, with its red "ornaments."

Modern Uses: At the end of the 19th century, the Yew made a comeback sculpted as topiary or maze hedge in formal gardens. The wood has been used to make billiard balls, eggcups, and even toothpicks. Today, the wood is prized for furniture, veneer and fine woodturning. Modern healthcare applications include experiments in treating epilepsy, diphtheria, rheumatism, arthritis, and tonsillitis, and for inducing miscarriage. Wild Yew trees have been over-harvested for tumor-active compounds (paclitaxel, later docetaxel), to produce anti-cancer drugs (Taxol®, Taxotere®). Now, for this purpose, farmed trees are mostly trimmed for their foliage, instead of chopping down the entire tree for the bark.

Threats and Conservation: European Yews never fully regenerated after hundreds of years of deforestation. Many old trees are neglected. Although not officially listed as threatened, the Yew has become extinct in many areas, and is a protected species in Germany.

Bettws Newydd (2009)

This face of this European Yew tree stands on one side as an unwelcoming tempestuous cauldron of fire. Still, after asking permission, I climbed up on the other side and tried to find a place to fit amidst the messy forest of epicormic spouts shooting up all over its trunk. Along with many windows in the trunk on both sides, there is an entrance to a cave inside—a portal I chose not to explore. Born from its internal tendrils, a new trunk has grown within its 7.6-foot-diameter (2.31 m) cavern. The entire tree measures over 32 ft. (10 m) in circumference and, based on its form, is probably 3,000+ years old. This Yew lives in the graveyard of the Church of Bettws Newydd, in the town of the same name in Monmouthshire, Wales.

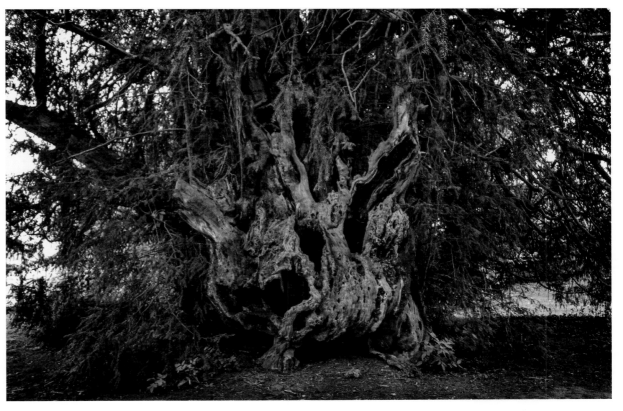

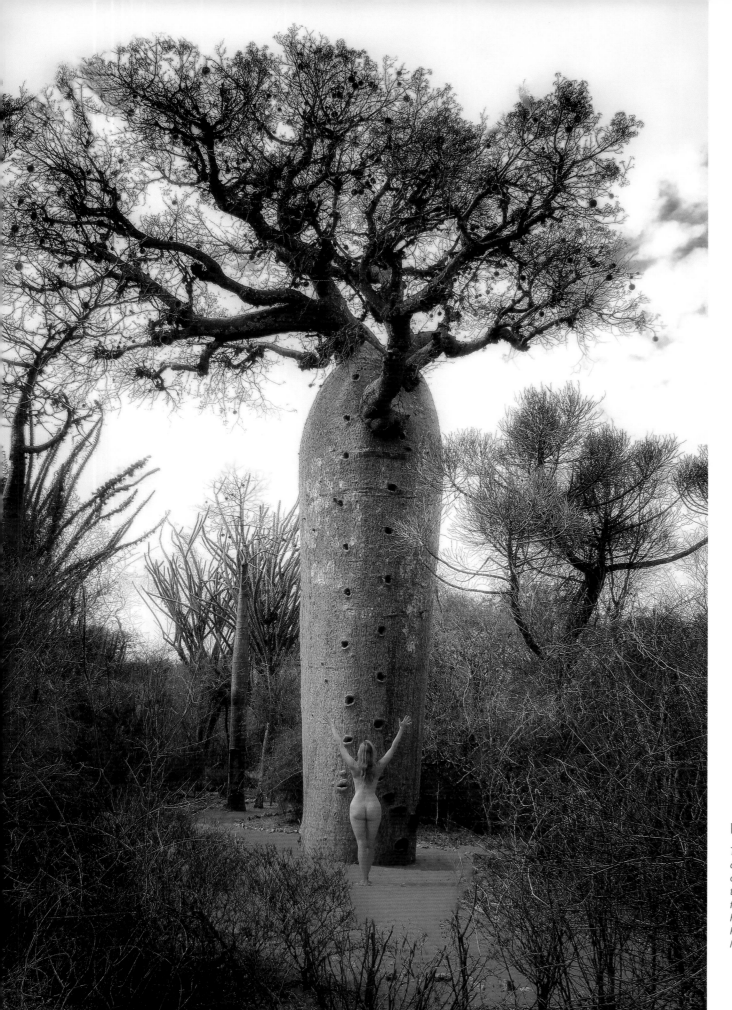

Baobab Climbing Holes (2008)

This outrageous-looking Fony Baobab tree stands out against the orange earth in the spiny forest, which is full of other fantastic and animated endemic species. What are those holes going up the tree, people often ask? They are for putting sticks into the trunk, to enable climbing up to harvest the fruit pods. This tree is protected in the Renialia Private Reserve (Renialia is a local word for Baobab), Ifaty-Mangily, west coast, Madagascar.

Fony Baobab

Adansonia rubrostipa

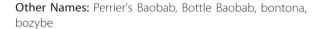

Other Names: Perrier's Baobab, Bottle Baobab, bontona, bozybe

Distinctive Characteristics: This crazy-looking deciduous tree has a distinctively cylindrical shape that narrows considerably at the top like a bottle, where it sprouts an extremely sparse crown. Trees may be tall or squat and stocky, some resembling a giant teapot or potato.

Distribution: Endemic to northwestern and southwestern Madagascar. Elevation: sea level to 2,600 ft. (800 m).

Ecosystem: The Fony Baobab is found in arid scrublands, in deciduous thorn thickets and spiny forests, and in moist montane forests. Found with another species of Baobab, *Adansonia za*, in the Kirindy region.

Maximum Age: Undetermined. Like all species of Baobab, its indeterminate rings make accurately dating the tree impossible. They appear older than they are because they hold water.

Maximum Height and Girth: 82 ft. (25 m) in height; 31 ft. (9.4 m) in circumference.

Animal Community: This tree provides food and habitat for various animals, including fruit bats and lemurs such as sifakas, fat-tailed dwarf lemurs, brown lemurs and possibly mouse lemurs. It is pollinated by the nocturnal fork-marked lemur. Birds, such as the greater and lesser Vasa parrots, feed on the flowers and fruits.

Traditional Uses:
Medicine: The bark, roots, leaves, fruits, and seeds have been widely used for medicinal applications, for both people and animals.

Food: The large gourd-like seedpods contain a dry, sour center high in vitamin C that has been made into candy, porridge, or a lemony drink; the seeds have been roasted; the young leaves and seedling roots have been eaten. The seeds can be pressed for cooking oil.

Fiber: The inner bark has been harvested to make very strong rope and cordage.

Art and Ceremony: The Baobab is the focus of much taboo, folklore, and spiritual reverence.

Shelter: Sheets of dried wood have been used as roofing material.

Threats and Conservation: Although not listed as threatened, the Fony Baobab is being over-exploited north of Toliara for local use for hut construction and rope making. It is also in danger from agriculture, livestock grazing, and lack of regeneration. It is protected in a number of reserves.

A Chunky Baobab (2008)

Another fantastical specimen of Fony Baobab, looking like a giant potato that has started to spout.

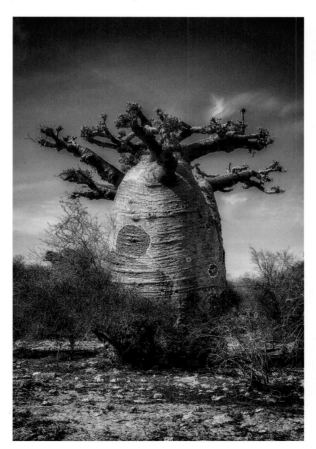

Brown Lemur (2008)

An inquisitive Brown Lemur, who is known to eat the fruit.

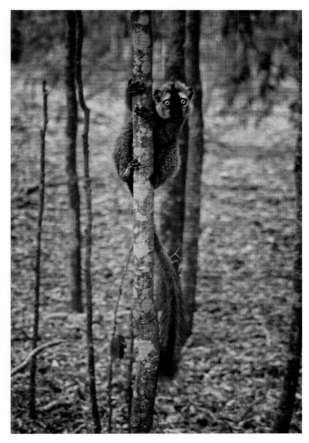

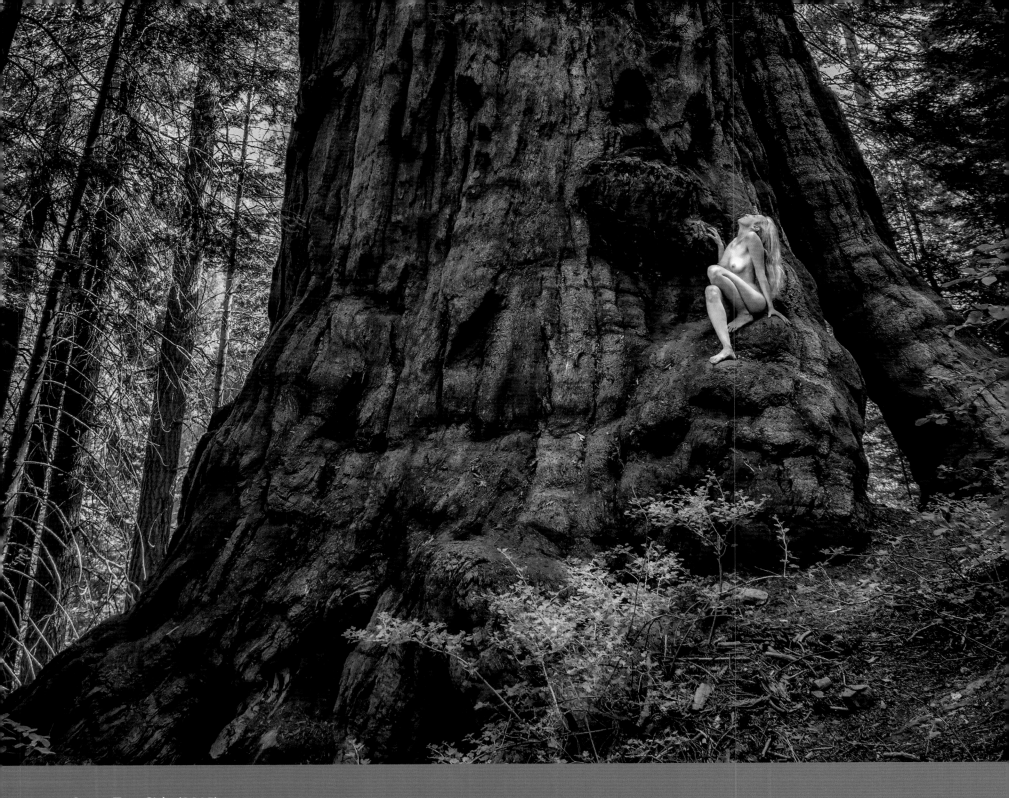

Stagg Tree Side (2015)

The Amos Alanzo Stagg Tree is named after a philanthropist football coach from the 1930s. How the sixth-largest Sequoia (by volume) got to be named after him is a mystery to me, but perhaps he was a tree lover. Half the tree's base engulfs the steep hillside, making it a challenge to set up my tripod and keep it steady. My treegirl friend and I enjoyed some quality time basking in the gracefulness of this giant beauty, who provided a small platform for lying and canopy-gazing. The Stagg Tree was measured in the 1990s at 71.62 ft. (21.82 m) in circumference (probably taken at DBH at the most level spot on the trail as opposed to its base at

the bottom of the hillside) and 109 ft. (33.22 m) in total perimeter. This mammoth tree is 243 ft. (74.06 m) in height, with a volume estimated at 42,557 cu ft. (1,205 m³).

During a snowstorm, tree climbers ascending the tree discovered a room burned out at the top, complete with an entrance, a window, and a flue. How lucky for them! I wonder who lived up there. Another mystery. This is the only really large Sequoia that still exists on private land, but there is easy public access on a marked trail. You can find it in Alder Creek Grove, within

the Sequoia Crest residential development in Sequoia National Forest, south of Kings Canyon-Giant Sequoia National Park, California, USA.

Giant Sequoia

Sequoiadendron giganteum

Common Names: Giant Redwood, Sierra Redwood, Wawona (Miwok), Toos-pung-ish (Tule River Tribe), and Hea-mi-withic (Tule River Tribe)

Distinctive Characteristics: Giant Sequoia is the most massive tree on Earth in terms of wood volume. *Sequoiadendron* is also one of the oldest-aged tree genera—its ancestors have been on the planet for 200 million years. Giant Sequoia is also one of the longest living trees in the world. Its longevity is in part due to its rot resistance, much like that of its Coast Redwood relative. Sequoia's rich, rusty-colored bark is fibrous, furrowed, and in mature trees it can be up to 3 ft. (1 m) thick. The insulating bark does not become covered by mosses, lichens, or other epiphytes found on other trees, and its thickness, as well as the lack of flammable terpenes in its sap (such is found in pine trees), provide significant fire resistance for protecting the trees. Fire is a necessary element for the Giant Sequoias' survival; they are dependent upon it for germination of the seeds. Also, in the absence of fire, other shade-loving tree species will crowd out young Giant Sequoia seedlings. Although a tree can reach hundreds of feet tall, its roots are shallow, only 4–5 ft. (1.5 m) deep, covering an area of 200–300 ft. (60–90 m) in diameter. Also, like Coast Redwoods, their rate of wood production increases as the tree ages (until near death), meaning, the oldest trees are making the most wood (because they have more leaves).

Distribution: Endemic to western slopes of Sierra Nevada range in California. There are between 65–75 groves (public and private). Elevation: 2,800–8,600 ft. (850–2,600 m), but typically 4,600–7,000 ft. (1,400–2,100 m).

Ecosystem: Usually in groves or mixed with Sugar Pine, White Fir, Incense Cedar, California Black Oak, Ponderosa Pine, and Pacific Dogwood, in a climate characterized by dry summers and wet, snowy winters. They do not grow solitary. Old growth Giant Sequoias are also ecosystems of themselves, with epiphytic plants and trees growing in the tops of canopies.

Maximum Age: Approximately 3,200 years.

Maximum Height and Girth: The current tallest Giant Sequoia is 315.91 ft. (96.29 m) in height. Historically, claims have been made as high as 321 ft. (97.8 m) and 341 ft. (104 m). The current largest girthed tree, the General Grant tree, is 91.1 ft. (27.7 m) in circumference. Since Giant Sequoias are buttressed at their base, and sometimes grow on slopes, their base diameters can be much larger, the current widest being an unnamed tree at 57 ft. (17.7 m). The General Sherman tree, although not the largest in terms of height or girth, is currently the largest living Giant Sequoia and largest living tree on Earth in terms of total wood volume, which was estimated at over 55,000 cu ft. (1,557 m³).

Animal Community: In addition to fire, two animal agents, the longhorn beetle and the Douglas squirrel, assist in the Giant Sequoia's seed release. The beetle's larvae bore holes in the seeds, allowing the cones to dry and open for the seeds to fall, while the Douglas squirrel dislodges and drops seeds as the green cones are eaten. Otherwise, the cones may stay on the tree for up to twenty years before falling.

Modern Uses: Even though their brittle wood usually shattered upon falling, making much of it unusable as timber, the trees were logged to make shingles, fence posts, and, horrifyingly, matchsticks! Giant Sequoia has since become a popular horticultural tree in landscapes and botanical gardens in many parts of the world.

Threats and Conservation: Upon their discovery in the 1800s, Giant Sequoias were logged nearly to extinction. Giant Sequoias are currently classified as a Threatened species by the U.S. Endangered Species List and classified as Endangered by the IUCN Red List. Although most of the ancient trees are protected from logging, preserved in Yosemite, Sequoia and Kings Canyon National Parks, as well as a few state parks and on private land, they are still at risk on National Forest lands. The historical absence of wildfires has contributed greatly to their slow regeneration.

Sequoia Cones

The chicken egg sized cones of the Sequoia contain between 150–250 seeds, but only a very few will germinate with the aid of fire (or extreme heat) and much fewer will become trees.

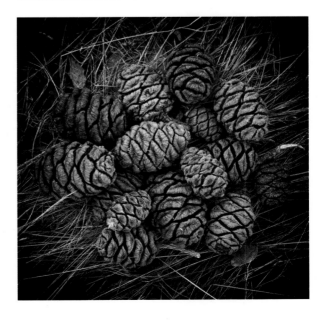

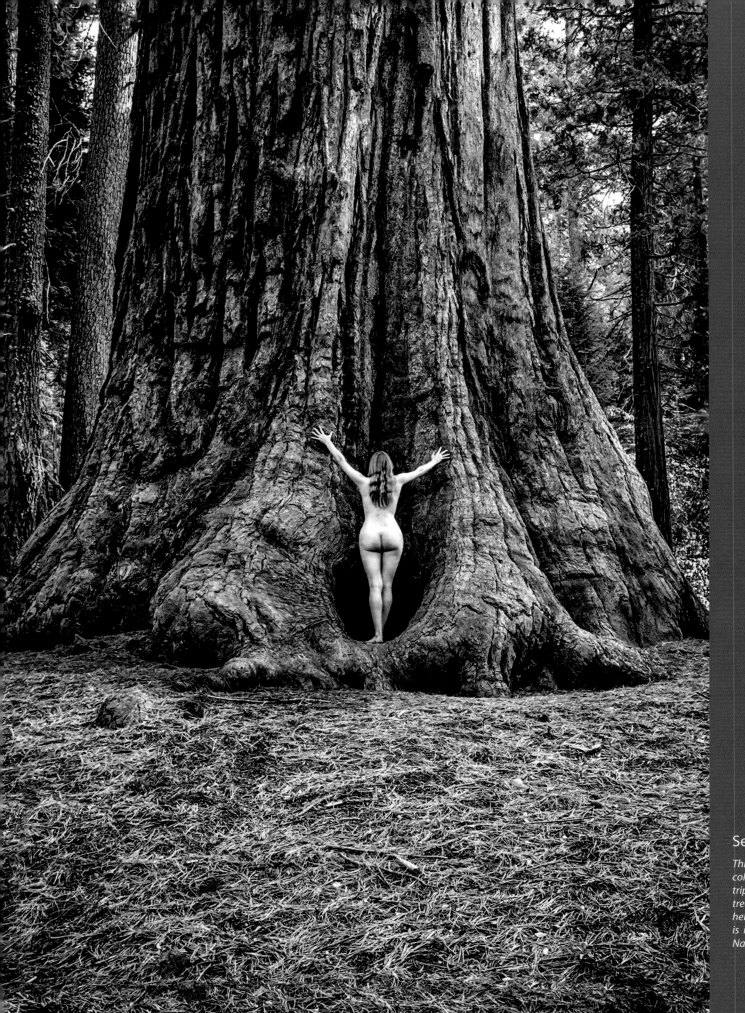

Sequoia Meditation IV (2001)

This favorite TreeGirl image was created spontaneously in collaboration with Devin Fleurdujon, during a camping trip for an interdisciplinary college course I taught on trees. It was one of my first and best tree encounters, and it helped marry me to my soul's purpose. This Giant Sequoia is in the less-visited, walk-in Merced Grove in Yosemite National Park, California, USA.

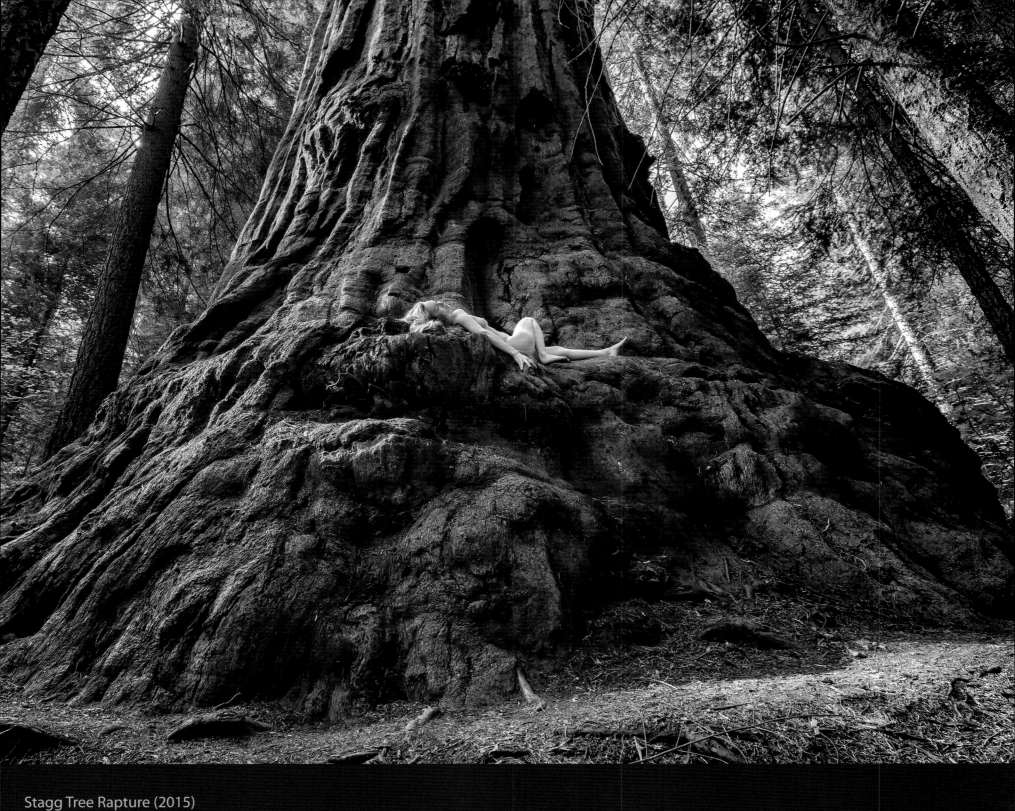

Stagg Tree Rapture (2015)

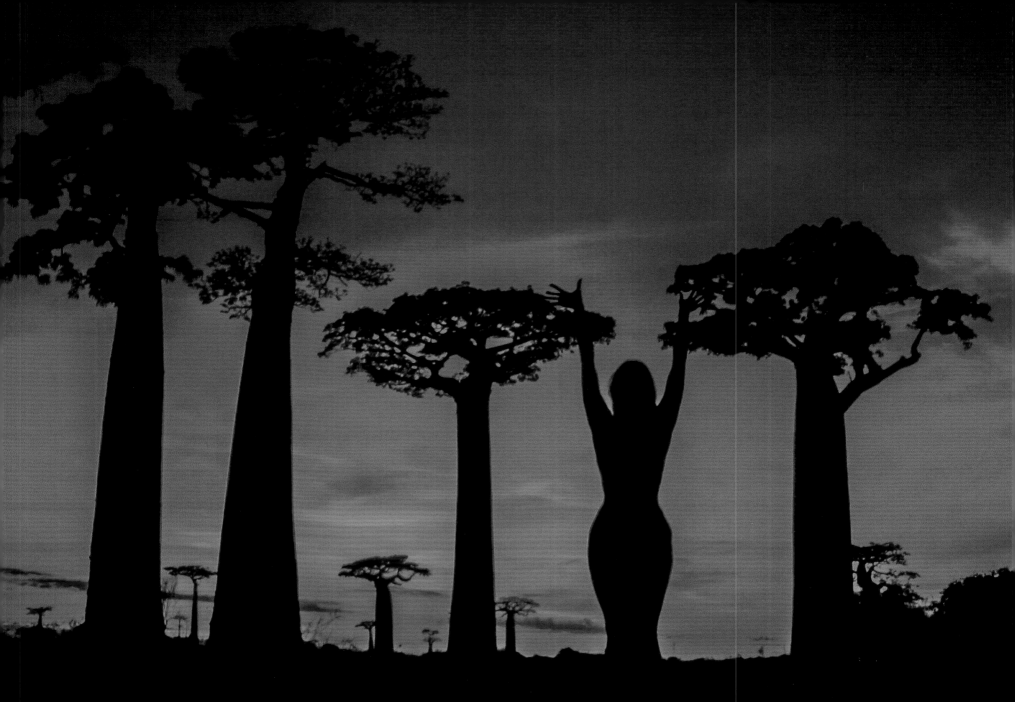

Morondava Sunrise Silhouette (2008)

The famous, picturesque Avenue of the Baobabs is what remains of a former, more dense, dry deciduous forest. At sunrise, with the aid of my guide Manda, I had only seconds to take this magical shot without being seen by the villagers, who would soon come down the avenue to start their day. The avenue is 15 kilometres northeast of Morondava, in Madagascar.

Grandidier's Baobab

Adansonia grandidieri

Other Names: the upside-down tree, renala or reniala (mother of the forest), bontouna

Distinctive Characteristics: These iconic trees, forming a distinctively cylindrical trunk, are the tallest of all the Baobabs. They are deciduous, have smooth reddish-grey bark, and end in a concentration of short, crown-like, stocky branches at the top. Like all Baobabs, the fibrous wood stores water, and thus the diameter of the trunk fluctuates depending on rainfall. They have white, showy flowers that only bloom for one night.

Distribution: Endemic to the southwestern coast and inland of Madagascar, between Lac Ihotry and Bereboka. Elevation: sea level to 200 ft. (60 m).

Ecosystem: Once part of dense dry deciduous tropical forest, they now grow solitary or in groves.

Maximum Age: Estimated at 800 years, they do not reach the millennial age or wide girth of its mainland cousin, *Adansonia digitata*. They appear older than they are because they hold water. Like all species of Baobab, its indeterminate rings make accurately dating the tree impossible.

Maximum Height and Girth: 98 ft. (30 m) in height; 30.7 ft. (9.3 m) in circumference.

Animal Community: These Baobabs are an important habitat tree to a variety of invertebrates, birds, reptiles, and mammals. Red-tailed sportive lemurs and Coquerel's dwarf lemurs feed on its leaves and flowers; non-native bush pigs and rats eat the fruit. They are pollinated by nocturnal pale fork-marked lemurs and Madagascar straw-colored fruit bats. Birds, such as green sunbirds and the Souimangas sunbird, enjoy the nectar; parrots and guinea fowl eat the seeds.

Traditional Uses:
Medicine: The bark, roots, leaves, fruits, and seeds have been widely used for medicinal applications for both people and animals.

Food: The large gourd-like seedpods contain a dry, sour pulp high in vitamin C that has been made into candy, porridge, or a lemony drink. The seeds have been roasted or pressed for cooking oil. The young leaves and tuberous taproots have been eaten.

Fiber: The inner bark is harvested to make very strong rope and cordage.

Art and Ceremony: The Baobab is the focus of much taboo, folklore, and spiritual reverence. Some individual trees are said to be the spiritual home of ancestors.

Threats and Conservation: This species of Baobab is listed as Endangered on the IUNC Red List. They have been put at risk by over-harvesting the bark and fruits, slash-and-burn agriculture, zebu cattle grazing, groundwater consumption, flooding and effluent from nearby rice fields, and lack of regeneration. This species is protected in the Reserve Speciale d'Andranomena, but the famous Avenue of the Baobabs—elder trees near Morondava—is not. Local organizations are working to promote conservation.

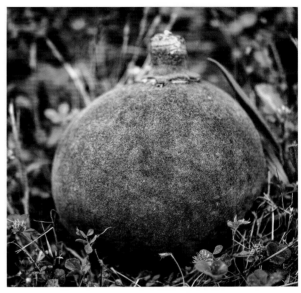

Baobab Seed Pod (2016) ▸

▾ Avenue of the Baobabs (2008)

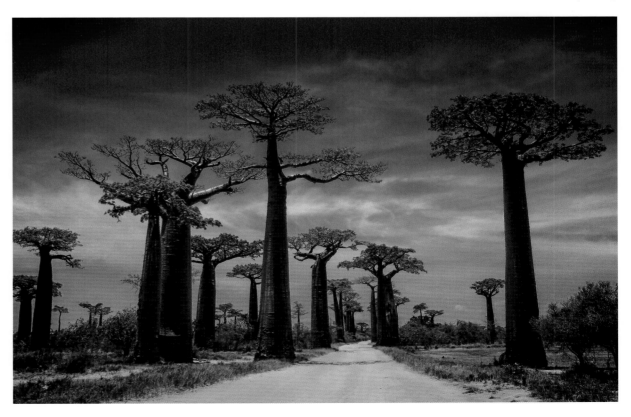

69

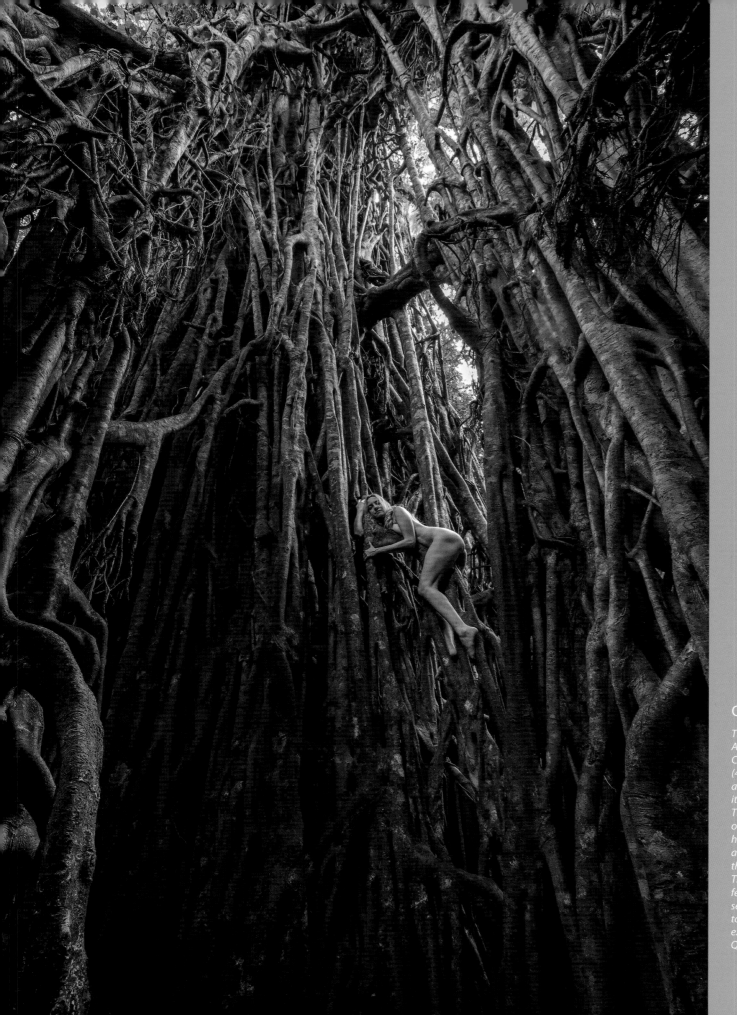

Cathedral Fig Dryad (2015)

This spectacular jungle of a Green Fig tree is famous in Australia for its size and unusual shape, along with the Curtain Fig. It is 144 ft. (44 m) in circumference, 157 ft. (48 m) high, and estimated to be about 500 years old. Like all strangler figs, when the host tree is finally overtaken, it will decompose, leaving an empty space in the middle. That is where you can walk into the narrow, tall cathedral of the Cathedral Fig, and look up at the aerial roots that have come down and fused together. Because these roots are intertwined, they create a ladder to climb up into the tree, with many places for a treegirl to hang on to. This tree is a whole ecosystem in itself. I could hear and feel the presence of many beings living in it, and had a sense that the entrance to the cathedral was accessible to visiting humans, but the rest of the tree was off limits, except to those who lived there. This tree is in Yungaburra, Queensland, Australia.

Green Fig

Ficus virens

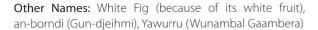

Other Names: White Fig (because of its white fruit), an-borndi (Gun-djeihmi), Yawurru (Wunambal Gaambera)

Distinctive Characteristics: *Ficus virens* is one of the rare deciduous trees in the rainforest. It can stand alone as a single-trunk tree, or grow as a parasitic strangler fig with many aerials. Strangler figs begin as epiphytes at the top of a host tree when a seed is pooped out from a bird or bat. They then germinate and send aerial roots down to the earth to bring up nutrients. Over time, the aerial roots encircle and smother the host tree, out-competing it for nutrients and light, eventually killing it.

Distribution and Range: Native to northern Australia, India, China, Japan, Southeast Asia, Malaysia, New Guinea, and the Solomon Islands. Elevation: 900–9,000 ft. (300–2,700 m).

Ecosystem: Monsoon forests, tropical lowland rainforests, and along streams. In Queensland, Australia, some trees share their home in the endangered mabi rainforest with Brush Mahogany, Black Bean, Candlenut, Queensland Maple, Red Cedar, and Red Tulip Oak. Epiphytes, such as a variety of ferns, make their home high in the tree's canopy.

Maximum Age: Over 500 years.

Maximum Height and Girth: 164 ft. (50 m) in height; the true circumference is tricky to determine on strangler figs because of the arial roots, however, one Green Fig in New South Wales, Australia, appears to be solid at 101.7 ft. (31 m). Single-trunked trees have been measured at 62.3 ft. (19 m) in circumference.

Animal Community: A wide range of birds, insects, bats, and other mammals eat the fruit. In Australia this includes Papuan frogmouths, rose-crowned fruit doves, emerald fruit doves, grey-headed robins, orange-footed scrubfowls, leaf-tailed geckos, Lumholtz's tree kangaroos, common brushtail possums, and green ringtail possums. Like all species of figs, there is one species of wasp that will fertilize the flowers.

Traditional Uses: There were most likely many traditional uses of this tree.

Medicine: A decoction of the bark was used as an injection in the treatment of leucorrhoea, as a wash on skin ulcers, and as a gargle, while a bark extract mixed with other plants was given for bone fractures.

Food: The fruits are edible, although not favorable. Young shoots and young leaves are eaten raw or cooked. The leaves, boiled and used in curries, are known in Thai cuisine as *phak lueat*.

Fiber: The inner bark was used to make rope.

Modern Uses: The latex is used for caulking boats and waterproofing. The wood, including the aerial roots, is used locally in light construction and tool making.

Threats and Conservation: Not threatened.

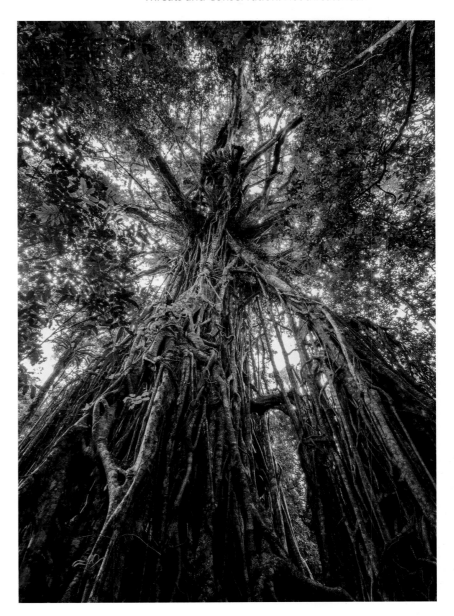

Cathedral Fig Up (2015)

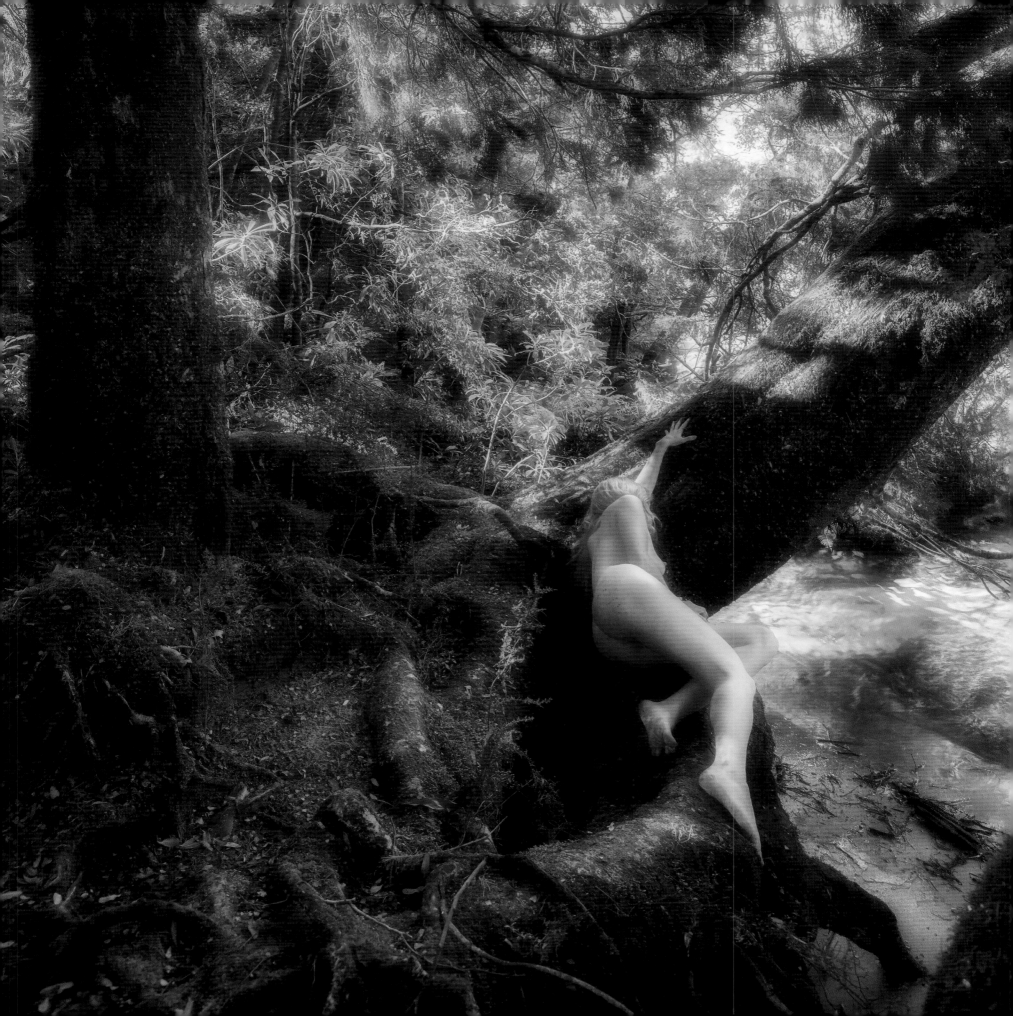

Huon Pine

Lagarostrobos franklinii

Other Names: Macquarie Pine

Distinctive Characteristics: The Huon Pine is a conifer— not a pine, but of the Podocarpaceae Family, having finely overlapping, tightly packed scales, not needles. One of Australia's longest-lived species, fossils have been dated at 10–20 million years old. It is also one of the few conifers that can form clonal thickets by vegetative reproduction. For example, a branch touching the ground, or one that has broken off, can form roots. It can also reproduce by its tiny, 3 mm-long cones. Because they are able to clone themselves genetically, much like aspens, they can reproduce themselves continually over thousands of years. One Huon pine stand—considered one organism, interconnected underground—has been dated at 10,500–15,000 years old! Since they are also slow-growing (only 1 mm per year), the wood grain is dense with narrow growth rings; the girth of the trunk is no indication of its age. Extremely fragrant and rot-resistant, because of the oil methyl eugenol, its wood is so durable that logs buried for thousands of years can still be milled.

Distribution: Native to 10,500 ha. of remnant riparian and subalpine forest in southwest Tasmania, Australia. Mostly in reserves, including Mt. Read, Tahune Forest Reserve, Teepookana Forest Reserve, Arthur-Pieman River State Reserve, and Franklin-Gordon Wild Rivers National Park. Elevation: 1,640–3,300 ft. (500–1,000 m).

◄ Huon Pine Hiding (2015)

The rare Huon Pine is not a true pine; it was probably so named because of the wonderful piney scent given off by the fragrant oils that preserve its wood. This pair of Huon Pines, though not wide in girth due to their slow growth, are about 400 years old. They were some of the sweetest trees I had ever met: gracious, welcoming, and I would even say loving in their elderhood. The giant root, or perhaps fallen limb, clings to the riverbank, with its toes dipped into the water where it regenerates itself, growing into the sandy soil. I lay down on the cold, damp yet cozy, moss-covered tree, blissed out with a peaceful view of the riverbank that the tree has witnessed for hundreds of years. Both the trees and I were hiding from the tourists, who primarily come to the park to walk the suspended bridge and the giant catwalk that ends in a cantilever 164 ft. (50 m) above the forest floor. Who comes to just see the hidden Huon Pines, let alone get naked with them? Tree nerds like me. These Huon Pines, noted as the most accessible ones in the world, are down off the boardwalk on the Huon Pine Loop Walk, at Tehune Airwalk, outside Geeveston, Tasmania, Australia.

Ecosystem: Cool, temperate rainforest (nanophyll moss forest) and subalpine forest, along the banks of rivers or near swampy flats or lakes, often with its roots in water. Grows with Myrtle Beech, Southern Sassafras, Leatherwood, King Billy Pine, Pencil Pine, Cheshunt Pine, Creeping Pine, Toothed Orites, Tanglefoot Beech, Celery Top Pine, Blackwood, and Laurel.

Maximum Age: Individuals estimated at 3,000–5,000 years old; stands of genetically cloned trees are estimated at 10,000–15,000 years old.

Maximum Height and Girth: Up to 125 ft. (38 m) in height; over 18.5 ft. (5.6 m) in circumference.

Traditional Uses:
Medicine: Because of its antimicrobial properties, the oil has been used for dressing wounds, treating toothaches, and as an insecticide.

Modern Uses: The wood has been highly prized—more than any other Australian wood—for its beautiful, smooth, tight grain, its creamy golden color, and its lightness, fragrance, and rot-resistance. The wood is valued for making boats, furniture, and fine artisan crafts.

Threats and Conservation: While 90% of the remaining stands are protected in reserves, including a world heritage site, the species itself is not considered threatened. Living trees can no longer legally be cut, but there is still wood available from logs destroyed by mining, fire, flooding from hydroelectric construction, and from the original logging period.

Huon Pine Wood

The beautiful wood is prized by carvers and furniture makers.

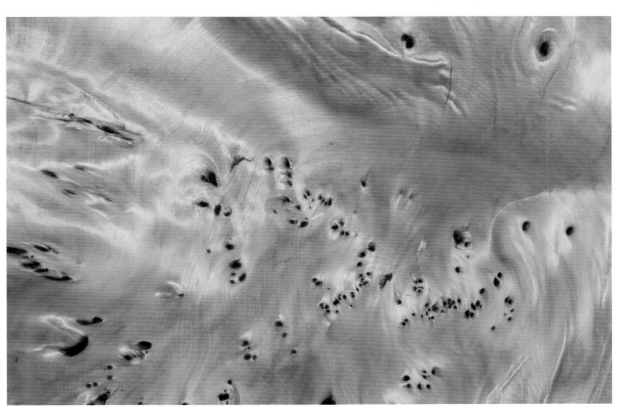

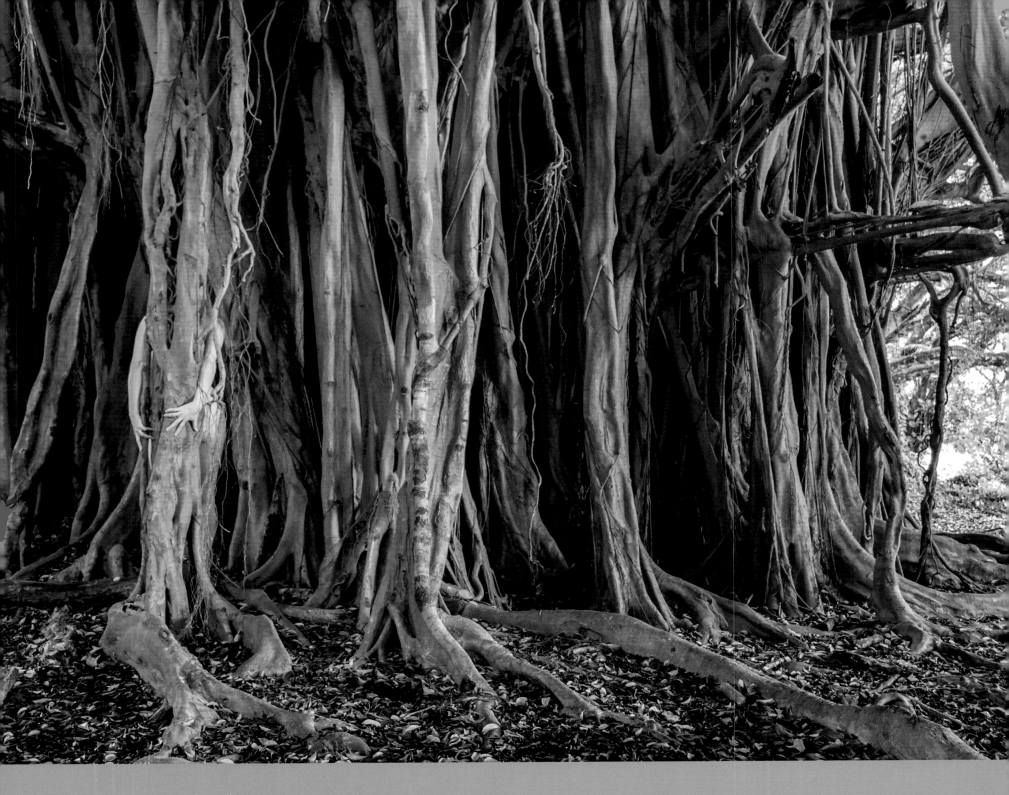

InterBanyanBeing (2013)

Most tourists come to see the waterfall near this Indian Banyan tree, taking only a minute or two to photograph it with their phones. I could have easily spent a couple of days shooting this tree. Banyans are not native to Hawaii, but they thrive in this hot, tropical climate. Big Banyans are perfect for a treegirl, because they are relatively easy to climb and there are so many photo possibilities. Banyans all over the world can conceal many creeping creatures that one needs to watch out for. (They can easily conceal many treegirls as well.)

However, the biggest hazard at this tree were the pair of gawking male voyeurs—an extremely rare occurrence, in which TreeGirl was actually seen in action. This tree is at Rainbow Falls, Hilo, Big Island, Hawaii, USA.

Indian Banyan

Ficus benghalensis

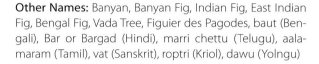

Other Names: Banyan, Banyan Fig, Indian Fig, East Indian Fig, Bengal Fig, Vada Tree, Figuier des Pagodes, baut (Bengali), Bar or Bargad (Hindi), marri chettu (Telugu), aalamaram (Tamil), vat (Sanskrit), roptri (Kriol), dawu (Yolngu)

Distinctive Characteristics: The Indian Banyan is known for its multiple aerial roots that propagate downward from the canopy, penetrating the ground and growing into multiple woody, pillar-like trunks. This tree, with a thick canopy of evergreen leaves, can spread out laterally indefinitely, covering a very wide area to form its own "forest."

The Banyan can grow as a single-trunk tree, but more likely it grows as a parasitic 'strangler fig': born from a seed dispersed by a bird or bat atop a tree, it starts life as an epiphyte, absorbing nutrients and water from the air, dropping down multiple aerial roots from the host's branches that eventually take root in the earth. Over time, roots aggressively intertwine the trunk and fuse together, creating a latticework. Out-competing the host for space, nutrients and light, eventually the new tree becomes freestanding. After it establishes other trunk-like props, the tree can survive even if its original trunk is removed or decayed. In terms of canopy coverage, Banyans are the largest trees in the world.

Distribution: Native to India and Pakistan, throughout the subcontinent. Elevation: in India, sea level to 3,900 ft. (1,200 m).

Ecosystem: Grows in tropical and subtropical forests.

Maximum Age: 300 years.

Maximum Height and Girth: 98 ft. (30 m) in height; 28 ft. (8.5 m) in circumference. One individual, named Thimmamma Marrimanu, in Andhra Pradesh, India, covers 5 acres (2 hectares). A tree known as The Great Banyan, in Kolkotta, has more than 3,300 aerial roots touching the ground, and covers an area of 4 acres (1.5 hectares).

Animal Community: Many small animals and birds find refuge in the tree. The figs are eaten by birds such as Indian myna birds. All *Ficus* species are dependent upon one species of wasp to pollinate them; the Indian Banyan relies on *Eupristina masoni* for reproduction.

Traditional Uses:

Medicine: The milky latex sap was applied externally to treat pains and bruises; ingested to treat nausea, inflammation, or rheumatism; or used as a toothache remedy or an aphrodisiac. The aerial root was used to treat bleeding, gum infections, nausea, hemorrhoids, venereal diseases, female sterility, and inflammation of the liver. The root and leaves were used to treat diarrhea. The leaves were used to treat ulcers. A paste made of the leaves was applied to relieve skin disorders. The bark was used to treat diabetes. String made of the bark was tied to heads to cure headaches and to make tourniquets.

Food: The small, scarlet-red fruits can be eaten fresh or dried, although they are not favored. The young leaves and shoots are also eaten as food in times of famine.

Fiber: The bark has been used for making paper and ropes.

Tools and Objects: The aerial roots have been used for making poles and cart yokes. String was made from the bark for fishing lines and nets, bags, fixing spearheads to shafts, baskets, ornaments, and slings.

Art and Ceremony: Venerated in Hindu culture as "the wish-fulfilling tree," this was one of the trees of religious importance during the Vedic period. It is the national tree of the Republic of India. Parts of the tree were used to create a 'Morning Star Pole' and arm and headbands (Bardi). Bark or roots were made into necklaces (Iwaidja). Latex sap was used to adhere feathers to performers' bodies (Kurlama), and put into beards (Tiwi).

Shelter: While the wood itself can be harvested for construction, a spreading tree's canopy and entire structure can provide shelter from winds and storms.

Modern Uses: A type of shellac is produced from a resinous secretion called lac, created by various insects that live on the tree, including *Laccifer lacca*. The shellac has many industrial uses, and is an ingredient in beauty products. The trunk wood is used to make furniture.

Threats and Conservation: Not threatened.

InterBanyanItself

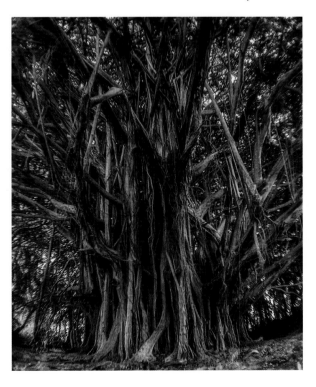

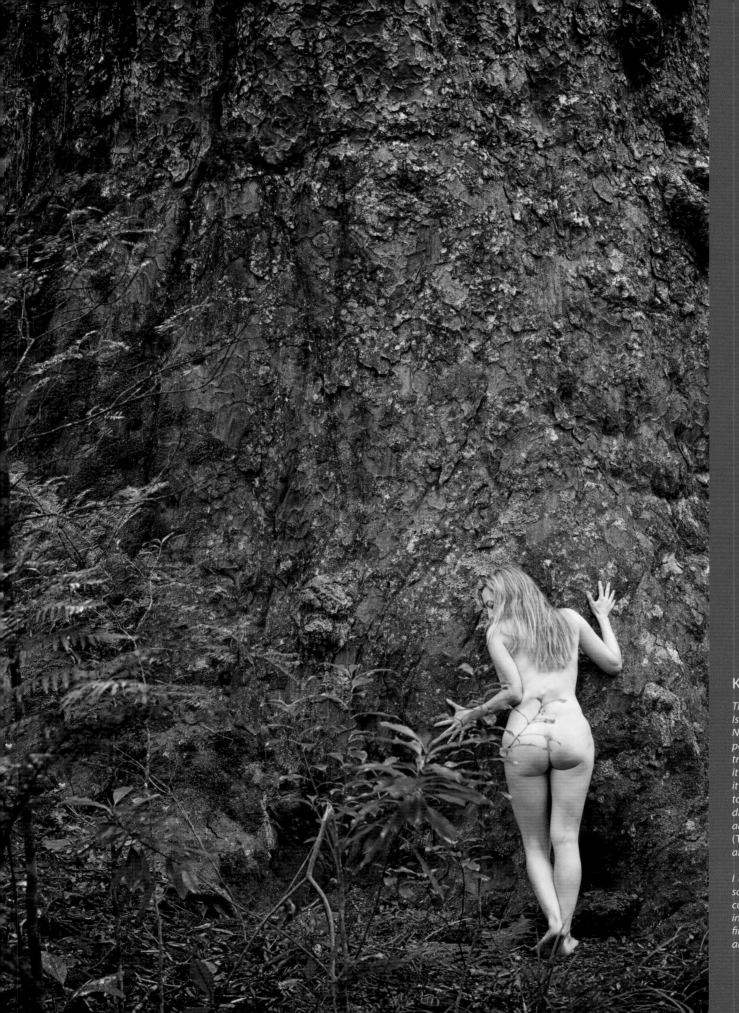

Kauri Love (2015)

The Waipoua Forest, on the far north of the North Island, is the largest remaining tract of native trees in New Zealand. This is one of only two big Kauri trees that people can get anywhere near. I missed the forty-minute trail to this tree the first time, so I briskly walked back to it on the wet, winding, wooden boardwalk, only to find it too dark to shoot when I got there. On my way back to the parking lot in pitch darkness, with my headlamp dimming, I literally ran into a non-native, but quite adorable, common brushtail possum mum with her baby (Trichosurus vulpecula) on her back, foraging for food and aloof to me.

I only had one more chance to photograph the Kauris, so I spent that night—long, cold, and uncomfortable—curled up in the back seat of my illegally parked rental car, in order to return to the trail once more to shoot the tree first thing in the morning, before the tourists arrived. I did, and it was well worth it.

Kauri

Agathis australis

Other Names: agathis

Distinctive Characteristics: Kauris are conifers, with small, leathery evergreen leaves that are elliptical in whorls, with shedding, flakey bark. Mature Kauri trees have characteristically large, massive, columnar trunks with little or no branching below the crown, which is characteristically narrow. Kauris ooze a sticky, resinous gum that protects them from fungi and wood-boring insects. Over time, the gum hardens, falls to the ground, and gets buried. The fossilized gum, or resin, called *kâpia*, has hardened over millions of years into various grades of beautiful copal, or amber.

Kauris are an ancient species, with ancestors from 190 million years ago. Prehistoric Kauri trees, known as swamp or ancient Kauri, have been preserved in salt marshes and dated to over 45,000 years old, and are still a viable source of preserved wood. Today, the Kauri forest is considered a long-term carbon sink. The total carbon content in living, above-ground biomass and dead biomass of mature Kauri forest is estimated to be nearly 1,000 tons per hectare—the second highest of any forest type recorded anywhere in the world, after Mountain Ash (*Eucalyptus regnans*).

Distribution: Native to the northwest corner of the north island of New Zealand, north of 38°S latitude from the Kāwhia Harbour in the west to the eastern Kaimai Range, with some patches on Great Barrier Island. Elevation: sea level to 2,000 ft. (600 m).

Ecosystem: Subtropical rainforest, with other trees including Rimu, Tōtara, Taraire, Kohekohe, Miro, Hard Beech, Kahikatea, and Pūriri. Thirty-eight species of ferns, orchids, and other plants were found in just one tree. On the forest floor grow astelia grass and ghania sedge.

Maximum Age: Estimated at 2,000 years.

Maximum Height and Girth: Over 185 ft. (56 m) in height; over 88 ft. (26.83 m) in circumference.

Animal Community: Branches provide habitat for short-tailed bats and birds, including kererū, kākā, kākāriki, and eastern rosellas. Mice eat the seeds. Insects such as the kauri leaf miner moth caterpillar mines the leaves; looper caterpillars, kauri weevils, and native longhorn beetles can be found on the wood.

Traditional Uses:
Medicine: The gum was used to as an insecticide in kumara plantations, as well as for dressing wounds, repelling leeches, and treating asthma and arthritis (Maori).

Food: The fresh gum was used for chewing (Maori).

Tools and Objects: Branch wood was used for making mallets, weapons, and spades. Scrapings of harder gum were used to kindle fire (Maori).

Art and Ceremony: The wood was prized for large carvings. Soot of the burnt gum was used as a tattooing pigment (Maori).

Transportation: The wood was prized for making canoes (Maori).

Modern Uses: Nineteenth-century Europeans exploited most of the accessible old trees, valuing the timber for its branch-free trunk (with tight grain and no knots), using it widely for making ships, houses, carts, furniture, wood paneling, fences, bridges, dams, church pews, post office counters, mine braces, and railroad ties. Stumps and perfectly preserved ancient swamp kauri are used for furniture, fine wood turning, carvings, and tourist trinkets. The fossilized amber- and milky yellow-colored resin was mined extensively and was a valuable commodity (used as money, it was said to have built cities). It was sold in chunks and carved and polished to make beautiful collectibles, and is still sold as jewelry. The resin was also used to make varnish, linoleum, waxes, candles, and molding material.

Threats and Conservation: It is estimated that, today, only 4% of the original, uncut forest remains in small pockets. Even though it was almost logged to extinction, and very few large trees are left, *Agathis australis*'s status is not threatened, while other species in the same genus are. However, Kauri dieback or Kauri collar rot (a *Phytophthora* disease), is a threat.

▼ Kauri bark (2015)

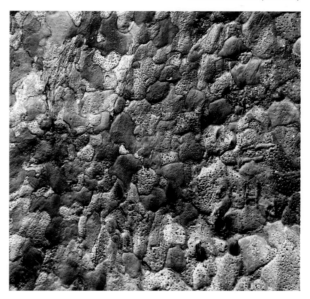

Te Matua Ngahere (2015) ▸

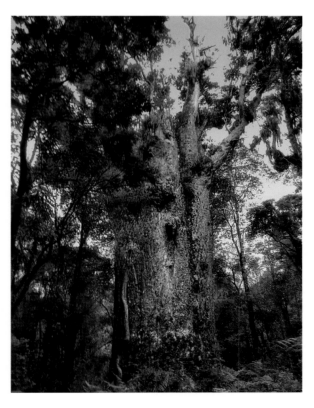

The 'Father of the Forest', revered by the Maori, has the largest circumference of any Kauri remaining, at 52 ft. (16 m), and has the second biggest volume. It has survived a storm in 2007 that tore off a major limb. It is estimated at 2000 years old.

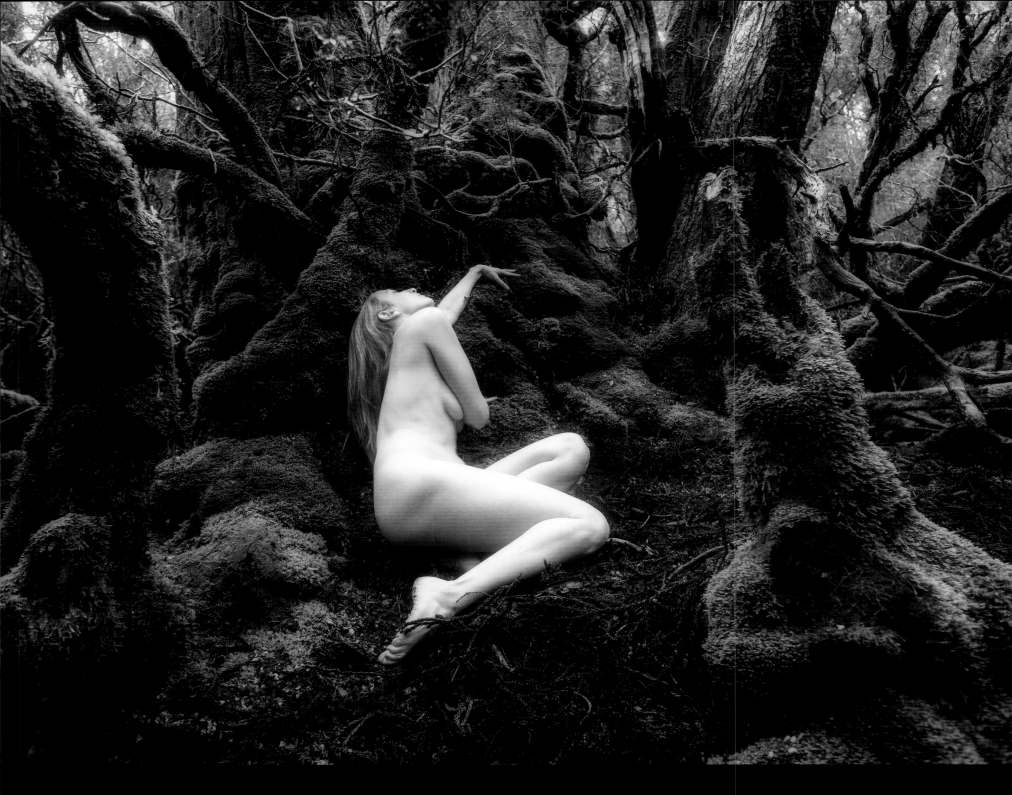

Waldheim Enchantment (2015)

This was, by far, the most enchanting forest TreeGirl has yet visited. It sits atop a cloud-covered plateau in the perpetually clouded-over Cradle Mountain-Lake St. Clair National Park in Tasmania, Australia—one of those special places on Earth that changes your life. On the previous day, I had driven all day over a horrible 4-wheel-drive road in my 2-wheel-drive rental car to get to Cradle Mountain, in hopes of seeing wombats in the wild. I woke up early, with only a couple of hours before I had to get to the airport to fly out. In the parking lot at the alpine trailhead, I was boldy greeted by a Black Currawong—the yellow-eyed Trickster Raven

of Tasmania. I entered the Waldheim track, a moss-covered dark tunnel of craggy trees, feeling tingles all over my body. I knew I had entered a truly magical world: moss-covered everything, the rich fragrant earth, and rocky trickles of creeks. If ever there were fairies and gnomes, forest sprites and spirits, they surely lived in abundance here. My intuition told me to go back on the trail to see where it had originally split. It was there, along the overgrown, rocky passage of an old trail, that I saw him: the King of the Forest. In absolute silence, the mythical elder Wombat walked away from me, hobbling over an old creek bed, disappearing into the fog like

a holy apparition. He had let me see him, if only for a few seconds. That was one of the most sacred nature experiences of my life. But I continued onward, with just a few moments left to photograph myself with the other ancient king of the forest: a King Billy Pine. After asking permission, I lay down for a few minutes on the cold, moss-covered roots, which were entangled with other trees. I felt as if I were lying on the belly of a giant, as the entire forest felt like one giant breathing organism, humming and vibrating with life-force energy.

King Billy Pine

Athrotaxis selaginoides

Other Names: King William Pine, Tasmanian Pencil Cedar, Red Pine

Distinctive Characteristics: Not a true pine, this conifer is in the Cypress family. It is extremely slow-growing, taking about forty years for a seedling to grow to one meter tall in the wild. It grows with a single or multiple trunks; in subalpine areas it occurs dwarfed, twisted, and gnarly. The soft, spongy bark is slightly furrowed and exfoliates in long strips.

Distribution: Endemic to the western and southwestern plains and mountains of Tasmania, Australia. Elevation: 1300–3600 ft. (400–1,120 m).

Ecosystem: Occurs generally in dense, shady, cool and wet climates including heavy snowfalls, in temperate rainforest, subalpine scrub, coniferous heathland, riparian or valley bottoms, and less commonly on rocky slopes and exposed ridges. Associated with Myrtle Beech, a Top Pine, Southern Sassafras, Cider Gum, Tasmanian Snow Gum, Huon Pine, Pencil Pine, Leatherwood, Deciduous Beech, Cheshunt Pine and various species of Eucalypts, Atherosperma, and Melaleucas.

Maximum Age: Estimated to 1,200 years old.

Maximum Height and Girth: 131 ft. (40 m) in height; 22.6 ft. (6.9 m) in circumference.

Animal Community: Common wombats, Bennett's wallabies, black currawongs, and various song birds.

Modern Uses: King Billy Pine has been popular for use in joinery, furniture, pianos, violins, drawing boards, carriage works, wooden pipes, venetian blinds, cabinetry, boatbuilding, doors, windows, and roofing shingles. Scavenged wood is still prized for fine wood-turning, and for making water craft and stringed instruments.

Threats and Conservation: Because of its slow-growing nature, regeneration is extremely poor. The species has declined about 40% over the last 200 years due to logging and fire. Although 84% of its forests are now in protected areas, fires are still a potential hazard. They are also in danger from climate change. The status is listed as Vulnerable on the IUNC Redlist; it is on the Index of Threatened Australian Plant Species; and it is classified as Endangered under Australia's Environment Protection and Biodiversity Conservation Act.

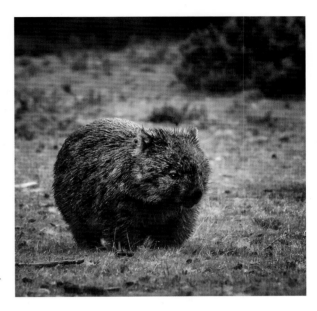

Wombat (2015) ▸

Black Currawong (2015) ▾

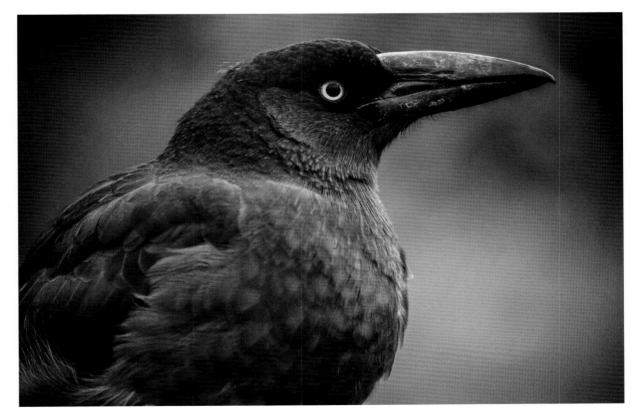

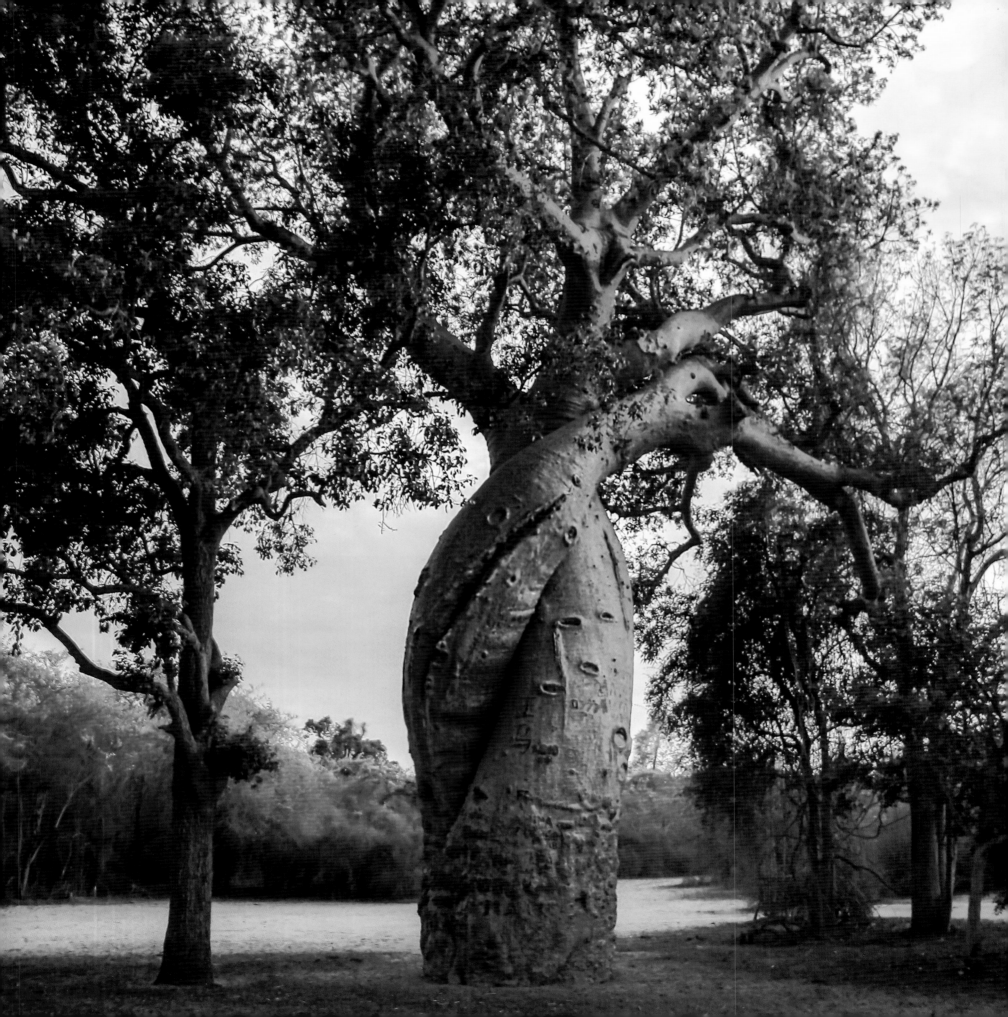

Trees make love quietly.

-Alex Shigo

Chapter 3
Touch Trees

I have always felt a strong connection with trees. In my childhood experience, Nature was filled with marvelous animate beings, like characters in a book that came to life with magic and mystery. Trees, especially, had wisdom to share, protection to offer, and company to give. I had a relationship with them. As a young adult, I moved to California and discovered the majesty of the native Redwoods, Giant Sequoias, Bristlecone Pines, and many species of Oaks. After spending a lot of time with trees, I began to really look at trees and experience them in a much deeper way. I became certified as an arborist because I had an affinity with trees and wanted to better understand the science of how they live and what causes them to die. I began taking notice of sad-looking trees who were not thriving, mostly in urban and suburban settings, and I wanted to be of service to them.

Now, with more training and years of observation, it seems that all I see in urban settings are trees that are isolated, unhealthy, ignored, butchered, or decapitated. Even in wilder areas, we are witnessing an increase in tree sickness and mortality from the rippling effects of global climate change, loss of ecological community, acid rain and other pollution, and an overall lack of the kind of land management indigenous peoples performed for thousands of years. Trees, like animals, retain genetic memory of their wild ancestry, memories of how the landscape was before modern human settlement—before gravel, asphalt, and concrete became dominant ground coverings. Trees want to be wild.

Regardless of humankind's well-intentioned attempts to control nature, a tree's branches will grow haphazardly. Branches, leaves, fruits, and seeds fall wherever gravity pulls them, conspiring with the wind to scatter messily on the ground or on the cars beneath them, sometimes irritating humans. In 2009, 63 healthy and mature holly oak trees (*Quercus ilex*) lining the shopping district in downtown Palo Alto, California, were cut down because city planners, as well as some shop owners and shoppers, found the abundance of acorns dropping on the sidewalks every autumn to be a safety hazard and a nuisance![1] But in the wildness of a county, state, or national park, those same city planners, shopkeepers, and shoppers might bask in the same tree's fertile abundance. When the trees were removed, there was public outrage at the unsightliness of the "improved" public space. Those trees were lost, in part, because of poor initial choices in urban landscape design (by perhaps the same city planners), as well as due to a culture-wide disconnect with Nature and her bountiful gifts. For thousands of years, in their native region those same acorns would have been respected and coveted as a food source. This urban tragedy was not an isolated incident; urban deforestation such as this occurs commonly on public lands and private lands, not just in wild natural areas. Trees, like most other species, have become expendable and disposable if they are in our way or not doing what we'd like them to. But a tree is not civilized, nor will it ever be—even when shaped as topiary. It will rebel. Wildness is the true nature of trees, and that is what we instinctively love about them.

In the urban landscape, where the botanical realm is forced to conform to imposed restrictions, even an untrained eye can see the inherent desire for freedom in tamed trees who are limbed up tidily, pruned like manicured lollipops, or topped in careless butchery, surrendering in competition with electrical power lines. Even an uneducated eye can discern what looks unnatural, and we feel empathy and compassion for such an unhealthy looking tree.[2] Call this intuition, conscious awareness, spiritual attunement, or simply common sense. We are able to recognize a tree's true character and sense who it could be if it lived under optimal conditions, free of pollutants and surrounded by its wild ecological community—not denied its true identity. Like people, trees need neighbors. But in most urban and landscaped situations they are *missing* the other trees, plants, mycorrhizal fungi, soil organisms, insects, birds, and mammals that make up a full ecosystem trees depend upon. A tree is not just a tree; it is a complex living system, responding to its community. Nothing lives in isolation.

Many species of California oak trees (*Quercus* sp.), for example, can be a host for 300 distinct organisms[3] that either call the tree home or rely on it as an important place to visit daily or seasonally. These relationships determine all organisms' identity; each one's existence depends upon the other, and such is the web of life. So where does a tree *end* and an insect, animal, plant, fungus, bacteria, or even water or land itself *begin*? Where does a hollow cavity in a trunk end and an owl's home begin? Where does a tree's bark end and epiphytic moss or lichen begin? Is there a distinct border between a tree and the wind, sunlight, and water it interacts with? Where is the line between a tree's root hair, the mycorrhizal fungi, and the mycelium network that assists the tree's cycling of nutrients deep below the surface? Rainwater absorbed into well-aerated soil is drawn up by these root hairs, travels against gravity upward into the phloem layer of the trunk, and branches out into thousands of leaf tips, before excess water is transpired into the surrounding air. Trees and forests create their own microclimates and help to regulate the surrounding temperature, not only through creating shade but also through recycling moisture. Where does a forest end and climate change begin?

Trees are one of the most vital ecological parts of our globally interconnected living system. They are our partners on this planet, shaping climate, regulating weather and the hydrologic cycle, producing oxygen, and sequestering carbon dioxide. They absorb atmospheric pollutants, prevent soil erosion, fix nitrogen, cycle nutrients, and provide habitat for countless organisms above and below ground. Trees have been on the planet for 360 million years, long before we have. With the exception of some cultures whose harsh ecosystems are treeless, humans have always relied on trees directly for subsistence and existence. Trees made human cultures possible, granting us food, shelter, windbreaks, wood, fiber, fuel, fertilizer, and countless other physical and cultural gifts. Wherever there were people and trees, there was traditional knowledge, understanding, and wisdom, based on generations of acute observation and intentional interaction. Where does a tree end and human culture begin?

Wherever the oak tree was a dominant species, acorns were the staple food for many cultures worldwide that developed agroforestry systems, before and even after the wide-scale production of agricultural grains. For thousands of years, native Californians gathered acorns, tending to more than a dozen species of oak trees, as well as other trees and plants. Acorns were a principle food source, sustaining ¾ of all native Californians by providing nutrition and calories (up to 45% of their daily diet). In some regions, a family could harvest, store, and eat up to 500 pounds of acorns a year! They used management techniques such as low, slow-burning, controlled fire to keep away diseases and damaging insects, to prohibit catastrophic fires, and to add nutrients to the soil. They also managed oak trees by pruning, propagating, and knocking on tree limbs.[4] They sung songs and did ceremonies for the oaks to maintain their relationship. They were nourished both physically and culturally by their intimate relationships with these trees.[5]

What happened to all these acorn-eating *balanocultures*[6]? With some exceptions, the processing of nutrient- and carbohydrate-rich acorn food has become a lost art. Some practitioners and researchers of traditional ecological knowledge believe that the rampant spread of disease in some species of oaks and other acorn-producing trees[7] is due to not only to the rippling effects of global climate change and environmental unbalance, but also stems from this loss of intimate interaction—a cultural reverence and reciprocal dependence. I suspect this is true as well for many other tree and plant species that are no longer being utilized and appreciated as they were traditionally. In the ethnobotanical co-evolution of plants and people, the phase of development in which a cultivated crop is no longer tended and goes feral is called "cultural abandonment."[8]

"When people don't use plants they get scarce. You must use them so that they come up again. All plants are like that. If they're not gathered from, or talked to, or cared about, they'll die."[9]

Today, we don't think much about oak woodlands being orchards, nor do we think of forests, grasslands, wetlands, deserts, or coastal, riparian, or tundra ecosystems as farms that require horticultural experts to grow and harvest crops, or to be managed for the future. Historically, everywhere people lived, the wild was tended and cared for because it was considered to be home, and other species were part of the community. Of course, the wild is still our home, and the biodiversity of these communities are worth preserving and reviving. In fact, at this point in the ecological crisis, the wild is depending upon us to lend a hand—to know when to listen, when to get involved, and when to step back and get out of the way.

So as the resurgence of traditional ecological knowledge continues parallel to the science of ecological restoration, and as the industrial-chemical, fossil fuel-dependent systems of agriculture, food production, and transportation continue to collapse, it is obvious that the science of environmental conservation includes reviving traditional cultural skills—what Dennis Martinez calls "eco-cultural restoration." The resilience of forests, woodlands, and everywhere else depends on this relationship.

Tree literacy is crucial for understanding this resilience, and arboriculture is the modern science of caring for individual trees, from birth to death.[10] Arborists are the trees' caretakers, or "doctors," and an effective arborist will have a good understanding of the sciences of botany, ecology, and biology, as well as an affinity, instinctual curiosity, and deep respect for trees. The health of a tree is an indicator of the health of the entire ecosystem, and vice versa. Could we say the same about the health of the trees and the health of the human species?[11]

A tree's health is determined by both its expression in vigor (genetic capacity for resilience) and vitality (ability to grow, reproduce, and adapt to its surroundings). A tree could be high in vitality, but low in vigor and vice versa.[12] Like us humans and most organisms, trees are affected by their environment. Trees react to biotic ("living") stresses—invasions from animals, insects, fungi, and microscopic diseases, as well as compact and unfertile soils. They are also affected by abiotic ("non-living") environmental factors, such as water, lightning, chemicals (natural and manufactured), salts, elements, gases, minerals, acid rain, erosion, geologic forces such as seismic activity, as well as altitude, gravity, wind, temperature, and, of course, sunlight. We've all seen a shaded plant or tree reach for the light (heliotropism), but did you know that trees are susceptible to too much direct sunlight? Some species of trees, positioned directly next to cement or asphalt (which reflect excessive heat and light) can sunburn, resulting in bark chipping off, creating a vector for disease. Keep a look out for these casualties on the sides of sidewalks, streets, or parking lots.

Many human activities and products affect trees, including concrete, underground pipes, and jackhammers. Well-intentioned but inherently poor pruning and trimming, bracing, filling of cavities, or simple nail punctures all place stress on trees' cellular structure. If a tree is vigorous enough, the conditions favorable and the damage not too radical, it will survive. However, many times it will not. Vigor cannot be increased without genetic manipulation, but vitality can be improved by cultural practices such as fertilization, soil aeration, watering, and supporting a healthy surrounding ecosystem.

Trees respond unfavorably to being grown in containers, and many such trees are doomed. Root-bound trees are found all too often in even the best-intentioned nurseries. Trees in bondage will rarely grow normal root systems that are able to absorb water, nutrients, and oxygen (yes, trees need oxygen, too). Girdling often occurs as contorted roots encircle the trunk and strangle the tree. A young tree may be planted like this, and a few years later simply fall over or die prematurely. Indeed, urban street trees are notorious for having low average lifespans and high mortality rates due to poor planting, poor environments, and poor management. The average lifespan of a tree in an urban environment, with soil that is compacted or stressed from lack of water and nutrients has, sadly, been estimated at between a mere seven and fifteen years (depending upon the study and the tree).[13] Keep a look out for these sad-looking characters when you search to park your car under the shade of an unhealthy canopy. Can you read a tree's vitality by observing it or touching it? How about with simple intuition?

Suburban trees fare a bit better due to somewhat more favorable conditions. In many cases, if the same urban trees were planted in optimal suburban conditions and managed properly, they would have a much longer life expectancy of 60–200 years[14], depending on the species. Many of the trees along roads and sidewalks, in parks and yards—doing their job of sucking up carbon dioxide, beautifying the landscape, providing habitat, and offering temperature regulation—are sentenced to a shortened lifespan and will have to be replaced at some expense, only to be possibly replaced again later. To avoid this tragic waste of trees from biotic, abiotic, or caretaking stresses, we need not only better urban design (such as permeable pavement), but we also need a greater knowledge and understanding of a tree's biology.

We can start by learning the names of our botanical friends, as well as whether they are native or exotic to our region and whether they are endangered and by what. This can help us feel a bit more confident in the eco-literacy of our place. For a *place* is simply a community of beings (biotic and abiotic) working together as a system. Suddenly being in the natural world becomes a little less lonely and a bit more like home when we can acknowledge our neighbors around us, and even more so when we know a bit of their story. It becomes even more intimate if we help shape that story. We don't all need to become arborists, but it would help if we all had some simple knowledge and tactile experience with trees in order to understand and care for them, whether on our own properties or in our communities. In fact, we know a lot more now about trees than we did even a generation ago, and there is always more to discover.

The late, great Alex Shigo was the father of modern arboriculture. Trained as a plant pathologist, he was also a conservationist with an unprecedented breadth and depth of scientific exploration. He brought professionalism and integrity to the arboricultural field, as well as attention to trees as living beings who command respect. With his passion for promoting a true understanding of tree biology, Shigo catapulted tree care out of the Dark Ages, ending misguided practices such as filling a tree cavity with cement or bricks, or painting a branch's stub cut with tar. He proved that many such old-school practices promoted decay and even hastened a tree's death. The simple lack of knowledge and forethought had created a culture of ignorance and fostered a disregard for the complexity of trees as biological beings. Shigo noted that if there is a tree problem near a home, or anyplace where there are many people, over 90% of the time the cause of the tree problem will be the people and their activities.[15]

Shigo also taught that trees have dignity. "Plan before you plant" is one well-known "Shigoism." Trees should not be planted where they won't have room to grow, only to be mutilated with bad "pruning," topped or chopped down later because they are overcrowded or in an inconvenient place. Take a ride down any street lined with electrical lines and trees, and you'll observe "necessary tree trimming" that often looks and feels like butchery. That's because many people, even trained tree workers, have not been educated properly, nor are they sensitive to the long-term health of an individual organism or its role in its ecosystem.

Perhaps the most radical aspect of Shigo's work was his emphatic teaching that, to be more effective, arborists need to understand tree biology, to get to know trees intimately—to "Touch Trees," a slogan that became his trademark. For Shigo, an arborist's knowing touch was equivalent to the mastery of a well-experienced surgeon. Using a chainsaw, Shigo examined the interiors of thousands of trees, both decaying and healthy, to understand their intelligence in self-maintenance by compartmentalizing their own wounds in four ways to stop the spread of disease.[16] Discovering how a tree worked to prevent its own death gave Shigo the understanding of how humans could unintentionally cause it, and thus how to prevent it. He was not unlike Leonardo da Vinci, a radical artist and scientist in his time, who dissected the human body and other once-living things, to comprehend his rendered subjects more intimately and make his beautiful art more realistically accurate. Shigo believed that a good arborist combines science (knowledge) with art (skill) and common sense.

Alex Shigo cared deeply about trees and about people. He encouraged arborists to care for their patients by knowing them as individuals rather than making assumptions about a tree's health from a distance. His primary research led to new standards in tree care practice, changed the consciousness of arborists and foresters, and earned him great respect. He wrote over 270 publications and lectured with a daring, outspoken, heartfelt candor. With his chainsaw in hand, Shigo was also a tree-lover and tree-hugger at his core. He created a life of intimacy with trees. Shigo is my hero, for he said what other scientists never dared to say: "You have to touch a tree and feel it." Sexy!

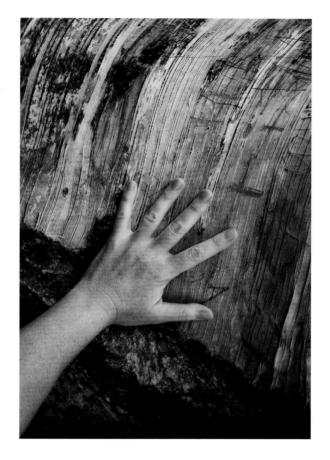

[1] *http://www.paloaltoonline.com/square/topic_print.php?t=9156* (accessed 8-13-13).

[2] Remember the scene in the 1970 film, *Harold and Maude*, in which the couple, out of love and empathy, steal a car, a policeman's motorbike, and a shovel in order to heroically transport and transplant a very sick-looking urban tree to its new home in a forest?

[3] *Oaks of California*. Bruce M. Pavlik, Pamela C Muick, Sharon G. Johnson and Marjorie Popper (Los Olivos: Cachuma Press, 1991).

[4] *Tending the Wild: Native American Knowledge and the Management of California's Natural Resources* (Berkeley: University of California Press, 2005). p. 285–290.

[5] Depending on the species, acorns can consist of up to 6% protein, 18% fat, and 68% carbohydrate. In comparison, modern varieties of wheat and corn consist of approximately 2% fat, 10% protein and 75% carbohydrate. *Oaks of California*, p. 96–99.

[6] 'Balanoculture' is a term coined by David Bainbridge in "*The Rise of Agriculture: a New Perspective*." Abstract: "Interest in and research into the origin and development of agriculture has increased sharply in the last twenty years, yet all of these studies have missed the common link between the areas where agriculture may have begun—the acorn. All three areas considered of significance to date—the Middle East, middle China, and Mexico—are, or were once, characterized by oak woodlands. The experience in California, where ethnographers and anthropologists have been able to study a fully developed balanoculture (from the Greek *balanos*—meaning acorn) reveals the primacy of acorn use and the complex interaction between people and oak woodlands. The California balanoculture was in fact a very successful agroforestry system that prospered for thousands of years. Balanoculture provided the stable communities necessary for agriculture to develop. The lower time and work cost associated with acorn use suggests agriculture may have evolved as acorns became more scarce from the decline in the oak woodlands brought about by the adverse human impacts resulting from overgrazing, fuel cutting and cutting for timber, and field burning, exacerbated by climatic fluctuation. A reevaluation of the record is in order: agriculture may perhaps be better considered a regressive rather than a progressive evolutionary event." *Ambio* Vol. 14, No. 3 (1985), p. 148–151. Published by Springer.

[7] The acorn-producing tree TanOak, *Notholithocarpus denisflorus*, is not a true oak but in the same family. TanOak acorns were and still are a favorite of indigenous tribes, and are severely threatened by the disease sudden oak death, caused by the pathogen *Phytophthora ramorum*. Fires would have kept these pathogens at bay.

[8] In the web article "Co-evolution: Man and Plants" by Brian Altonen, the author explains the four phases of development of domesticated crops. While oak trees were not a 'domesticated' agricultural plant per se, they were tended to assure healthy crops of acorns as a primary food source for thousands of years. I argue that while oak trees were not significantly engineered or altered like some plants, a relationship of harvesting and tending significantly affected their production. "There are four fairly well defined stages in ethnobotany and economic botany that can be defined based on the historical events which take place whenever a particular plant species or its value is discovered…The fourth or final phase in species development is the abandonment stage, when for some reason the plant is no longer useful and passively returns to the situation it was first surviving in within the natural ecosystem. The disadvantage this change has on the plant is that since the plant has been significantly modified, it is also more susceptible to losing the battles it faces related to the natural selection processes…This may be worsened by environmental changes taking place due to the human cohabitation process itself. The simple act of reducing woodlands and natural grasslands and swamplands size and amounts reduces the likelihood the plant's original habitat will be found. "Co-evolution: Man and Plants," Brian Altonen, *http://brianaltonenmph.com/natural-sciences/4-projects/plan-tae-the-evolution-of-plant-chemicals/co-evolution-man-and-plants/* (accessed August 12, 2014)

[9] Mable McKay, Pomo Elder, in "*News from Native California*", Vol. 7, No. 3 (Summer 1993).

[10] Arboriculture usually refers to the science of urban trees, as opposed to forestry or silviculture (growing trees for harvesting).

[11] See *Forests, Trees and Human Health*. Kjell Nilsson, Marcus Sangster, Christos Gallis, Terry Hartig, Sjerp de de Vries, Klaus Seeland and Jasper Schipperijn, Ed. (New York: Springer). Kindle Edition.

[12] Alex L. Shigo. *A New Tree Biology: Terms, Topics, and the Treatments for Trees and Their Problems and Proper Care* (Durham: Shigo and Trees, Associates, 1986).

[13] Lara A. Roman and Frederick N. Scatena "Street tree survival rates: Meta-analysis of pervious studies and application to a field survey in Philadelphia, PA, USA." *Urban Forestry & Urban Greening* 10 (2011) 269–274.

[14] *http://www.powertrees.com/urbanforestry.htm* (accessed 8-12-13).

[15] Alex L. Shigo. *A New Tree Biology: Terms, Topics, and the Treatments for Trees and Their Problems and Proper Care* (Durham: Shigo and Trees, Associates, 1986).

[16] Alex L. Shigo. *A New Tree Biology: Terms, Topics, and the Treatments for Trees and Their Problems and Proper Care* (Durham: Shigo and Trees, Associates, 1986).

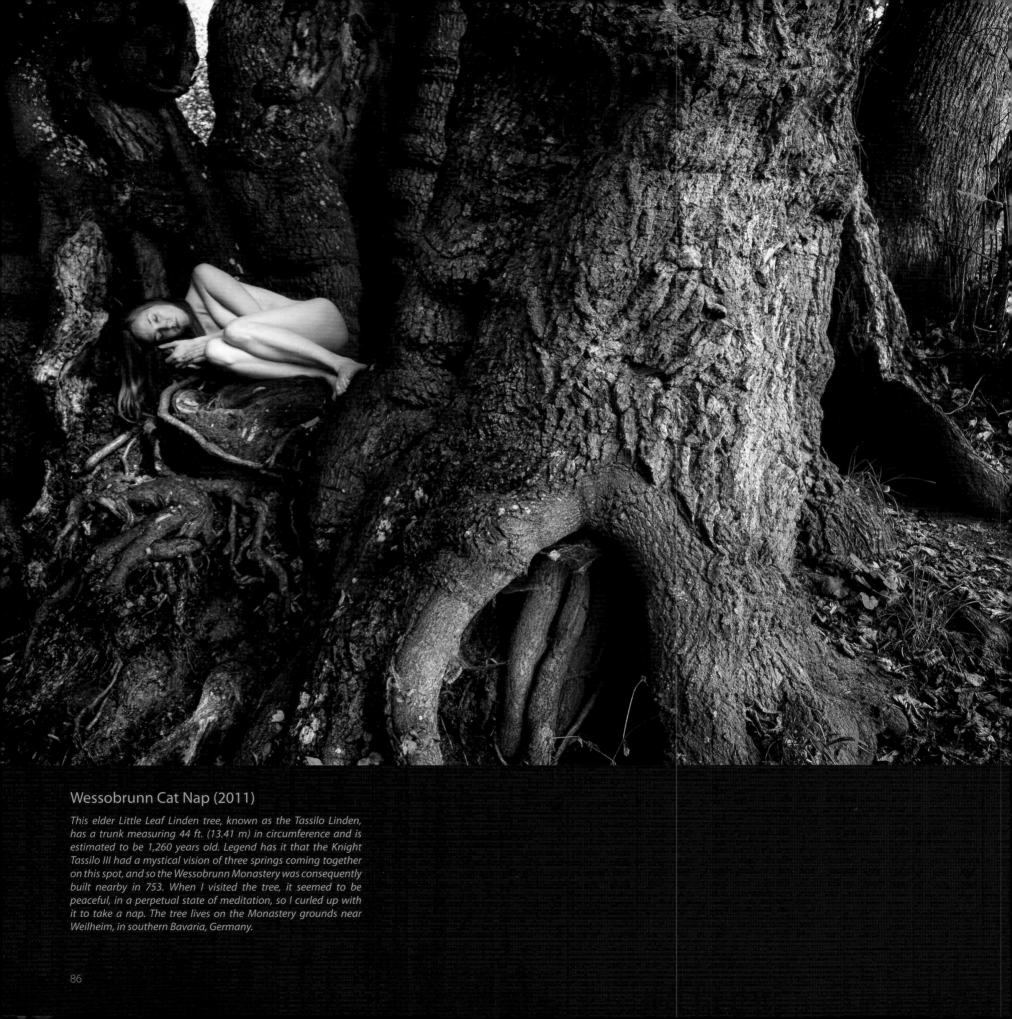

Wessobrunn Cat Nap (2011)

This elder Little Leaf Linden tree, known as the Tassilo Linden, has a trunk measuring 44 ft. (13.41 m) in circumference and is estimated to be 1,260 years old. Legend has it that the Knight Tassilo III had a mystical vision of three springs coming together on this spot, and so the Wessobrunn Monastery was consequently built nearby in 753. When I visited the tree, it seemed to be peaceful, in a perpetual state of meditation, so I curled up with it to take a nap. The tree lives on the Monastery grounds near Weilheim, in southern Bavaria, Germany.

Little Leaf Linden

Tilia cordata

Other Names: Small-leaved Lime, Small-leaved Linden, Little-leaf Linden, Lime

Distinctive Characteristics: Both the Little-leaf Linden and its relative, the Big-leaf Linden (*Tilia platyphyllos*), are handsome, fragrant, densely foliated, deciduous trees with heart-shaped leaves. Like yews and beeches, ancient Lindens can hollow and deteriorate into what seem like separate trees. Aphids like to eat the flowers and leaves, dropping a sticky substance, called honeydew. They are known for their longevity, and responding well to coppicing.

Distribution: Native to Britain, most of Europe, Turkey, and Western Asia. Elevation: 6–3,400 ft. (2–1033 m).

Ecosystem: Moist deciduous forests with English Oak, Sessile Oak, European Beech, Oriental Beech, Sweet Cherry, European Ash, Scots Elm, White Fir, European Larch, Silver Birch.

Maximum Age: Estimated at 1,300 years.

Maximum Height and Girth: 130 ft. (40 m) in height; 59.7 ft. (18 m) in circumference.

Animal Community: Lindens provide habitat for a number of bird and insect species.

Traditional Uses:
Medicine: The flowers, leaves, wood, and charcoal have all been prepared for medicinal purposes including relief from colds, cough, tonsillitis, diarrhea, fever, infection, rheumatism, and inflammatory skin disorders, including burns and hemorrhoids. Preparations were also used as a treatment for high blood pressure, vomiting, indigestion, as well as a nervine for calming. The flowers in water were used for a cold face wash. The German verb for alleviate is *lindern*.

Food: Young leaves are edible.

Fiber: The fibrous inner bark (bass or bast) was used to make cordage, ropes, fishing nets, paper and shoes.

Art and Ceremony: Culturally, the living trees themselves were often the center of social life in many villages, as well as judicial meetings. Special *Tanzlinden* in Germany were social centerpieces that served as settings for the village dance floor for ceremonial dances (*Tanz* means *dance* in German). Many of these special trees still are used today, some recognizable by their circular scaffolding built to encircle the dance ring and hold up the big branches. The wood was a favorite for carving musical instruments, religious sculptures and small canvas on which to paint.

Modern Uses: The wood has been used for beehives, spoons and light furniture. Sweet and fragrant *Lindenblüten* (Linden blossoms) are still an everyday popular herbal tea in Europe.

Threats and Conservation: Not threatened. It is protected in Great Britain, as populations are rare.

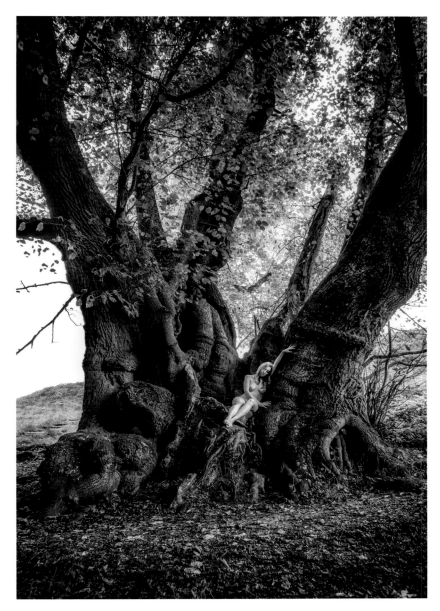

Wessobrunn Dream (2011)

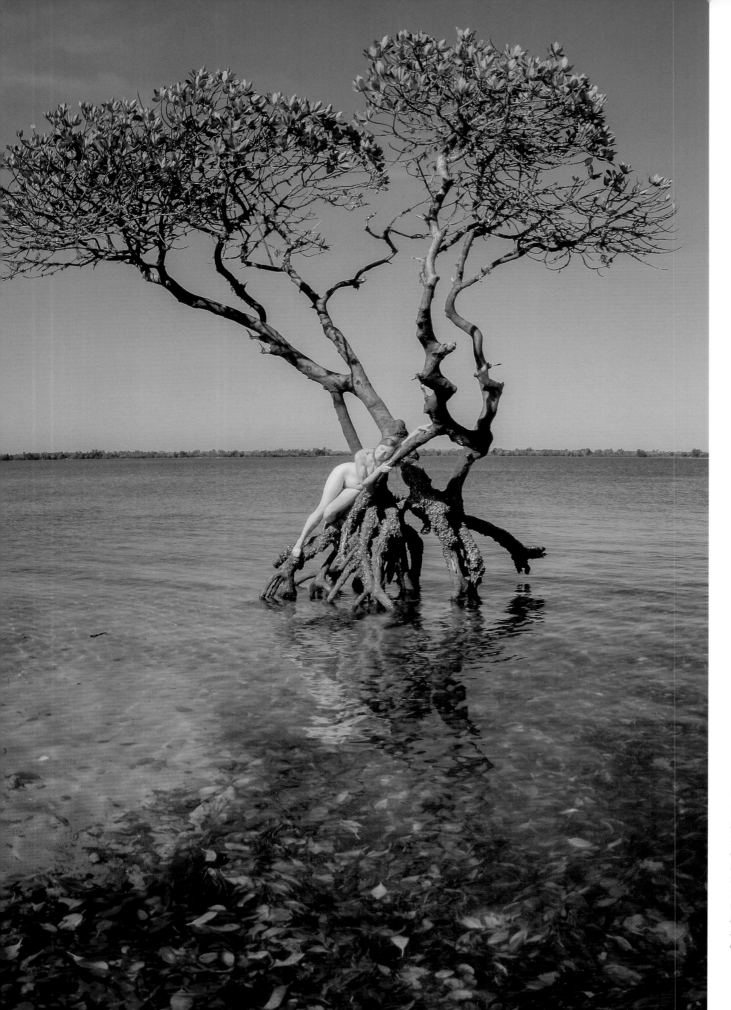

Mangrove with TreeGirl (2008)

This lone, unidentified Mangrove species is located just off the warm, pristine waters of the southwestern coastal town of Ifaty, in Madagascar. The stilts I lay down on were encrusted in barnacles and a bit slippery, but I managed to find where I fit in order to get the photo. The fallen leaves of the Mangrove had turned the water red with their tannins. Wading into the warm water changing colors was like walking into a brilliant watercolor painting. Once again, I only had seconds to take this shot, and accidentally left some clothing on the beach in order to make my stealthy departure.

Mangrove

unknown species

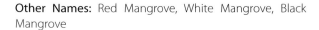

Other Names: Red Mangrove, White Mangrove, Black Mangrove

Distinctive Characteristics: "Mangrove" refers to a variety of evergreen tree and shrub species that is specially adapted to salty waters, having evolved physiologically to overcome high salinity and frequent tidal inundations. They are an extremely important type of forest ecosystem for the planet. Because they live in a coastal zone, between fresh and salt water, they prevent soil erosion and build up land by catching sediment and debris between their roots, also protecting coral reefs from suffocating in silt. They are able to 'breathe' in oxygen through their bark.

What makes mangroves aesthetically beautiful is the exposure of their sturdy and delicately balanced root-stilted systems during low tides. Their long, thick roots plant themselves firmly into the sandy soil, creating an underwater landscape, while their aerial roots act as snorkels, exposed to the air to absorb massive amounts of CO_2. Masses of propped roots break the waves and dissipate their force, protecting coasts even from tsunamis. Their seeds are adapted to their moving waterscape; the *Rhizophora* species have seeds that 'pre-germinate', growing while still in their pod on the tree, so that when they hit the mud, they are immediately able to root themselves.

Distribution: Mangrove forests are on tropical coastal regions throughout the world. This unidentified species is native to the Western coast of Madagascar. Elevation: sea level.

Ecosystem: Tropical and subtropical waters, where fresh water from rivers flows into the ocean. In Madagascar, most of the mangrove stands contain six species in four families: Rhizophoraceae (*Rhizophora mucronata*, *Bruguiera gymnorrhiza*, and *Ceriops tagal*), Acanthaceae (*Avicennia marina*), Lythraceae (*Sonneratia alba*), and Combretaceae (*Lumnitzera racemosa*).

Maximum Age: Some species are estimated at over 100 years old.

Maximum Height and Girth: Red Mangroves are the tallest, growing to 80 ft. (25 m); trunk circumference is not substantial.

Animal Community: Mangrove habitats provide a unique sheltered habitat for a great diversity of coastal and estuary species such as mussels, crustaceans, and barnacles; endemic and migratory birds; as well as insects, fish, turtles, and even crocodiles. In this way they bridge the food webs of both the ocean and land.

Threats and Conservation: Because mangroves reside in sensitive and vulnerable ecotones, they are in danger everywhere across the planet. Mangroves are increasingly threatened by pollution from development of urban areas, over-fishing, over-harvesting of wood, and erosion caused by deforestation. In Madagascar the massive deforestation of the island has led to a constant flow of red, salty soil into the sea. Rising sea levels due to climate change, are also a threat. Globally, the greatest threat is the over-conversion of mangrove areas into economic use for rice farming, salt harvesting and production, beach resort developments, shrimp aquaculture, and even dumps.

Mangrove Seeds

The seeds of the Rhizophora *mangrove germinate on the tree before they are dispersed. When the seedling drops from the parent tree, it uses the already grown stem as an anchor in the mud or as a floating device.*

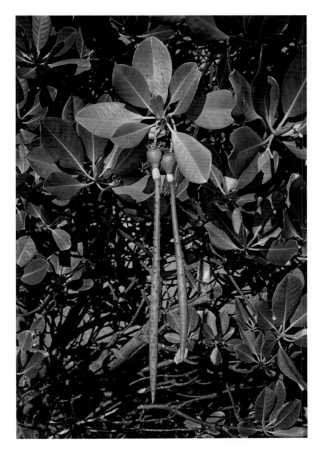

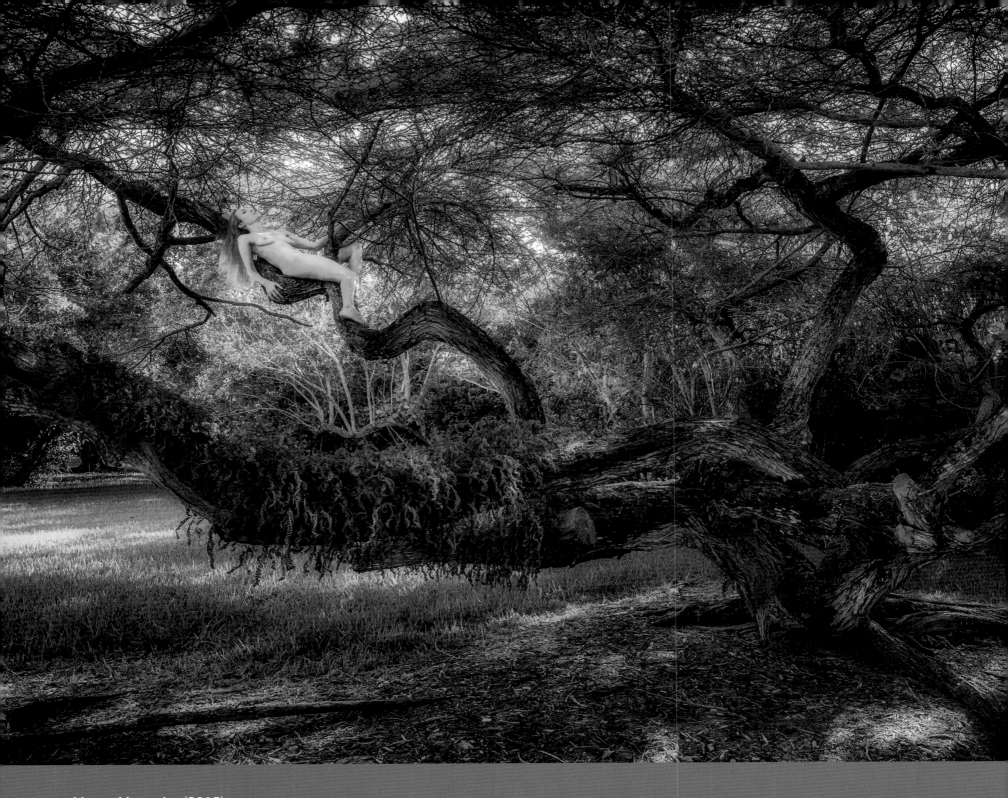

Mossy Mesquite (2015)

Sometimes a tree does not have to be big to be a perfect treegirl tree; it just has to have character and a comfortable place for a girl to rest. I know it's a perfect tree when I see instantly where my body would fit. If I fit, it's a good shot; if I don't fit, there's no shot worth taking. The perfect treegirl tree will also be in a very private place. Too often, a tree is right on a trail. But part of the thrill of being a treegirl is the ability to shift into invisible stealth mode to avoid unsuspecting passersby. In this case, landscape workers on golf carts are not looking for a naked woman up in a tree, so they simply do not see her. This octopus-shaped, shaggy Mesquite

tree was unlike any other I had ever met. After sussing out the tree, asking its permission, and setting up the shot, I put on my invisibility cloak and quickly got into place. Having someone to scout is extremely helpful as well. And the entire day was made possible only with the aid of a dear friend and photographer, The Legend. This Mesquite lives a humid life in Fairchild Botanical Gardens, Miami, Florida, USA.

Mesquite

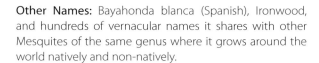

Prosopis juliflora

Other Names: Bayahonda blanca (Spanish), Ironwood, and hundreds of vernacular names it shares with other Mesquites of the same genus where it grows around the world natively and non-natively.

Distinctive Characteristics: Drought-tolerant, nitrogen-fixing, semi-deciduous, single- or multiple-trunked shrub or tree with twisted, cracked, rough bark. Branches have thorns up to 2 in. long. Characteristic leaves are bi-pinnate, compounded with 12–20 small leaflets. The seedpods, characteristic of leguminous trees, are 8–12 in. (20–30 cm) long and contain 10–30 edible seeds that remain viable for up 10 years. Mesquites have some of the deepest known roots, found at depths up to 175 ft. (53 m) in search of water.

Distribution: Native to Mexico, Central and northern South America, and the Caribbean. The *Prosopis* genus is an established and aggressive invasive species in the southwest U.S. and other countries outside of its native range, including Africa, Asia, the Middle East, and South America. Elevation: sea level to 6000 ft. (1800 m).

Ecosystem: Plains, canyons, and hillsides in arid and semi-arid regions, near watercourses such as desert washes, streams, arroyos, and creeks.

Maximum Age: Estimated at over 250 years old.

Maximum Height and Girth: Height up to 39 ft. (12 m); circumference up to 12 ft. (3.7 m).

Animal Community: Wild and domestic animals eat the leaves and seeds. Bees make honey from the nectar.

Traditional Uses:
Medicine: The soothing, astringent, antibiotic, and antiseptic properties of mesquite have long been known and utilized by many native tribes throughout the southwest United States and northern Mexico, including the Pima, Seri, Papapago, Paipai, Paiute, Tewa, Mayo, and Yaqui. The branches, stems, and inner yellow bark were used as purgatives; the stems to treat fever; the bark for bladder infection, measles, or fever. A poultice of leaves was used to treat red ant stings. The pods were used to make eyewashes, to treat sunburn, and as a drink for animal stings. The gum was used externally as an eyewash and to treat skin infection, wounds, burns, chapped fingers and lips, sunburn, hemorrhoids, and lice. Internally, the gum was used to treat sore throat, laryngitis, cough, fever, painful gums, diarrhea, and digestive inflammation, and as a purgative. The leaves were prepared to treat pinkeye, diarrhea, headaches, bladder infections, and painful gums, and to cleanse the digestive tract.

Food: The seeds can be made into a tea, fermented into an alcoholic beverage, boiled into a sugary syrup, or ground as a highly nutritious flour for baking.

Fiber: The bark has been used to make cloth, baskets, and rope.

Tools and Objects: The wood was used for small tools and construction.

Art and Ceremony: The gum was used to make face paint, hair dye, and pottery paint, and as a glue for mending pottery.

Modern Uses: The small wood is used for small lumber, posts, and tools. Agriculturally, the tree is used for erosion control, windbreaks, and shade, while the leaves and pods are used as livestock fodder. It is also popular as firewood.

Threats and Conservation: Not threatened.

Honey Bees on Mesquite Flowers

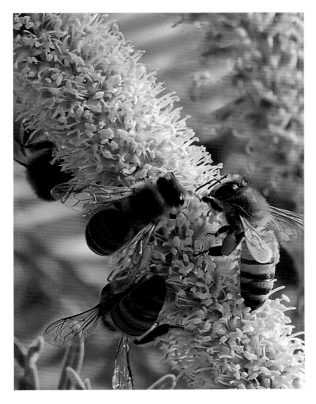

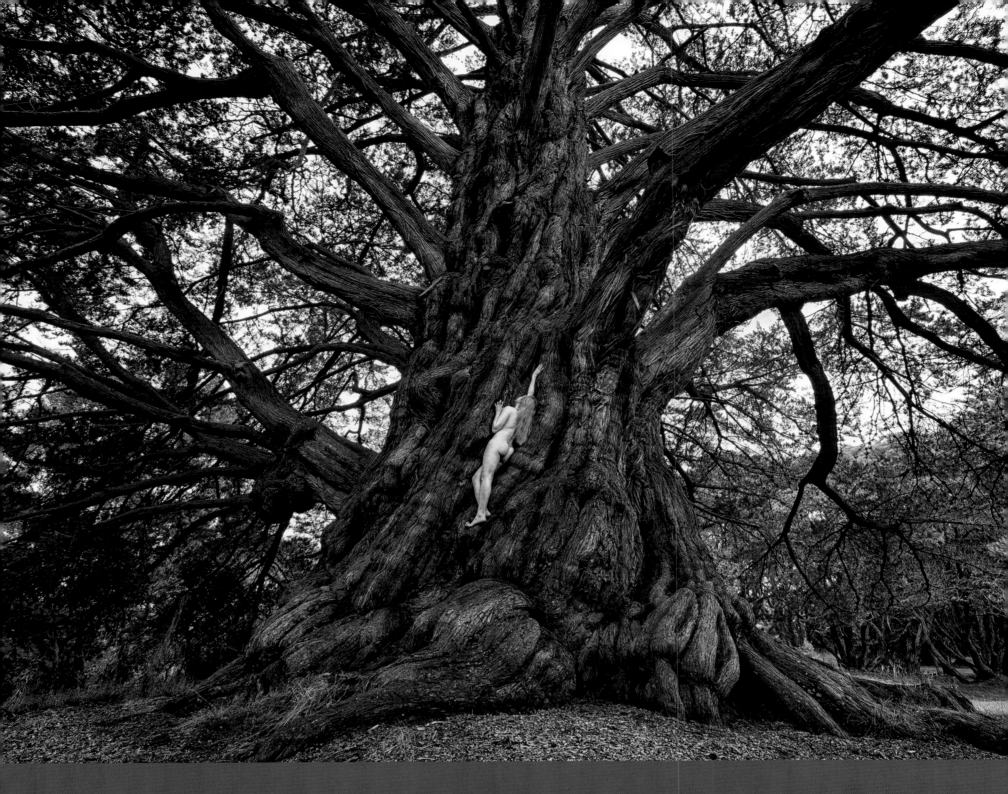

Monterey Cypressness (2015)

This magnificent Monterey Cypress is a giant, twisting upward like a vortex into another dimension. It is one of my favorite individual trees in the world. Trees of this species rarely get this big outside of their native region in California. But they seem to thrive in countries like New Zealand, where they have been planted with many other California trees, replacing the native trees that were deforested.

I had two glorious photo shoots with this tree: one in the evening at dusk when it started to rain, and one sunny morning on my

last day in New Zealand—the final day of an epic forty-seven-day TreeGirl expedition. I could have not asked for a more perfect ending. This tree is located in the Awhitu Park, outside of Auckland. It is 78.74 ft. (23.77 m) tall, with a circumference of 47.47 ft. (14.47 m).

Monterey Cypress

Cupressus macrocarpa

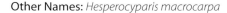

Other Names: *Hesperocyparis macrocarpa*

Distinctive Characteristics: This conifer can be dramatically sculpted into twisted and distorted forms when exposed to strong ocean winds, salt spray, and poor soils on the coast. The crown is widespread and sparse with foliage. The trunk can be deeply fluted and the base buttressed. When cultivated outside its native range, it is a relatively fast-growing, erect tree. While it is semi-dependent upon fire for reproduction, fire can easily destroy trees.

Distribution: Native to a narrow coastal strip in only two small populations, between Cypress Point and Pescadero Point on the north side of Carmel Bay, and near Point Lobos State Natural Reserve, on the Monterey Peninsula, California. It has been planted widely outside its native range in the U.S. and internationally. Elevation: sea level to 100 ft. (30 m).

Ecosystem: Along foggy, ocean salt-sprayed coasts on sandy soil, on rocky cliffs, rock crevices, slopes and headlands, exposed to nearly constant onshore winds. Forms in pure stands, or with Monterey Pine and Gowen Cypress trees, and associated with understory including California sage brush, pearly everlasting, coyote brush, ceanothus, bracken fern, chamise, Hooker manzanita, chaparral broom, blue blossom, liveforever, seaside daisy, golden yarrow, lizard tail, salal, Douglas iris, bush monkeyflower, Pacific bayberry, skunkweed, poison oak, and California huckleberry.

Maximum Age: Estimated to 300 years.

Maximum Height and Girth: Height up to 158 ft. (48 m); up to 25 ft. (7.6 m) in circumference. (None of the largest trees occur in their natural range.)

Animal Community: Rodents and deer consume seedlings.

Traditional Uses:
Medicine: A decoction of the foliage was used for rheumatism (Costanoan).

Modern Uses: The wood is used for building furniture, fence posts, boats, and fine crafts. It is planted in Australia and New Zealand as a hearty windbreak tree on farms.

Threats and Conservation: Its status is "Vulnerable" on the IUNC Redlist, due to its limited natural distribution, semi-dependence upon fire for regeneration, and climate change.

Monterey Cypress Wind (2015)

Monterey Cypress (Cupressus macrocarpa), Slope Point, South Island, New Zealand.

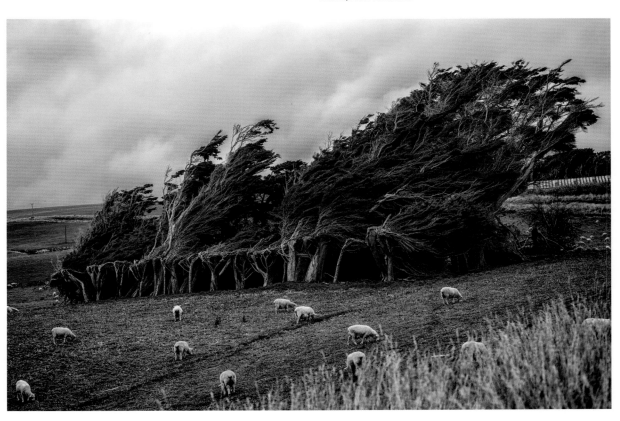

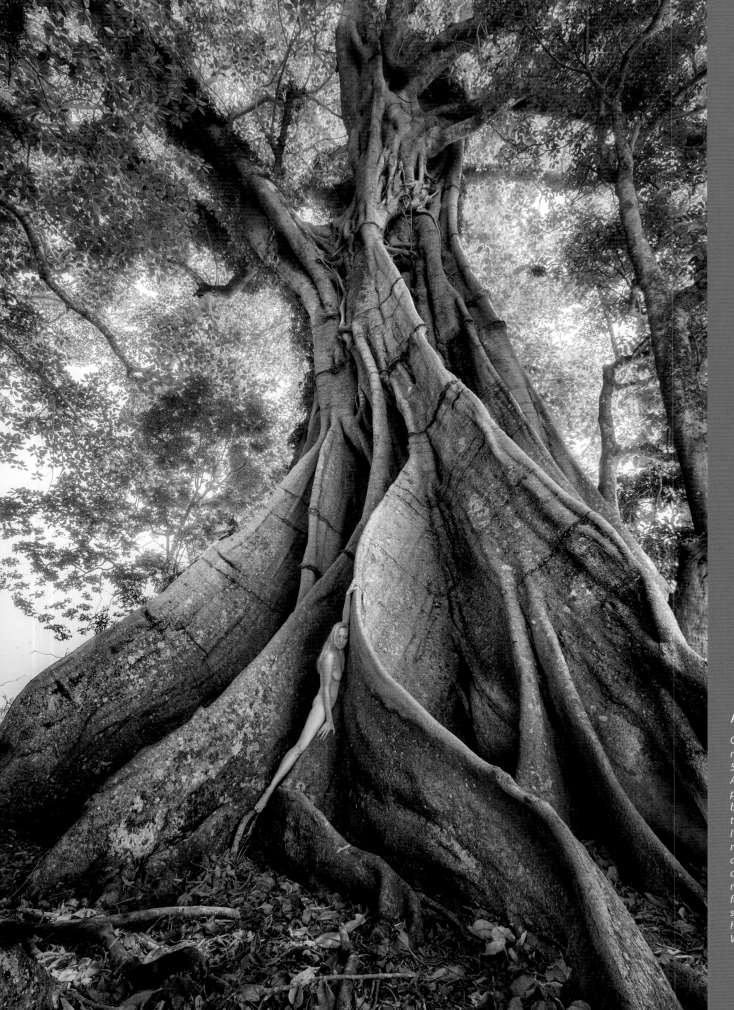

Avatar Tree (2015)

Occasionally I meet a tree whose presence astounds me so much I have to name it. This is my Avatar Tree, measuring 59 ft. (18 m) in circumference and 164 ft. (50 m) in height. Are they siblings, mother and baby, lovers, friends, or partners of some kind? Their giant, ribbony, rippling, buttressed roots almost touched each other; entering the space between them felt like a doorway into figness. I entered their world and spent three hours with them, mounting and conforming to the flared base of their wild architecture, which felt like separate rooms of a treehouse or folds of a moving organ in a giant's body. Holding myself still against buttresses of the larger tree, I could feel it moving as if it were growing and expanding in super slow motion as it anchored in place. This tree is a bit hidden off the road, in a field near Bellingen, New South Wales, Australia.

Moreton Bay Fig

Ficus macrophylla

Other Names: Australian Banyan, Figwood, Black Fig, kar-reuaira (Dharawal), guwanggaga (Gumbaynggirr)

Distinguishing Characteristics: Like many Fig trees, the Moreton Bay Fig develops a wide trunk with tall, ribbony, buttressed roots. It forms a dense canopy of glossy evergreen leaves with rust-colored undersides. It can grow as a single-trunk tree, but more likely it grows as a parasitic 'strangler fig': born from a seed dispersed by a bird or bat atop a tree, it starts life as an epiphyte, absorbing nutrients and water from the air, dropping down aerial roots from the host's branches that eventually take root in the earth. Over time, roots aggressively intertwine the trunk and fuse together, creating a latticework. Out-competing the host for space, nutrients and light, eventually the new tree becomes freestanding. The leaves and branches bleed a milky sap if cut. The fruits are small figs that turn from green to purple with spots when ripe.

Distribution: Native to the east coast of Australia, from the Atherton Tableland, Queensland, to north of Illawarra, New South Wales. Elevation: sea level to 3,000 ft. (900 m).

Ecosystem: Found in subtropical, warm temperate, and dry tropical rainforests, from river valleys to slopes, often along watercourses in association with White Booyong, White Walnut, Giant Stinging Tree, Lacebark, Red Cedar, Hoop Pine, and Green-Leaved Fig.

Maximum Age: Estimated at 200 years.

Maximum Height and Girth: 200 ft. (60 m) in height; 60 ft. (18 m) in circumference.

Animal Community: The figs are an important food source to such birds as the wompoo fruit-dove, topknot pigeon, rose-crowned fruit-dove, Lewin's honeyeater, yellow-eyed cuckoo-shike, pied currawong, Australasian figbird, green catbird, regent bowerbird, and satin bowerbird. It is also consumed by rodents such as the Sydney bush rat, and bats such as the grey-headed flying fox. All *Ficus* species are dependent upon one species of wasp to pollinate them; the Moreton Bay Fig relies on *Pleistodontes froggatti* for reproduction.

Traditional Uses:
Medicine: The sap was used to treat wounds.

Food: Although not very palatable, the small fruits were eaten.

Fiber: String fiber from the bark was made into fishing nets, dilly bags, and cloth.

Tools and Objects: The root buttresses were made into shields.

Transportation: Bark and branches were made into canoes.

Modern Uses: The Moreton Bay Fig is used as paneling, in cabinetwork and to make packing cases.

Threats and Conservation: Not threatened.

Becoming Figness (2015)

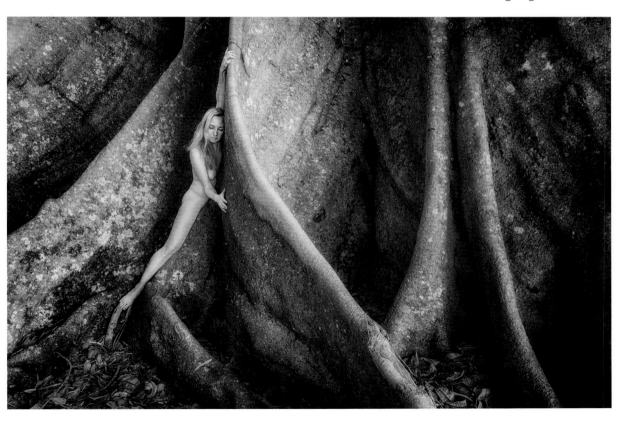

95

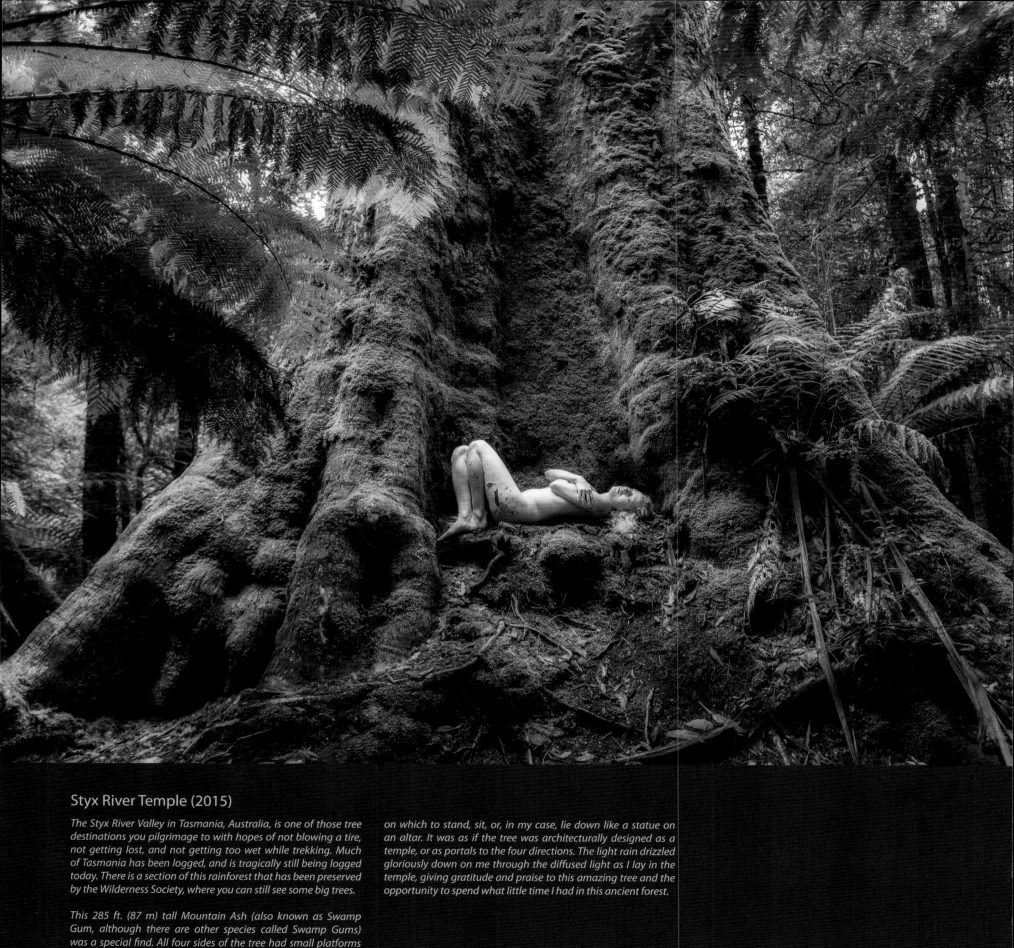

Styx River Temple (2015)

The Styx River Valley in Tasmania, Australia, is one of those tree destinations you pilgrimage to with hopes of not blowing a tire, not getting lost, and not getting too wet while trekking. Much of Tasmania has been logged, and is tragically still being logged today. There is a section of this rainforest that has been preserved by the Wilderness Society, where you can still see some big trees.

This 285 ft. (87 m) tall Mountain Ash (also known as Swamp Gum, although there are other species called Swamp Gums) was a special find. All four sides of the tree had small platforms

on which to stand, sit, or, in my case, lie down like a statue on an altar. It was as if the tree was architecturally designed as a temple, or as portals to the four directions. The light rain drizzled gloriously down on me through the diffused light as I lay in the temple, giving gratitude and praise to this amazing tree and the opportunity to spend what little time I had in this ancient forest.

Mountain Ash

Eucalyptus regnans

Other Names: Mountain-ash, Swamp Gum, Stringy Gum, Giant Ash, Tasmanian Gum, Tasmanian Oak, White Mountain Ash, Victorian Ash, Australian Oak, yowork, warreeha, yowat (Karnathun)

Distinctive Characteristics: This evergreen eucalypt is one of the tallest and fastest-growing tree species in the world; it can reach a towering 213 ft. (65 m) in just fifty years. Mature trees are buttressed at the base, with the lower 50 ft. (15 m) covered in long, peeling strips of bark up to 98 ft. (30 m) long, with the upper bark being smooth and grey. According to calculations, Mountain Ash forests are some of the best for carbon storage.

Distribution: Native to Tasmania and Victoria, Australia. Elevation: up to 3,300 ft. (1,000 m).

Ecosystem: Grows in cool, temperate rainforests, in pure stands or with Blackwood, Grey Gum, Victorian Blue Gum, Manna Gum, Shining Gum, Myrtle Beech, Silver Wattle, Southern Sassafras, Celery Top Pine, Leatherwood, Alpine Ash, Messmate, and Red Stringybark. It hybridizes with the last two of these. At least nine species of epiphytes grow on the Mountain Ash.

Maximum Age: Estimated at 500 years.

Maximum Height and Girth: Mountain Ash is considered the tallest flowering plant (angiosperm) in the world and the second tallest tree in the world, next to Coast Redwood (which is a gymnosperm, having cones). Historically, the Thorpdale Tree (also known as the Cornthwaite Tree) was measured at 375 ft. (114.3 m). Currently, the tallest is the Centurion, at a height of 327 ft. (99.6 m) and a circumference of 25 ft. (7.6 m).

Animal Community: The tiny Leadbeater's possum uses the hollows in old Mountain Ash trees for nesting, shelter, and foraging of insects; it also feeds on the sap, along with the yellow-bellied gliders. Koalas eat the foliage, although it is not their preferred species. Yellow-tailed black cockatoos nest in the hollows of old trees, and the Tasmanian wedge-tailed eagle nests in the tops.

Traditional Uses:
Transportation: Canoes were made from the a large sheet of bark.

Modern Uses: Mountain Ash has been valued as timber for general construction, as well as for making furniture, flooring, paneling, veneer, newsprint, plywood, and wood chips.

Threats and Conservation: Although this species is not listed as threatened, old growth forests of Mountain Ash are in danger. Much of the acreage has been clear-cut, and most of what is left is controlled by the Department of State Forestry. This species is also very susceptible to lack of regeneration after fire.

Mistaken Zig Zag Tree (2015)

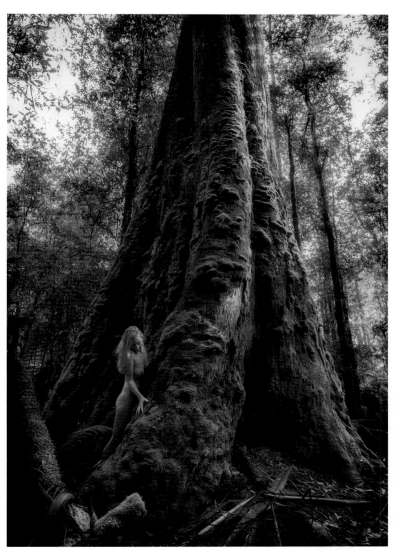

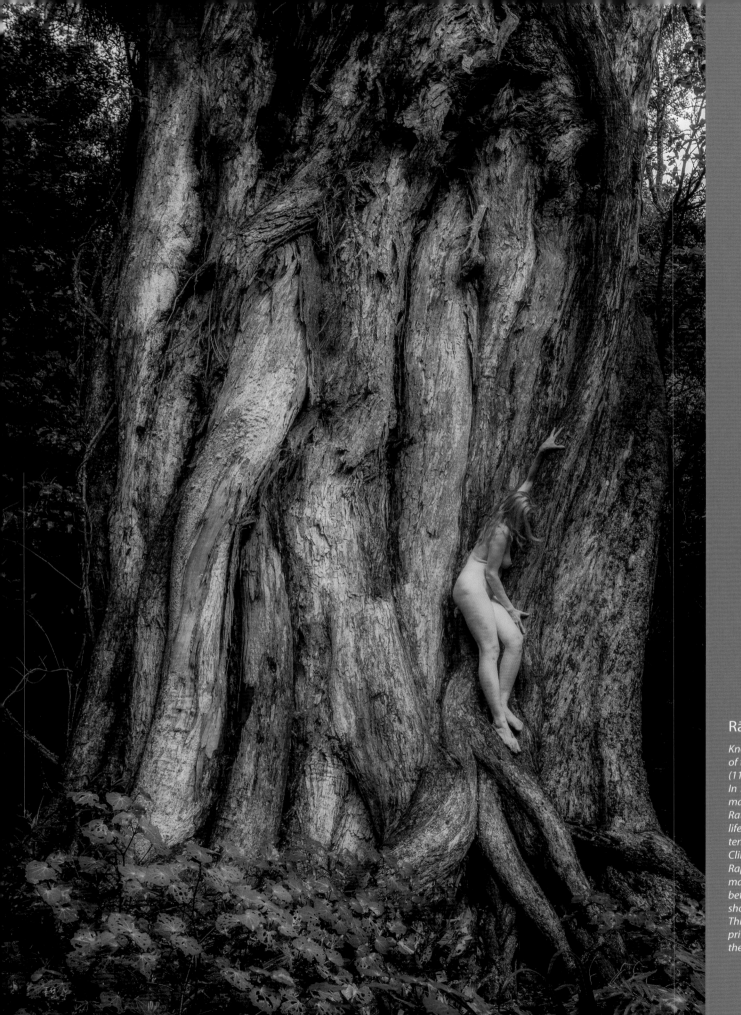

Rātā Spirit (2015)

Known as Ratanui ("Big Rata"), this Northern Rata is one of the largest, if not the largest, left of its kind, at 38.02 ft. (11.59 m) in circumference and 143 ft. (43.58 m) in height. In Maori mythology, Tane (the god of forests and birds) mated with female deities such as Rere-noa to produce the Rata and other epiphytic plants. Indeed, the Rata starts life up in other trees, such as the Rimu, and sends down tendrils like a strangler fig until it forms itself into a tree. Climbing up on this tree, I felt like a tiny insect climbing Rapunzel's giant braid of luscious hair. As happens with many of my photo shoots, I arrived at my destination just before dusk, with minutes of daylight in which to get my shot, without my secret TreeGirl purpose being revealed. This tree lives north of Wanganui, at Bushy Park Stables, a private 87-hectare park that includes rare virgin bush, on the North Island, New Zealand.

Northern Rātā

Metrosideros robusta

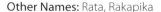

Other Names: Rata, Rakapika

Distinctive Characteristics: The evergreen Northern Rātā often begins life as an epiphyte started from a windblown seed; over hundreds of years, it sends down vine-like roots to the ground, which fuse together. The host dies (or some say it is killed) and the Rātā is left hollow. The grain of the wood is therefore twisted and gnarled. The Southern Rātā grows like a regular tree, and not to as great a height. They have masses of bristly flowers that range from dark orange to red to dark red. As the Latin species name *robusta* implies, it is sturdy, hard, and durable, as a tree and as timber. The genus name *Metrosideros* means "iron-hearted."

Distribution: Native to both the North and South Islands of New Zealand; from Te Paki to Wellington on the North Island, and from Nelson to Greymouth and Hokitika on the South Island. Elevation: up to 3,000 ft. (900 m).

Ecosystem: In the lowlands up to montane forests and along the coasts, often associated with such tree species as Rimu, Rewarewa, Tawa, Hīnau, Kānuka, Kahikatea, Kāmahi, Kohekohe, Pukatea, and Māhoe.

Maximum Age: Estimated to 1,000 years old.

Maximum Height and Girth: 141 ft. (43 m) in height; 40 ft. (12 m) in circumference.

Animal Community: Short-tailed bats, lizards, bees, as well such as birds as tūi, bellbirds, stichbirds, silvereyes, kākā, kea, and kākāpō drink the nectar. Kākāriki birds eat the flowers and buds. Native stick insects, the common forest looper, and the moth caterpillar eat the leaves. Introduced species, such as goats and deer, will eat young trees, while brushtail possums can destroy it.

Traditional Uses:
Medicine: The nectar was used as a remedy for sore throats. An infusion of inner bark was used as a remedy for diarrhea. The bark was boiled and taken internally to treat colds, as well as being used externally as an astringent to treat ringworm, pains, bruises, and wounds. Young leaves were chewed for toothaches (Maori).

Food: The nectar was eaten (Maori).

Tools and Objects: The wood was made into weapons, paddles, and mauls (Maori).

Art and Ceremony: The wood was made into flutes (Maori).

Modern Uses: Its hardness, strength, and durability later made it useful for ship and bridge construction, machine bearings, cartwheels, carving-chisel handles, telephone pole cross-arms, and woodturning. The outer bark has been used for tanning leather. The bark can be made into a dye for wool.

Threats and Conservation: The invasive brushtail possums love to eat the Rātā to death. They are also threatened by forest clearing, cutting for firewood, and hybridization with the Pōhutukawa tree. However, its status is not considered threatened.

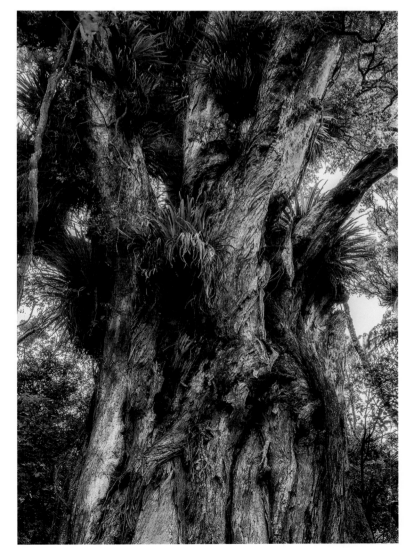

Rātā Tree Epiphytes (2015)

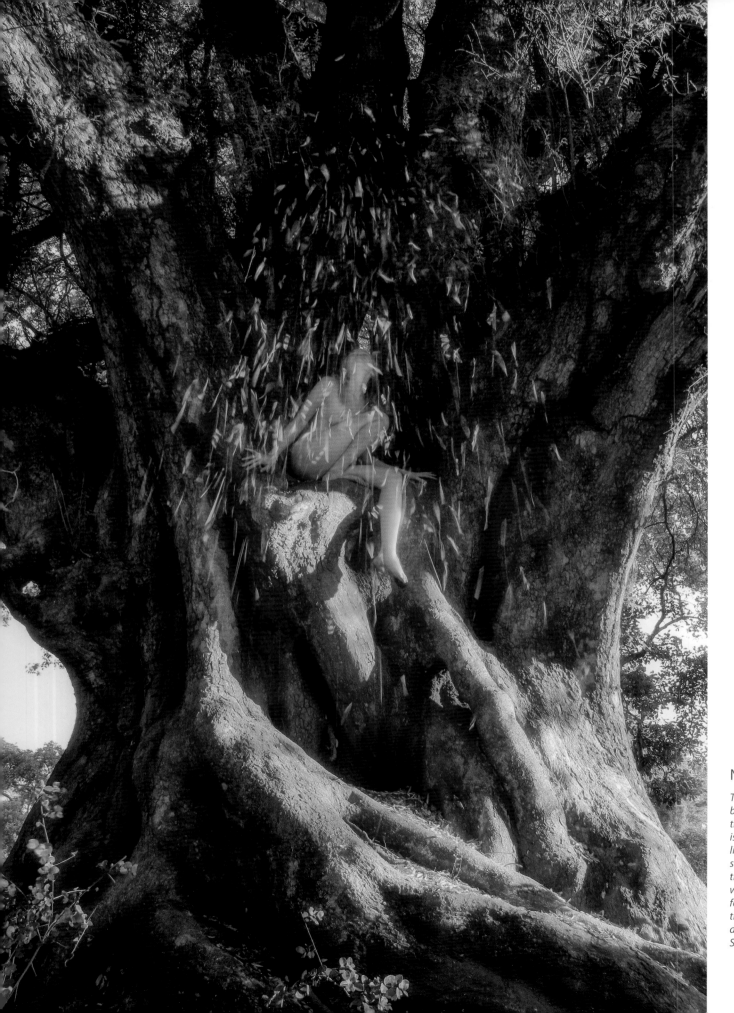

Nyala Falling Leaves 8 (2008)

The Nyala tree is named after the nyala, the handsome bearded ungulate with delicate stripes and spiral horns that eats this tree's fruits. The wild South African landscape is full of such animals that feast off the vegetation; I felt like one myself, dining off the trees with all my senses. My safari guide, Brett Greenaway, took me to this elder Nyala tree, struck with the radiant sunset lighting up the leaves it was shedding. I climbed up on it as it if it were a throne fit for a tree fairy, and tossed the leaves into the air, capturing them in the photo like sparks of flames or fairies raining down on me. This magical tree is in the northern tip of South Africa, near Pafuri, within Kruger National Park.

Nyala

Xanthocercis zambesiaca

Other Names: Nyala Berry, Muchetuchetu (Shona), Musharo (Shona)

Distinctive Characteristics: The Nyala is a common leguminous evergreen tree of the bushveld in Southern Africa. It is recognizable by a characteristically large-grooved, gnarly single or multiple trunks and large, dense canopy. The ripe berry-like fruits are yellow-brown.

Distribution: Native to northeastern South Africa, Botswana, Zambia, Zimbabwe, Malawi, and Mozambique.

Ecosystem: Lives solitary in the dry woodlands near floodplains and riparian zones with Fever Trees, Natal Mahogany, understory shrubs, and grassland. Elevation: sea level to 5,000 ft. (1,524 m).

Maximum Age: Estimated at 600 years.

Maximum Height and Girth: 98 ft. (30 m) in height; 31 ft. (9.4 m) in circumference.

Animal Community: Nyala Trees are often found in conjunction with giant termite mounds, which do not damage the trees. The tree is popular with many animals who seek its shade and fruit, including the nyala antelope, impala, kudu, elephants, giraffes, baboons, mongoose, dassies, vervet monkeys and birds. Insects are attracted to the sweet-smelling blossoms.

Traditional Uses:
Food: The fruit may be eaten fresh, or dry, grounded into a pulp for porridge.

Threats and Conservation: Not threatened.

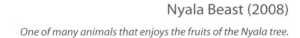
Nyala Beast (2008)
One of many animals that enjoys the fruits of the Nyala tree.

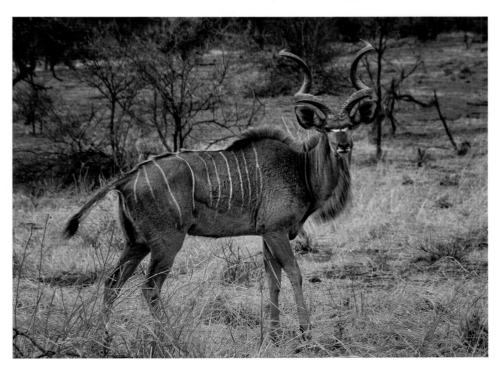

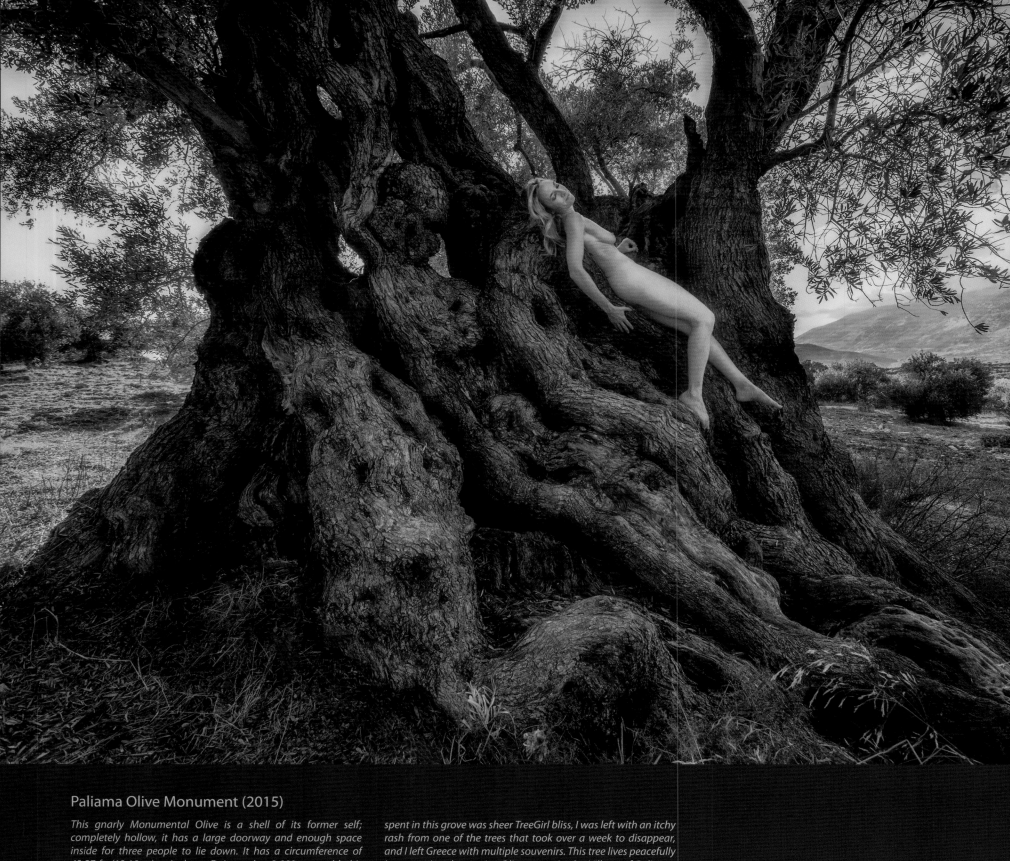

Paliama Olive Monument (2015)

This gnarly Monumental Olive is a shell of its former self; completely hollow, it has a large doorway and enough space inside for three people to lie down. It has a circumference of 43.27 ft. (13.19 m) at its base. Estimated at 3,000 years old, this tree—a variety known as chontroelia—was grafted onto a wild Olive and planted without any awareness that it would become such a monument thousands of years later.

I found myself drawn to cozy up to the "backside" of this majestic, gnarly tree that had many places to perch. Although my time

spent in this grove was sheer TreeGirl bliss, I was left with an itchy rash from one of the trees that took over a week to disappear, and I left Greece with multiple souvenirs. This tree lives peacefully in an ancient but active Olive grove in the Village of Paliama, Rethynmnon, Crete, Greece.

Olive

Olea europaea

Other Names: European Olive, African Olive, olivo (Spanish), elaía (Greek), zayit (Hebrew), and many others

Distinctive Characteristics: The iconic Olive tree is the culturally invaluable botanical elder of the Mediterranean landscape. The trunk and limbs are characteristically gnarled, fissured, and twisted, with extremely hard wood. The distinctive evergreen, lance-shaped leaves are grey-green on the upper side and silvery-white on the underside. The well-known small fruit is a fleshy drupe encasing a hard stone. Purple-black when ripe, the olive fruit varies from 0.19–1.5 in. (0.5–4 cm) in length. The hardy tree responds well to pollarding and pruning. The tree puts out shoots that can form a new tree even after the trunk has died, burned, or been cut down. Drought tolerant, the Olive may send down roots up to 20 ft. (6 m) in search of water. Slow-growing, the tree becomes more gnarled as it ages. Ancient trees can hollow out completely and survive, producing olives even after a thousand years. Wild Olives are known as the species *Olea oleaster*. To cultivate trees for producing edible olives, saplings must be grafted onto the stumps of old trees. There are perhaps 700 cultivars, or varieties, of *Olea europaea*, with their fruit varying widely in color, size, shape, and quality.

Distribution: Native to the Mediterranean, Asia, and northern Africa. Elevation: sea level to 2,900 ft. (900 m).

Ecosystem: Farmed in groves in dry Mediterranean ecosystems.

Maximum Age: Estimated at 3,000 years.

Maximum Height and Girth: 49 ft. (15 m) in height; 34 ft. (10.5 m) in circumference.

Animal Community: Rabbits, voles, and mice eat the bark. Many species of birds eat the fruit and disperse the seeds, extending the trees' range to the point of being invasive in some areas.

Traditional Uses: One of the world's oldest cultivated plants—and the Mediterranean's most important crop trees—the Olive has been cultivated for olive oil, the olive fruit, the leaves, and fine wood for an estimated 9,000 years.

Medicine: The oil has been used medicinally to treat skin ailments and ulcers, as a laxative, and as a cosmetic treatment, applied to the skin and hair for grooming and general health. The leaves were used as an antiseptic, antiviral, astringent, sedative, and fever reducer, and to treat malaria.

Food: The fruit and oil were eaten.

Tools and Objects: The oil was used for burning in lamps. The wood was used to make tools and furniture.

Art and Ceremony: The Olive tree, branch, leaf, fruit, and oil have long been considered sacred in many civilizations and religions. The tree has been a symbol of peace, wisdom, glory, fertility, abundance, power, and purity. The leafy branches were offered ritually to deities and powerful figures as symbols of blessing and purification. The oil was used to anoint kings, priests, athletes, the dying, and sacrificial offerings. In Egypt, the Olive was associated with the goddess Isis, and leafy branches were found carved into the pyramids and wrapped around the head of King Tutankhamun. Images of the trees appear in Minoan frescoes. In ancient Greece, the Olive tree was associated with the goddess Athena and the founding of Athens. The oil was burned in the sacred lamps of temples and was the "eternal flame" in the original Olympic games. Victorious athletes were crowned with its leaves and given bottles of sacred oil as prizes. The wood was used for carving statues of deities. In ancient Rome, the Olive was associated with the goddess Minerva. The Olive tree, olive oil, and olives played an important role in the Bible and the Quran.

Modern Uses: Olives are one of the most extensively cultivated fruit crops in the world, with Spain producing the most, followed by Italy and then Greece. Ninety percent of all harvested olives are turned into oil, which is mostly used for cooking; about 10% are used as table olives. The raw oil is considered a health food, and is also used as an ingredient in natural soaps. The wood, prized for its hardness, durability, color, and interesting grain patterns, is carved into kitchen utensils, wooden bowls, cutting boards, fine furniture, and decorative items. Dye can be made from the olive skins and leaves.

Threats and Conservation: Not threatened.

Olive Fruits (2015)

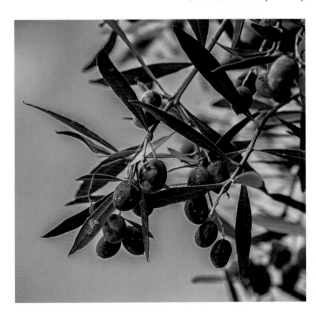

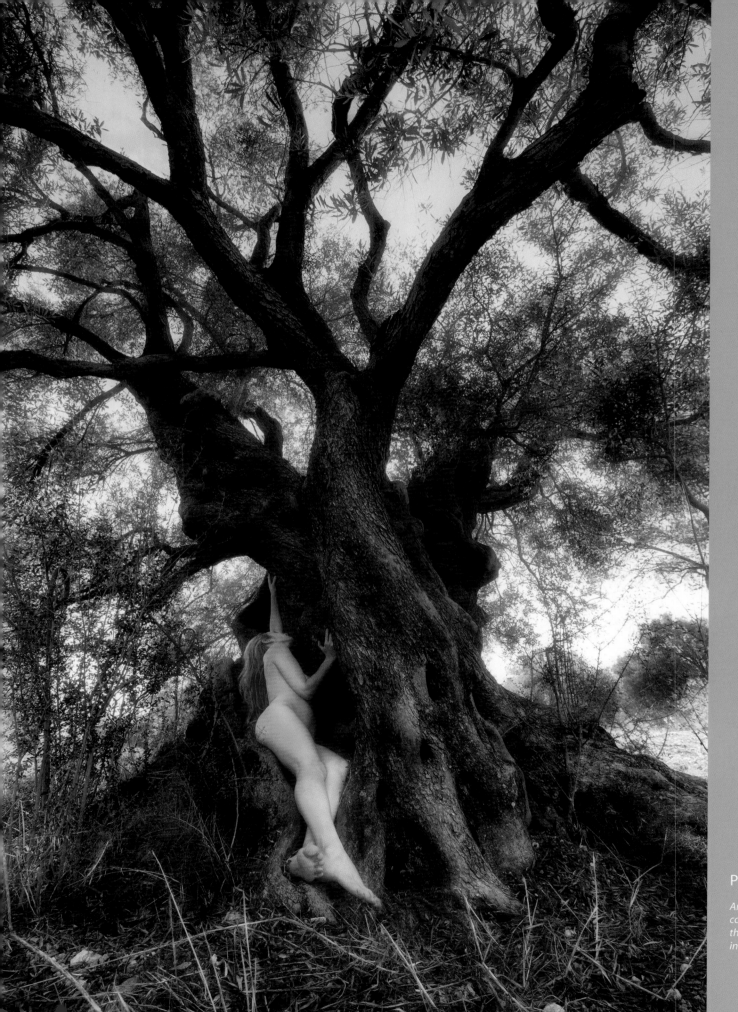

Paliama Olive Dryad (2015)

Another charismatic, gnarly, ancient Olive tree that I could camouflage myself within, this tree is adjacent to the giant Paliama Monumental Olive in an ancient grove in the Village of Paliama, Rethynmnon, Crete, Greece.

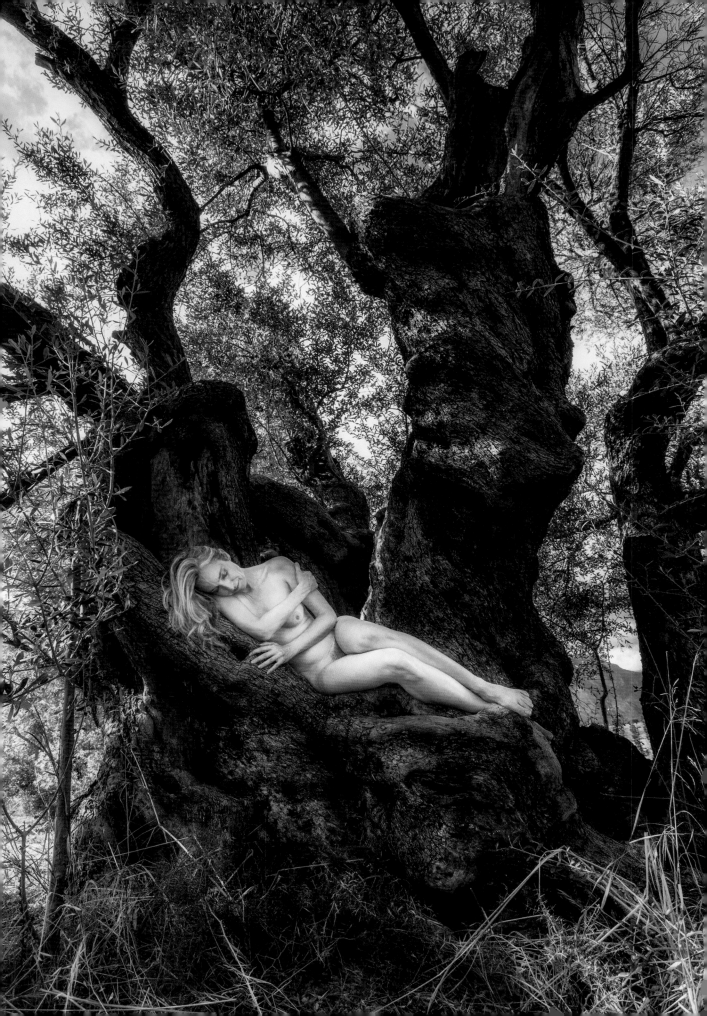

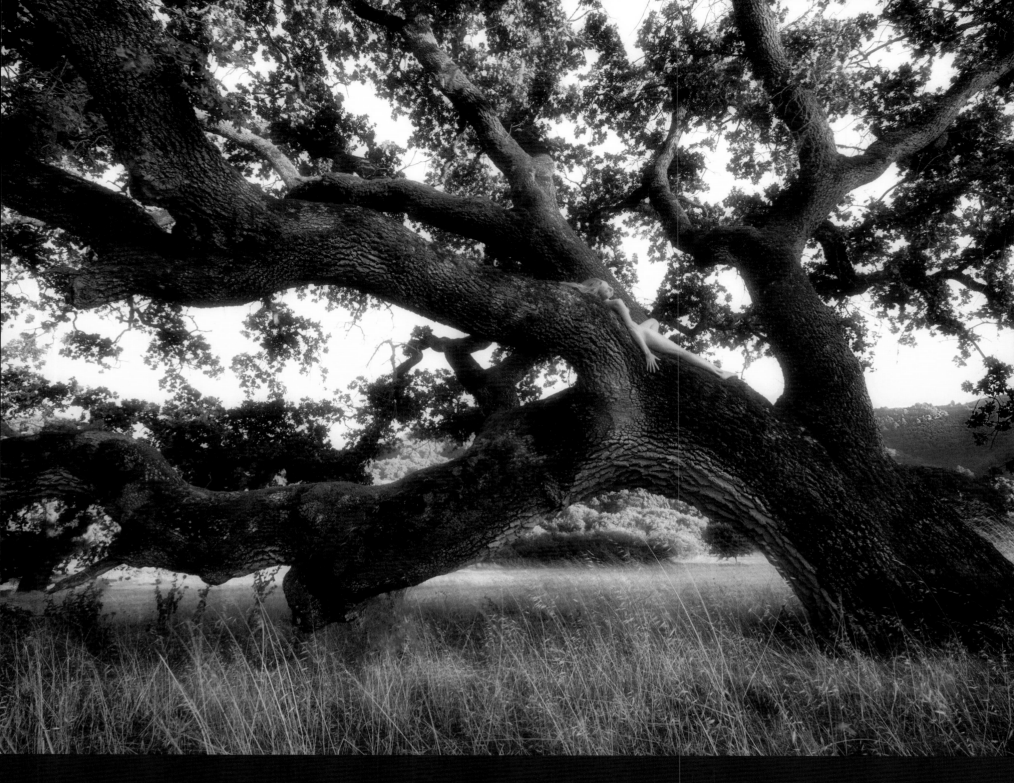

Sundrenched Stafford Oak II (2016)

There are some days when I can truly say, "I love my job." While I have travelled all over the world, some of my favorite species of trees are accessible right here in California. Looking at this arching oak tree, it seemed ideal to climb, when actually it proved to be one of my most physically challenging trees to mount. I had to work for it. Once I did so successfully, I found myself in a moment of pure bliss: a strong limb to cradle me, a bird's eye view, a summer sunset casting a golden glow on the landscape, deer browsing within my sight, my exposed skin warmed by the sun, *the brisk breeze of dusk tickling my bare body and whipping my long hair into a wild, uncontrolled mess—all with the knowledge that no one could see me. It was my secret pleasure. This bowing Oak lives with its cousins in Stafford Lake County Park, Novato, California, USA.*

Oregon White Oak

Quercus garryana

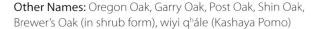

Other Names: Oregon Oak, Garry Oak, Post Oak, Shin Oak, Brewer's Oak (in shrub form), wiyi qʰále (Kashaya Pomo)

Distinctive Characteristics: The deciduous leaves, with deep rounded lobes, are dark and shiny on the upper side and lighter and slightly hairy on the underside. The mature bark is brownish-gray and deeply fissured. The acorns bulge out of their shallow, scaly cups. They hybridize with the similar-looking Valley Oak (which generally has slightly larger, shallower lobed leaves), Blue Oak, and several shrub oaks.

Distribution: Native to California, Oregon White Oak is the only native oak species in Oregon, Washington in the US, and southwestern British Columbia in Canada, where sparse populations are found on Vancouver Island and along the Fraser River. Elevation: sea level to 4,000 ft. (1,220 m), and up to 7,400 ft. (2,100 m) as the shrub × *breweri*.

Ecosystem: Grows in a variety of plant communities and climates, either solitary, in pure groves, in open woodlands, or in closed-canopy forests. They may be found on slopes, hilltops, valley bottoms, or riparian areas, with Valley Oak, Blue Oak, Black Oak, Canyon Live Oak, Coast Live Oak, Douglas Fir, Tanoak, Yellow Pine, California Buckeye, California Bay Laurel, Pacific Madrone, Water Birch, Lodgepole Pine, White Pine, Ponderosa Pine, Big Leaf Maple, Coast Redwood, Oregon Ash, Grand Fir, White Alder, White Fir, Red Fir, Vine Maple, Rocky Mountain Maple, White Alder, Rocky Mountain Juniper, Willow sp., Western Redcedar, Western Hemlock, manzanita sp., ceanothus sp., Oregon grape, hazelnut, bunch grasses, bracken fern, mountain mahogany, snowberry, oceanspray, hawthorn, serviceberry, mistletoe, California brome, balsamroot, and poison oak.

Maximum Age: Five hundred years.

Maximum Height and Girth: 120 ft. (36 m) in height; 25.3 ft. (7.7 m) in circumference.

Animal Community: Oregon White Oak woodlands are critical habitat for these rare species in Washington and British Columbia: Lewis woodpecker, propertius duskywing butterfly, leaf-mining moth, slender-billed nuthatch, sharp-tailed snake, Western gray squirrel, Western tanager, Western wood peewee, and Western bluebird. Downy woodpeckers, white-breasted nuthatches, and long-eared myotis (bat) use the tree cavities for nesting. Bear, deer, elk, grey squirrels, raccoons, acorn woodpeckers, band-tailed pigeons, Merriam's wild turkeys, rodents, and scrub jays eat the acorns.

Traditional Uses:

Medicine: The bark was used to treat tuberculosis. The interior of the bark behind knots was pounded and either put in warm water to drink or rubbed on a mother's body before childbirth.

Food: The acorns were prepared for food by many tribes, including: Chehalis, Cowlitz, Karok, Mendocino Indians, Nisqually, Paiute, Pomo, Kashaya Pomo, Kawaiisu, Coast Salish, Shasta, Squaxin.

Shelter: Oak wood was used to make poles and supports for sweathouses, earth lodges, dance houses, and other structures.

Modern Uses: The wood's hard, decay-resistant heartwood has been used for shipbuilding, railroad ties, wagon parts, fence posts, mine timbers, crates, caskets, small construction, wood turning, small furniture, veneer, cabinets, wine casks, flooring, wood pulp, and firewood.

Threats and Conservation: While common in California and Oregon, over 50% of its habitat in Washington, and up to 90% in British Columbia, has been destroyed due to land conversion and lack of fire. Conservation efforts are active in Tacoma, Washington, where an Oak Tree Park has been established; in the Scatter Creek Wildlife Area in Washington; and in Oak Bay, British Columbia, where a fine of up to $10,000 may be issued if a Garry Oak tree is cut or damaged.

Leaves and Acorns of Oregon White Oak

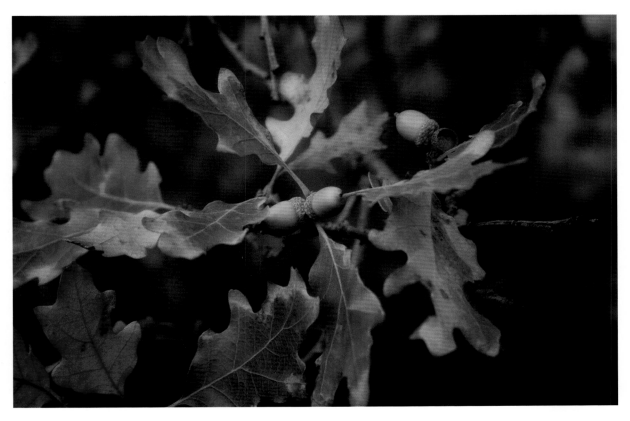

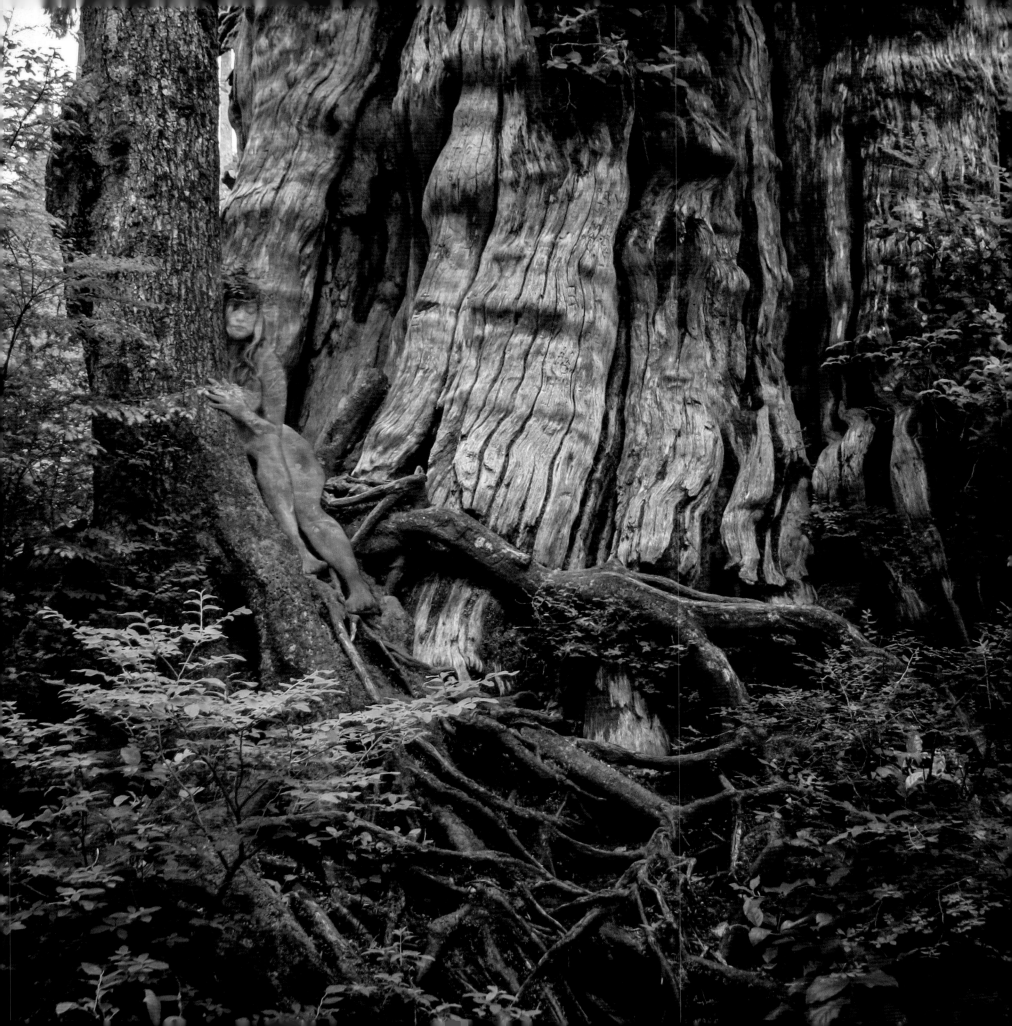

The Goddess and the Green Man

Once there was a girl who became so civilized, so consumed by busyness, and so full of self-doubt that she forgot her own divinity. One spring day, she left her village and went deep into the forest in search of companionship with nature. She was sitting among the branches of a tree, listening for birds, when she felt someone gazing intently at her. At first glance, she thought she saw a figure with skin as green as moss moving through the foliage. Indeed, there was a handsome man, with an intense gaze, completely covered with leaves and lichen. He had strong limbs like arms that swooped her up and held her gracefully off the ground. He fed and nourished her with food from the forest. The rustling leaves became his whisper as he told her of love and of her own earthly beauty. The girl had found her lover—in a tree. She felt so at home with him that she returned to the forest to be with her new beloved every day. Eventually, she forgot her own humanness and began to sprout leaves around the crown of her head and roots from her feet. Her arms split into the branches of a canopy and her silky torso hardened into bark. Feeling at home in her stillness and new vitality, she transformed completely into a beautifully majestic tree, rooted forever in the forest, covered with green leaves that turned golden every autumn.

Chapter 4

The Goddess and the Green Man

Magical and revered trees have long been the focus of folklore and myth, in nearly every culture all over the world. From the archaic to the modern, the World Tree is a potent, universal, cosmic archetype. It is a symbol of knowledge, strength, growth, life, sacrifice, fertility, transformation, immortality, the axis of the universe, and the journey from the center into the upper or lower realms. Through imagery and symbolism, the Tree holds the world in place for humanity's consciousness and unconsciousness, and therefore also holds the hope for the trees of the world to live as divine, not only in our collective hearts and minds, but also on the Earth itself. I wonder how many of us secretly would like to be a tree, or to embody what trees mean to us.

In India, the land of hundreds of gods and goddesses, there lives a glorious species of tree called the Banyan (*Ficus religiosa*). Banyan trees expand their surface area and reproduce by dropping tendrils from their branches that take root in the soil, creating additional "trunks" that continue to grow in girth. Each new tendril multiplies the perimeter of the canopy. The largest and most famous of these banyans is in the city formerly known as Calcutta (now Kolkata). At an estimated age of 250 years, its canopy has been measured at 1.5 hectares in circumference, forming a temple-like colonnade of 2,880 prop roots connected overhead, and able to provide shelter from the rain and sun for 20,000 people sitting beneath.[1] Only in India can we imagine 20,000 people crammed together in a quarter of a square mile under a single tree!

Previous p. 108:
Quinault Redcedar Greenwoman (2012)

The Quinault Lake Redcedar is the champion of all Western Redcedars. She is 61.26 ft. (18.67 m) in circumference, and 174 ft. (53 m) tall. She is completely hollow inside, with a large entrance. The womb-like cavern is filled with woody, stalactite-like knobs that hang down from the walls and ceiling (indicative of its age and decay), creating a very cave-like room. Even with her hollow, she has an estimated 17,650 cu ft. (500 m³) of wood, making her the largest of her species. Above seventy feet, she is topped by many massive staghorn reiterations, giving her a regal crown. On one side is a Western Hemlock tree with large, snake-like roots, being hosted on the Redcedar's base. Both inside, and outside of the tree, I felt myself transforming into the Green Woman. It is not known how old this ancient tree is, but it is probably at least 1,000 years old. She lives near Quinault Lake in Olympic National Park, in Washington State, USA, and is accessible by an arduous, steep uphill hike.

In the center of Auroville, a large spiritual ecovillage near Pondicherry in the state of Tamil Nadu, lives another banyan tree—smaller, but just as revered by its community. People gather under its canopy during the day, and on some nights they hold special festivals of light. I had the pleasure of encountering this tree when I lived in the village for three months. It was the most pleasurable of all the "temples" in India in which I had walked barefoot. On my last day in India, I visited the tree one last time to say goodbye to the one thing in that country that didn't give me culture shock. I had to express my gratitude for its grandeur. "You are magnificent," I thought as I gazed at the tree. "You are like a goddess." As if standing in front of the matriarch of all of India, I was briskly and unmistakably corrected:

YOU are the Goddess! If you do not acknowledge and honor the Divine Feminine in yourself, then you are dishonoring yourself, the miracle of your body, and the Divine Feminine everywhere, in all things, especially the Female. You, too, are the Goddess. Without you, there is no Goddess!

Me, "the Goddess"? I was startled. There must be some mistake. I'm obviously a mere mortal—a young white American at that—and not a very special woman, in the larger scheme of things. I thought humility was a virtue. What was the tree asking of me? How could I accept my own divinity—my own sacredness, my own "goddess-ness"[2]? It took me a while to comprehend that I was not being addressed as a divinity. Rather, I was being reminded that, as I was born into a female body, I have a responsibility to embody, respect, and manifest the self-love, pride, and creative feminine power that all goddesses represent—more important, that *all* women must personify. There is no room and no time for shame in being a woman, for being timid or docile, for being suppressed or oppressed. Collectively, we have a lot of work to do to help the planet and one another. We must use our creative power and energy fiercely and wisely.

When a woman becomes pregnant and gives birth, she has an instant spiritual connection to the divine expression of the Mother Goddess. Nurturing the miracle of Life inside her womb, the void of all things, and releasing it to the material world is pure magic. Embodying the Goddess, however, is not limited to motherhood, as goddesses take many forms and serve many functions. When I heard those words from the tree, I felt—from that moment on—that I was being continuously witnessed and watched over by goddess energy, no matter where I was, whether I ever gave birth to children or not. I was

given a purpose and an obligation: to forever live in my body from a place of love, honoring my feminine creative power; to stay connected to this divine source, without fear or excuse. And I had to find out exactly what that actually meant.

Perhaps she meant that I am one with Gaia, the Earth Goddess, as one conscious, living planet. Perhaps she meant that all females, including all goddesses throughout time, together co-create the sacred feminine—and I was no exception. To experience my mortal, human self as embodying the Goddess is a daily challenge. To gain strength in myself, I look toward nature and the Spirit of nature. I seek advice from trees. I get pure answers. The replies I "hear" are most assuredly about Love. Self-love. Loving another. Loving life. Protecting and serving life. Transforming, growing, and changing. Being true to my heart. What could be more divine?

It is no surprise that, as an arborist and a photographer of trees, I am attracted to artistic renditions of trees, leaves, and tree-related images, and that I create them myself. After all, art is a mirror of our outer reality, inner reality, and spirituality. One such image that I am drawn to is not feminine at all, but masculine: the Green Man.

The name "Green Man" was coined in 1933 by folklorist Julia Somerset, also known as Lady Raglan, the first scholar of his imagery.[3] But I wonder what was he called by the people who created these images and the generations of people who looked at him for centuries? Green Man imagery has been found in many cultures around the world.[4] While the earliest depiction is dated at 400 AD and found in many ancient cultures, the highest concentration is in European Romanesque and Gothic architecture from the sixth to the fourteenth century. The Green Man is represented in art as a face (and only a face, almost mask-like), surrounded by or composed of leaves. Sometimes sprouting branches and more leaves emerge out of the mouth or nostrils, like a corpse decomposing, transforming, and becoming fertilizer.

The leaves may be extremely stylized, extremely realistic, or somewhere in between. Some Green Men have been painted with vivid colors, while most are simply carved. Commonly used as a decorative architectural ornament in stone carving, and sometimes in wood, Green Men frequently appear inside and outside of churches and cathedrals (often in obscure spots, and sometimes prominently), as well as on civic and secular buildings, monuments, and other locations as ornamentation.

No two Green Men are exactly alike in their expression; multiple renditions of a foliate face may appear near one another. The artist's creativity and imagination determine what or whom the face looks like. Sometimes the Green Man appears very serious, and other times quite jovial. Sometimes he appears as a trickster, with grimacing or deviously grinning features. Generally, he is not a very compassionate-looking character.

Although still a mystery, the Green Man is believed by many to be related to the natural vegetative deities that sprang up independently in different cultures throughout the world, as symbols of growth, fertility, and the seasonal cycles of life, death, and rebirth and renewal. He is also seen as an animistic nature spirit with many faces, or a group of nature spirits. Perhaps because he is reborn, he can have many appearances. Some renditions of leaves are identifiable as a specific tree or plant, like oak or ivy, while others are extremely stylized. Some are combinations of plants and fruits. Perhaps people believed in an "oak man" or "oak spirit" and an "ivy man" or "ivy spirit." In the pagan tradition of medieval Europe, the seasonal cycles were mythologized in the story of the deciduous

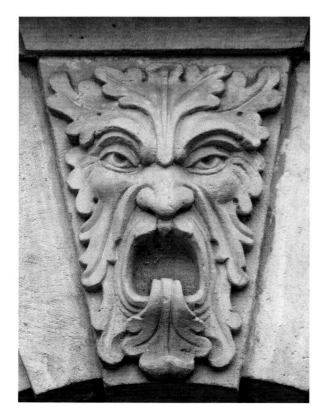

Oak King and the evergreen Holly King, who battled it out to rule half the year each, slaying each other annually, only to enter the womb of the goddess and be reborn anew.

The Green Man imagery that appears in the wood and stone carvings of Romanesque and Gothic architecture is thought to be carried on, by the artisans and masons who carefully crafted them, from earlier Pagan spiritual systems. Harmoniously, they are integrated into the same sacred space as Christian imagery in or on the outside of churches and cathedrals. So strong was his iconography, he often appears alongside or underneath sculptures of knights, saints, and even Mother Mary and Jesus, sometimes supporting them from below. In most instances, he is placed at the top of columns or ceilings, where he is almost hidden from view, or at the intersection of ribs on a vaulted ceiling, secretly looking down on the people below. Even if you can't see him, he is there—and he can see you. If one enters a church where he resides, one is sure to feel watched and may imagine him coming to life in the silence of a dark night.

Scattered about Europe and Britain with no particular pattern (in one church, but not in another), the Green Man was probably carved in stone so that he would not be forgotten. Sometimes he is carved into the baptismal font, or serves as part of an ornamental *tympanum* (arch above an entrance). Sometimes he is placed prominently in the center of town on civic or secular buildings, over a doorway or on a well or fountain with the water of life cascading from his mouth. An animistic symbol of nature unified with humanity, he may have been carved in the walls of public spaces and in the forests of stone columns inside the church to remind people not to forget nature. Perhaps, in their day, people knew what messages the Green Men had to share, just like they knew stories of the saints and characters in the Bible.

In addition to the face of the Green Man, many churches, civic structures, and secular buildings had carvings of other natural imagery, including the tree of life, leaves, flowers, and animals. Similarly, the sacred geometry of architectural elements often formed stylized imagery of nature, such as rose and cinquefoil windows. Consider how the inside of many cathedrals are likened to forests in their rows of tall columns, sometimes twisting, forming 'canopies' in the intricate vaulted ceilings.

But if Nature is often associated with the feminine, what about a Green Woman? While Green Women do exist less commonly, and even more rarely a Green Cat or other

Green creature, the Green Man in his masculinity may be seen as the male counterpart of, son of, or consort to the Mother Goddess, the source of all life, who appears in modern culture as a watered-down, faceless incarnation we refer to casually as "Mother Nature." Perhaps more Green Men were rendered than Green Women, especially in Europe, simply because those societies were patriarchal. The artisans were most likely male as well. Mother Mary was the only goddess allowed in the Christian Church, and she was demoted to the status of saint.[5]

Throughout time and across cultures, deities—as well as earthly, spiritual, and supernatural forces—have been rendered as anthropomorphic (human-like) or zoomorphic (animal-like), if not fully human. These figures were usually given faces in artistic expression, with personalities, eyes to witness, and mouths to speak and sing. The use of masks is universal in nearly all cultures as a ritual tool of transformation. Why? For one, in animistic and shamanic cultures, all aspects of nature have spirits. We also identify with faces; we relate to them, talk to them, listen to them, and feel witnessed by them. So it is no surprise that the Green Man is represented with a face. And how do we know that the people who carved those Green Men didn't believe they could see real Green Men in the forest? Perhaps they actually did see them. Or perhaps the Green Man was never seen in solid form in the forest, but rather appeared and disappeared in the leaves, changing shape with the movement of light and wind. Perhaps people simply felt watched in the forest. Perhaps they only heard his voice. What did he say?

The deep, dark woods were home to all sorts of animals, creatures, and spirits; the forests were very much alive. Some believed in a King, or Guardian, of the Forest. Perhaps they needed to ask permission from the King of the Forest to take anything from it, and leave an offering in thanks. The woods provided three life staples—food, medicine, and wood—and so much more. The Green Men were there through thousands of years when the woods were cut down for agriculture and settlement. The Green Men were there, as the woodlands and forests of Europe were cut down to fuel centuries of war. By the end of the Renaissance (fourteenth to seventeenth century), much of Europe and Britain had been deforested, and Green Man imagery in churches and civic buildings had devolved from representing a familiar character with intentionally rendered presence into a faint decorative ornament. Coincidence? Had he gone extinct? Or did people stop "believing" in him? Was the "Spirit of nature" being replaced with the Spirit of intellectualism, realism, and science? Who knows how much Green Man artwork has been lost forever—paintings, tapestries, manuscript illustrations, glass, ceramics, wood, stone, perhaps even masks—destroyed by time, or intentionally eliminated by those with opposing artistic styles or belief systems? The Green Man remains a mystery, and we do love a mystery, don't we? It is not a common occurrence to see real Green Men or Women today, perhaps because our modern Western culture doesn't believe in them anymore, just as it no longer believes in Demeter, Hades, or Pan. Perhaps these deities and spirits are just dormant. Perhaps they manifest themselves now in different ways. Perhaps they distrust an industrial world where the culture lauds the transformation of nature into manufactured objects and toxins, instead of honoring the transformative mystery of the life force.

The Green Man is an archetype our modern world needs to revive—and not simply on quaint garden statuary. We need to see him in person, often. He is surely not dead, for he symbolizes resurrection after death (a not-uncommon theme in world religion and spiritual systems), as well as the acknowledgment of an animate natural world. He has not really been forgotten. In fact, we have seen multiple revivals of him in artistic imagery. It began with a brief reprisal of his imagery in nineteenth-century British and U.S. architectural ornament, and modern artists continue to bring him back. People remember him not only as an archetypal symbol, but as a vital ecological life force as well. It is time he fully reawakens from his own cycle of dormancy—a very long winter, indeed. What does the Green Man have to say today?

There is no separation between yourself and Nature.

Certainly, a conscious awareness of the need to restore and heal the Earth has been growing in momentum steadily since the 1970s—alongside a parallel movement to reawaken and restore feminine power.[6] Of course, they are related: The Goddess and the Green Man, female and male, symbolize and embody complementary aspects of the life force. And they can both live within all of us. They connect us to our subconscious, the collective unconscious, and the entire human story. They are channels of divine Nature, elucidating our dependence on the natural world for survival and our biological unity with the cycle of life and death. They also remind us of our responsibility to all of life, and of our own potential for growth in conscious awareness and for inner and outer transformation. When we accept our responsibilities as a spirit with a physical body, we live through the cycles of growth and decay, regeneration and renewal, each time we choose to either

destroy life or serve it by healing, loving, making love, and accepting the shared divinity of all life. If we must destroy life to continue life, let it be in the respectful and humble context of the cycle of life and death. Both male and female divinities and spiritual beings of all cultures represent various aspects of the creation and destruction of life and our companionship with that cycle. Knowing that spirit endures reinforces the law of physics that states that energy cannot be created or destroyed, only transformed.

So here am I, a woman, embracing the divine feminine and in search of the divine masculine within the masculine archetype of the Green Man, my male consort. I have traveled to Europe to find historical evidence of his prominence. I have combed medieval churches and streets in anticipation of meeting him. I looked for his living face in the remnant groves and ancient, rooted trees. I want to know that he existed and that he still exists. Even when carved in stone, his many timeless faces appear very alive to me. I suspect he may continue to be there, carved in stone, long after we humans have died out as a species

and the Plant Kingdom reclaims and consumes the industrial corpse of the built world.

I look forward to returning into the soil when I have ended my own life cycle and the Goddess takes me back to my womb. May the roots and branches of the plant world grow up and out of my fertile remains. Not surprisingly, I hope to become a tree, in both physical form and in spirit. Then, I imagine, I will truly find peace. While doing what trees do all day and night—growing, shedding, breathing, producing food, providing homes for critters, waving in the breeze, and simply holding strong and still—I'll whisper to passing humans to remember their own divinity, as well as their mortality and their place in the family of nature. Until that day, I'll take my divine goddess-human self deep into the woods and stay on the lookout for the Green Man.

[1] Colin Tudge, *The Tree: A Natural History of What Trees Are, How They Live, and Why They Matter* (NY: Three Rivers Press, 2005) p. 191.

[2] According to the modern Goddess movement, "the Goddess" is not necessarily seen as monotheistic but is often understood to be an inclusive, encompassing term incorporating many goddesses in many cultures, known by many names and manifest in many forms. The Wiccan ecofeminist author and activist Starhawk speaks of the Goddess as both a psychological symbol and "manifest reality. She exists *and* we create Her." (*The Spiral Dance: A Rebirth of the Ancient Religion of the Great Goddess* (NY: Harper, 1979) p. 77.

[3] Clive Hicks, *The Green Man: A Field Guide* (Helhoughton: Compass Books, 2000) p. 1.

[4] A variety of artistic renditions and mythical incarnations of the Green Man have been found, predominantly throughout Europe, but also in Borneo, Nepal, China, India, Rajasthan, Mexico, Lebanon, and Iraq. Because of his association with vegetation and resurrections, scholars have associated him with deities such as Tammuz (Sumerian), Osiris (Egyptian), Viridios (Celtic), Cernunnos (Celtic), Sylvanus (Roman), Dionysus (Greek), and even Jesus. He is also related to the mythical nature spirits Derg Corra (Celtic "man in the tree"), as well as the woodwose or the wild man (European). He is also thought to be related to the English folkloric characters of the Holly King, the Oak King, the Green Knight, Green George, John Barleycorn, and Robin Hood, and has been an important feature of the May Day celebration.

[5] Joseph Campbell and Bill Moyers, *The Power of Myth* (NY: Doubleday, 1988) p. 176.

[6] Gus diZerega, *Faultlines: The Culture War and the Return of the Divine Feminine* (Wheaton, IL: Quest, 2013).

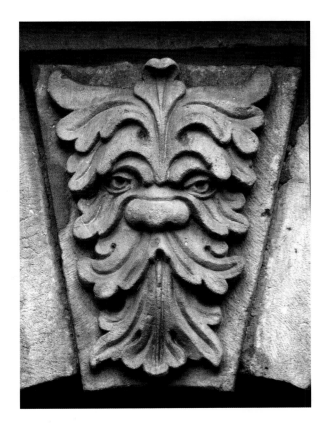

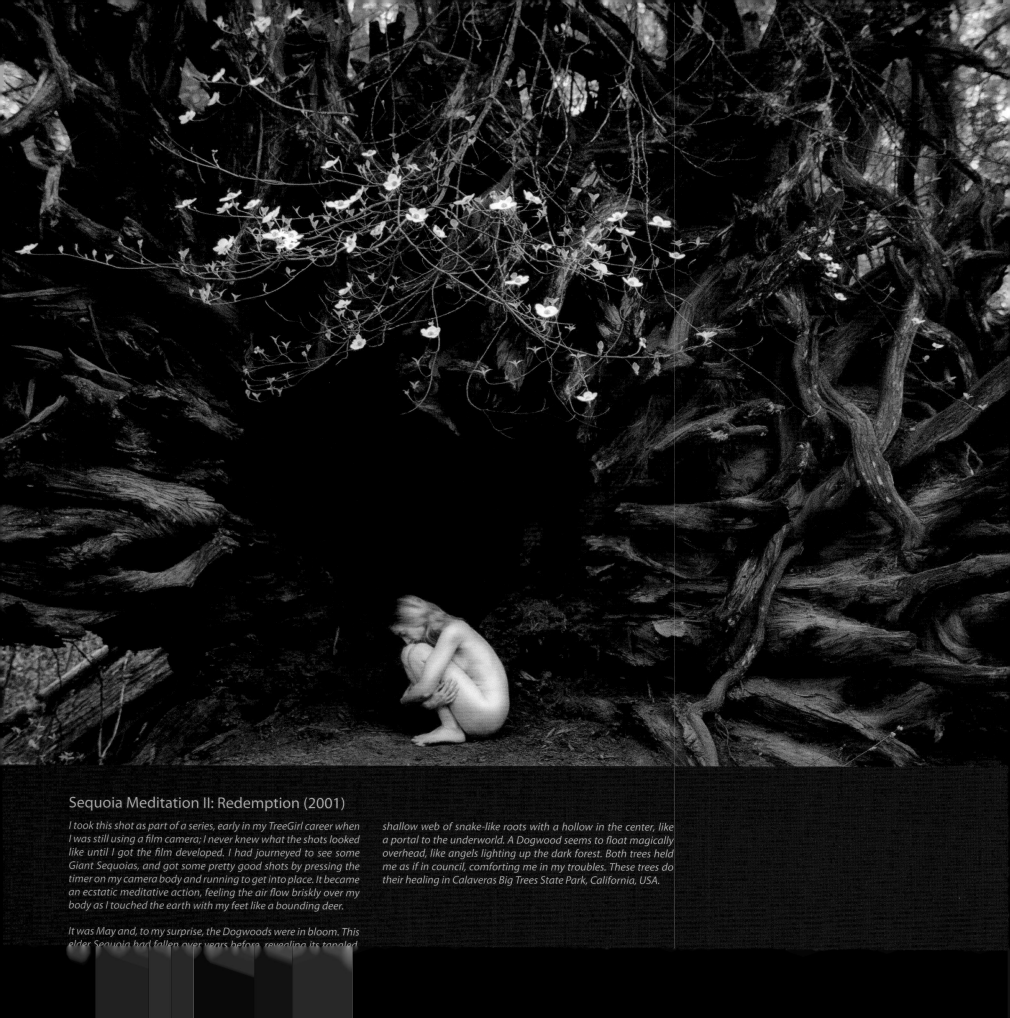

Sequoia Meditation II: Redemption (2001)

I took this shot as part of a series, early in my TreeGirl career when I was still using a film camera; I never knew what the shots looked like until I got the film developed. I had journeyed to see some Giant Sequoias, and got some pretty good shots by pressing the timer on my camera body and running to get into place. It became an ecstatic meditative action, feeling the air flow briskly over my body as I touched the earth with my feet like a bounding deer.

It was May and, to my surprise, the Dogwoods were in bloom. This elder Sequoia had fallen over years before, revealing its tangled

shallow web of snake-like roots with a hollow in the center, like a portal to the underworld. A Dogwood seems to float magically overhead, like angels lighting up the dark forest. Both trees held me as if in council, comforting me in my troubles. These trees do their healing in Calaveras Big Trees State Park, California, USA.

Pacific Dogwood

Cornus nuttallii

Other Names: Mountain Dogwood, Western Dogwood, Western Flowering Dogwood, California Dogwood, Nuttall's Dogwood, hayu qhále (Kashaya Pomo)

Distinctive Characteristics: This multi-branched deciduous shrub or tree is known for its showy white, creamy white, or greenish-white "flowers," which are actually composed of 4–6 bracts around a yellow-green button-like cluster of flowers that light up the forest understory. These "flowers" bloom from April to May, and occasionally again in September. The crowns of the trees are open and irregular, with arching branches. The leaves turn pinkish red. Twigs are sometimes reddish. Fruits are clustered into bright red drupes.

Distribution: Native to the lowlands of southern British Columbia, south through Washington and Oregon, to the mountains of southern California, with an inland population in central Idaho. Elevation up to 6,500 ft. (2,000 m).

Ecosystem: Pacific Dogwood is common along stream banks in moist, open, or dense coniferous, hardwood, and mixed coastal forests. Depending on the region, it is associated with various tree and understory plant communities:

British Columbia
- Douglas Fir and Western Hemlock.

Washington and Oregon
- Cascades: Douglas Fir, Western Hemlock, Pacific Silver Fir, Grand Fir, Noble Fir, Pacific Yew, Western Redcedar, Big Leaf Maple, Vine Maple, Oregon grape, salal, red huckleberry, and Pacific rhododendron.
- Mount Rainier National Park: Western Hemlock and salal.

Oregon
- Siskiyou: Tanoak, California Bay Laurel, Redwood, salal, evergreen huckleberry, Oregon grape, Pacific rhododendron, and poison oak.

Oregon and California
- Northern Coast Range: Vine Maple and salal.
- Southern Coast Range: California Bay Laurel, Tanoak, and evergreen huckleberry.
- Eastern Coast Range: Douglas Fir, Western Redcedar, Big Leaf Maple, Oregon Ash, and Bitter Cherry.

- Willamette Valley: Oregon White Oak, California Black Oak, Canyon Live Oak, Big Leaf Maple, Oregon Ash, Ponderosa Pine, Incense Cedar, Pacific Madrone, and Tanoak.
- Coast Range: Douglas Fir, Tanoak, Pacific Madrone.

California
- Redwood forests: Douglas Fir, Western Hemlock, Tanoak, California Hazel, salal, Pacific ninebark, and Pacific rhododendron.
- Klamath Range: Ponderosa Pine, Douglas Fir, White Fir, Sugar Pine, Port Orford Cedar, California Hazel, and Oregon grape.
- Ponderosa Pine forests: Ponderosa Pine, Sugar Pine, Grey Pine, White Fir, Incense Cedar, Oregon White Oak.
- Sierra Nevada: Ponderosa Pine, Tanoak, California Black Oak, Giant Sequoia, Canyon Live Oak, White Alder, California Hazel, Scouler Willow, manzanita, deerbrush, California coffeeberry, common snowberry, birchleaf mountain mahogany, poison oak, Sierra mountain misery, California wildrose, and Sierra gooseberry.

Idaho
- Western Redcedar, Western Hemlock, Douglas Fir, Grand Fir, Bitter Cherry, Rocky Mountain Maple, Scouler Willow, Red-Osier Dogwood, oceanspray, Saskatoon serviceberry, common snowberry, and thimbleberry.

Maximum Age: Estimated at 150 years.

Maximum Height and Girth: Up to 65 ft. (20 m) in height; 25 in. (61 cm) in circumference.

Animal Community: Deer and elk browse young Pacific dogwood sprouts. The fruit is eaten by deer mice, red tree voles, band-tailed pigeons, and pileated woodpeckers.

Traditional Uses:

Medicine: The bark was prepared as a lung strengthener and fever-reducer (Sierra Miwok), for treating malaria, ulcers, and stomach upset; part of the "10-bark medicine" (Saanich); as a laxative (Lummi), as a purgative (Plateau Indians), as a tonic (Hoh, Quiluete), and with cascara bark as a blood purifier (Thompson).

Fiber: Young shoots were used for basket-weaving. The long, slender branches were used for making baby baskets (Kashaya Pomo).

Tools and Objects: Wood was used to make bows and arrows.

Art and Ceremony: A dye was made from the bark (Nlaka'pamux). The boughs were burned for heat in a sweat lodge (Karok).

Modern Uses: The wood has been used commercially to make thread spindles, cabinets, piano keys, tool handles, and golf-club heads.

Threats and Conservation: Pacific Dogwood is very susceptible to Dogwood anthracnose, a disease caused by the fungus *Discula destructiva*, but it is not classified as a threatened species.

Dogwood Flowers

The white 'petals' of the Dogwood are called bracts, and are not actual flowers. The flowers are in the center.

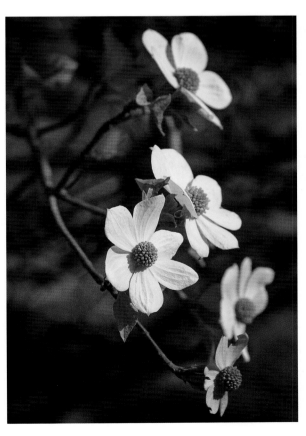

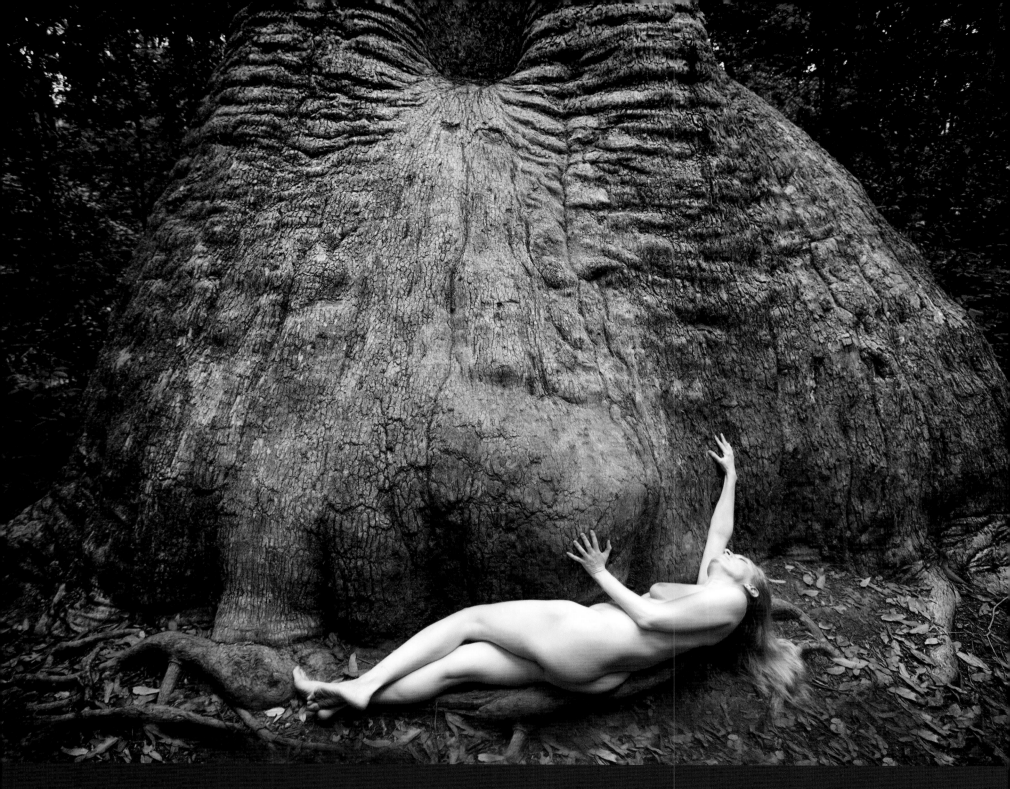

Bloodwood Bottle Bliss (2015)

This Red Bloodwood is known affectionately as "Old Bottle Butt"—not a very distinguished name for such an outstanding specimen of the species. It was probably not named by a distinguished gentleman or outstanding specimen of his species, either. But the tree is so well-known that it has its own parking lot, picnic area, and short trail with a boardwalk and fence to keep tree huggers from climbing all over the tree.

It was near dusk, and I had just made it to the tree after a dangerous, misguided detour down a 4-wheel-drive "road" in my 2-wheel drive campervan. I didn't have a ladder or the physical strength to climb up into Old Bottlebutt's belly-button nook, which was tempting me. So I lay down beside it and basked in its bulbous bottom, which looked like a giant, wrinkly elephant foot. This tree is measured at 53.47 ft. (16.30 m) in circumference and 169.94 ft. (51.79 m) tall. You can find it near Wauchope, in Burrawan State Forest, New South Wales, Australia—but don't miss the signs.

Red Bloodwood

Corymbia gummifera

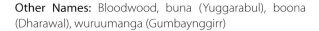

Other Names: Bloodwood, buna (Yuggarabul), boona (Dharawal), wuruumanga (Gumbaynggirr)

Distinctive Characteristics: The evergreen Red Bloodwood is a species of eucalypt so named because of its dark-pink to dark-red heartwood and its distinctive blood-red-colored sap. The bark is rough and fibrous, and the white flowers are showy. It usually grows as a tree, but may also take the form of a mallee (meaning that it reproduces from multiple stems and trunks that arise from underground lignotubers). Some mallee groves are dated to thousands of years old.

Distribution: Native to coastal Australia, from extreme eastern New South Wales into southeastern Queensland as far north as Fraser Island, with isolated populations inland in Wondai State Forest, Mt. Walsh National Park, and northeast of Toowoomba. Elevation: sea level to 2,600 ft. (800 m).

Ecosystem: Grows on flat areas, low hills, or sand dunes, in tall, open woodlands in pure stands or in community with many other eucalypts, including Blackbutt, Stringybark, Grey Gum, Ash, Silvertop, Scribbly Gum, Sydney Peppermint, and Angophora, as well as Banksia. Understory plants include eriostemon, spike waddle and gymea lilies.

Maximum Age: Estimated at 200 years.

Maximum Height and Girth: 197 ft. (60 m) in height; 53.5 ft. (16.3 m) in circumference.

Animal Community: Cockatoos eat the seeds. The sap is an important feeding source for sugar glider possums, who actively scar the tree bark to access it. Invertebrates consume the flower nectar. A range of species of birds and bats breed and roost in the hollows of older trees. The flowers attract parrots, galahs and other cockatoos.

Traditional Uses:

Medicine: A poultice of mud and leaves was used to stop bleeding. The sap was used as an antibacterial and antifungal (Gumbaynggirr). The sap was also used to treat venereal diseases, diarrhea, and inflammation of the bladder. Charcoal from the bark was used as an antiseptic. The sap was used for toothaches and as a mouthwash (Dharawal)

Food: The flowers were sucked or soaked in water to make a sweet drink.

Fiber: Fibers of balga grass were soaked in the gum before making cordage from them.

Tools and Objects: The gum was used to adhere tools and weapons. The gum was mixed with ironbark to tan string (Dharawal). The bark was used to tan cabbage tree palms for ropes and fishnets (Dharawal). The hollowed out branches or young trunks were made into didgeridoos.

Art and Ceremony: The sap was used for mixing with paints for decorating objects and art on cave walls (Dharawal).

Modern Uses: Its strong and durable heartwood has been used for fencing, poles, railroad ties, and mining timbers.

Threats and Conservation: While the species itself is not in danger, a grove of great ecological importance is at risk in the ancient sand dunes of Spring Gully, in and around the Royal National Park of Bundeena.

Old Bottlebutt (2015)

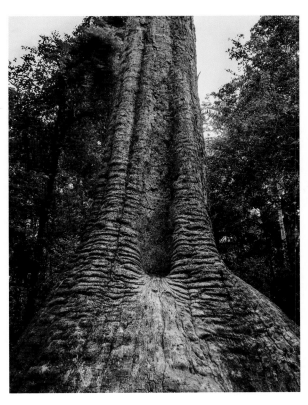

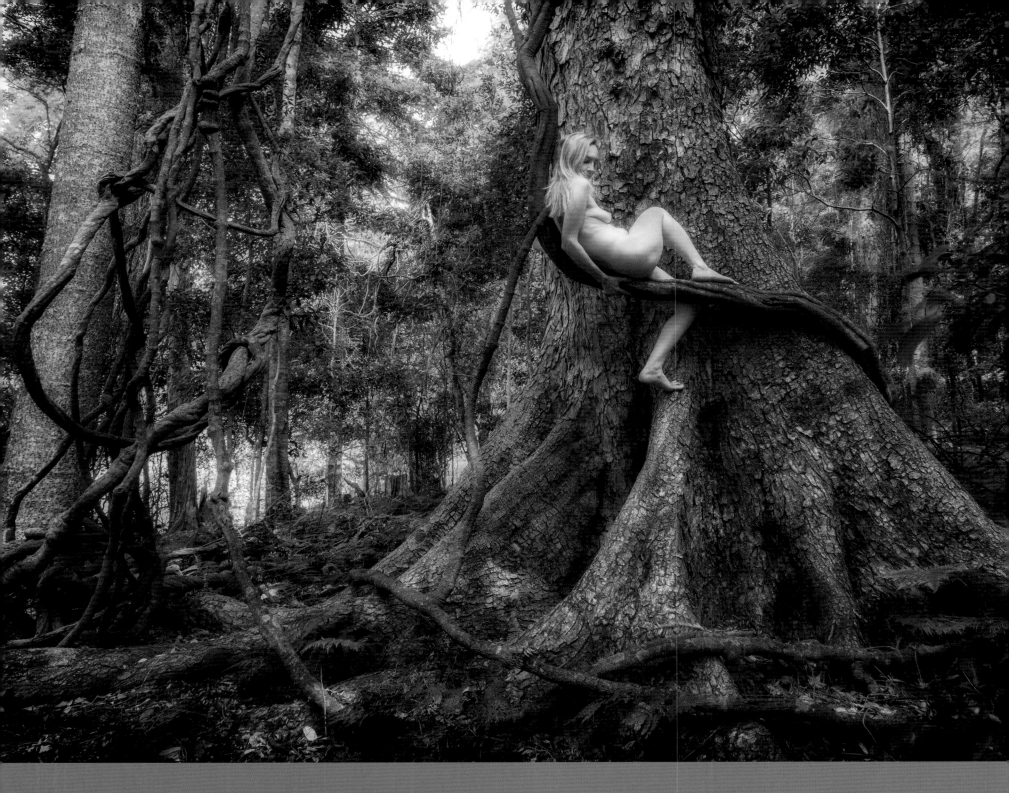

Red Cedar Escalade (2015)

I went to Bunya Mountains National Park for one day to see the famous Bunya Bunya trees, but was unable to photograph myself with them. Walking on the trail back to the center of the park, which was filled with cute wallabies grazing on the grass, I spotted this handsome Red Cedar tree tangled in vines. I couldn't help myself; I wanted to climb it, and hoped that the vines were not poisonous or too fragile to hold my weight. I climbed up, and soon found myself imagining what it would be like to be an aerial acrobat in the circus, like my cousin had once been. Safety, time, and nudity limited my acrobatic exploration, but I did get a thrill out of hanging on a swinging vine. This tree lives in Bunya Mountains National Park, New South Wales, Australia.

Red Cedar

Toona ciliata

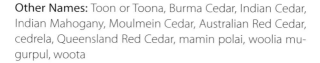

Other Names: Toon or Toona, Burma Cedar, Indian Cedar, Indian Mahogany, Moulmein Cedar, Australian Red Cedar, cedrela, Queensland Red Cedar, mamin polai, woolia mugurpul, woota

Distinctive Characteristics: This is one of Australia's few deciduous rainforest trees. New foliage is bright red and stands out distinctively. As it matures, the base of the trunk buttresses.

Distribution: It is not a true cedar (belonging to the Genus *Cedrus*), nor is it related to the Cedars found in Europe or North America. Native from northeast Queensland to southeastern New South Wales, Australia. Elevation: from sea level to 5,000 ft. (1,500 m). Also found in China, India, Pakistan, Malaysia, Bangladesh, Cambodia, Indonesia, Laos, Myanmar, Phillipines, Thailand, Vietnam, Nepal, and Papua New Guinea.

Ecosystem: Found in a range of rainforest types, from warm temperate to subtropical and tropical. Associated with Red Carabeen, Sassafras, Yellow Carabeen, Silver Quondong, Native Tamarind, Moreton Bay Fig, and occasionally Hoop Pine.

Maximum Age: Over 500 years old.

Maximum Height and Girth: 197 ft. (60 m) in height; 38 ft. (11.5 m) in circumference.

Traditional Uses: There were likely many traditional uses of this tree.

Medicine: The leaves have antibacterial, antifungal, and analgesic properties.

Modern Uses: Referred to as "red gold," the timber was highly valued and heavily exploited by early settlers, who logged it extensively for furniture, wood paneling, and construction, including shipbuilding.

Threats and Conservation: While mature trees are rare, the species is not threatened.

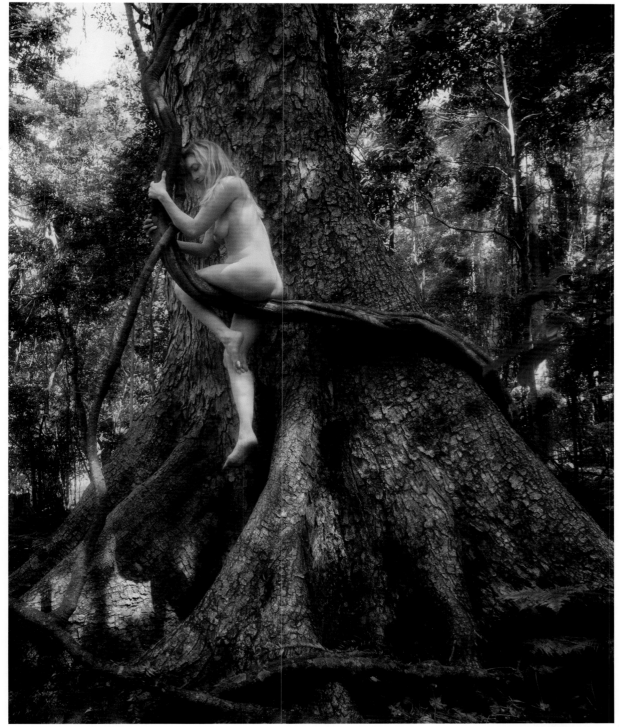

Red Cedar Aerial (2015) ▸

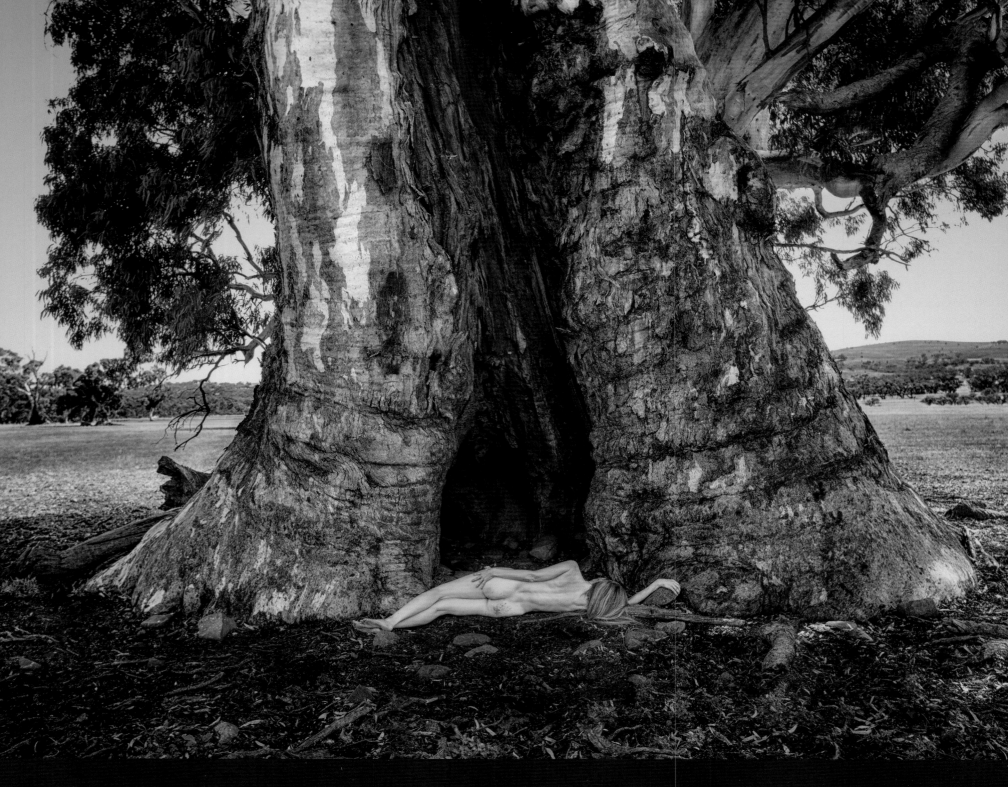

River Red Gumgantium (2015)

My return to the trees of Australia after twenty years started with this field of picturesque River Red Gums at Beautiful Valley Creek, near Wilmington in the Flinders Ranges, South Australia. I could never have found this mammoth tree on my own, so I enlisted the aid of the Eucalypt expert and arborist Dean Nicolle and his wife Annett Boerner, who graciously took me to this tree and others. The tree had been named Sheepy, and as we approached we had to break up a flock of surprised sheep that were congregating in its shade. After trying all sorts of treegirl positions, I lay down in front of the tree's cavern to show its girth, getting intimate with *what the sheep left behind on the earth. The second largest in girth, this River Red Gum, ssp. minima, has a circumference 46.09 ft. (14.05 m) and a height of 104.98 ft. (32 m).*

River Red Gum

Eucalyptus camaldulensis

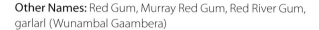

Other Names: Red Gum, Murray Red Gum, Red River Gum, garlarl (Wunambal Gaambera)

Distinctive Characteristics: The River Red Gum is named for its heartwood, which varies from pink to red. They can grow to be quite large in girth. Their crowns are densely foliated with evergreen leaves.

Distribution: Native to Australia, it is the most widely distributed of all the eucalypt species. Elevation: sea level to 4,900 ft. (1,500 m).

Ecosystem: Common along the banks and floodplains of watercourses.

Maximum Age: Over 700 years old.

Maximum Height and Girth: Height up to 157 ft. (48 m), 45 ft. (14 m) in circumference.

Animal Community: Birds such as rosellas, cockatoos, galahs, owls, kookaburrahs, and the threatened supurb parrots nest in hollows; corellas and many other birds rest on the branches. Koalas eat the leaves. Bees, butterflies, birds, possums, and flying foxes eat the nectar. Hollows create habitat for many species.

Traditional Uses:

Medicine: The sap was used as an ointment for burns and other skin problems and as a gargle for sore throats (Yarra). The sap was also used as an astringent to treat diarrhea. A leaf infusion was applied as a wash for fever or headache, and to treat coughs and colds. Steam from boiled leaves was inhaled.

Food: The white ash of the burnt bark was used to prepare the round yam, which is poisonous, before cooking (Wunambal Gaambera). The gum was eaten.

Tools and Objects: Weapons were made of the wood. The roots and bark were fashioned into containers. The bark was made into shields. Branches were used as digging sticks to find food and water. The wood was used in a children's disk-throwing game.

Art and Ceremony: The wood was used for making didgeridoos. Babies were smudged with the smoke to help them grow fit and strong.

Shelter: The bark was used to make shelters.

Transportation: The bark was made into canoes.

Modern Uses: Its durable wood is used for heavy construction, poles, railroad ties, fencing, flooring, craft furniture, particleboard, chipboard, pulpwood, and charcoal. It is a honey production tree.

Threats and Conservation: Not threatened, although grazing and logging have restricted its regeneration. Since they are often dependent on flooding for water supply, they could be affected by climate change.

River Red Sheep Soiree (2015)

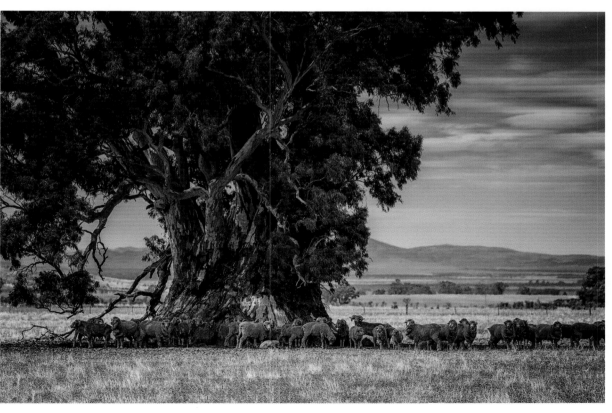

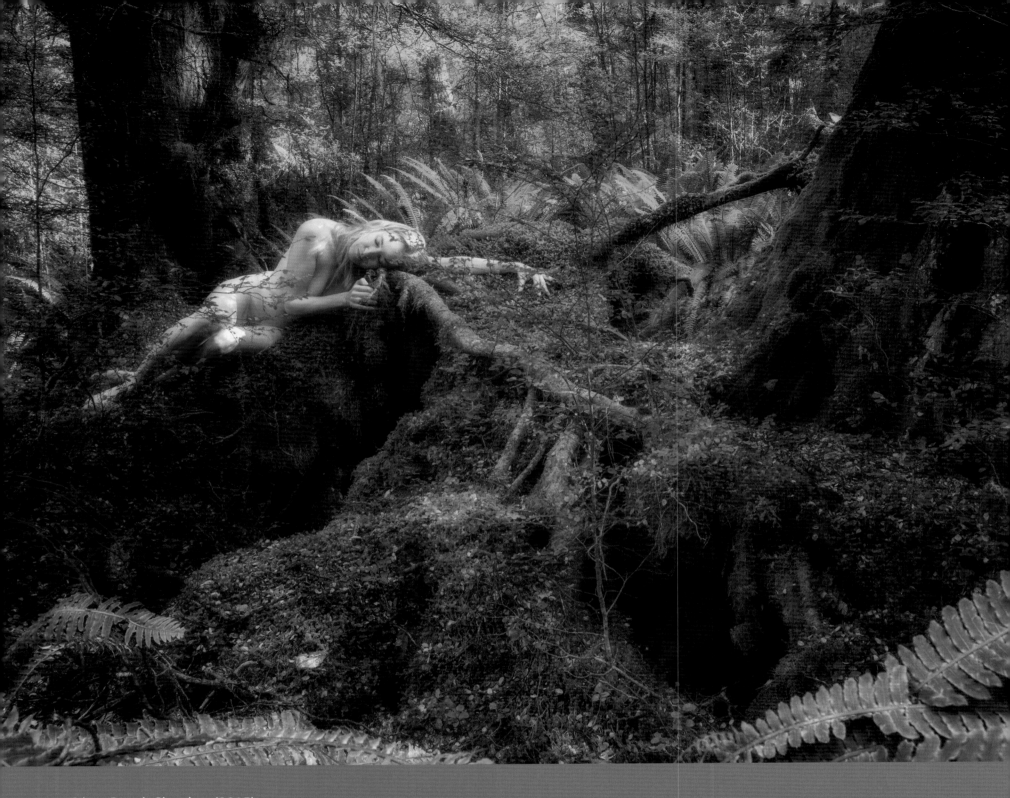

Silver Beech Slumber (2015)

The first thing I noticed about the wild forests of New Zealand—in comparison to the wild forests of Australia, from which I had just come—is the definitive lack of wildlife. New Zealand has no native mammals; strangely, you can feel their absence. In fact, there are also no poisonous critters to watch out for when enjoying oneself naked in New Zealand. On the Te Anu Track of Fiordland National Park, South Island, all I had to be aware of were unsuspecting trekkers. Luckily, I had help from a new friend named Tumbleweed who scouted for me, and could easily hide from passersby in a large hole where a former tree had once been rooted. This mound upon mound of forest was made up of roots and decomposing trees covered in moss, little plants, and the tiny leaves that had rained down off the Silver Beech trees. There were little holes and caverns within the mounds, which may have been the homes of the invasive (non-native) possums who feast off the eggs of the native bird population. Or perhaps the mounds were homes for fairies or gnomes (probably native). I did spot a healthy population of red amanita mushrooms on my way out. It was undoubtedly a magical place.

Silver Beech

Lophozonia menziesii

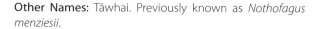

Other Names: Tāwhai. Previously known as *Nothofagus menziesii*.

Distinctive Characteristics: This evergreen tree, with small dark, double serrated shiny leaves sheds every 3–5 years throughout the growing season, leaving a carpet of lighter-colored dead leaves sprinkled on the forest floor. The silvery grey bark of older trees is rough and flaking, often covered in moss. On larger trees, the trunk is buttressed. This Beech is not related to the European or American Beech (*Fagus*).

Distribution: Native to New Zealand. Of the remaining native New Zealand vegetation (of which only approximately 10% is the original), there are two major forest types: the Kauri and the Beech forests. Beech forests are widespread. Elevation: on the South Island from sea level to 4,700 ft. (1,430 m), and on the North Island from sea level to 2,000 ft. (600 m).

Ecosystem: There are three species of Beech in New Zealand: Red Beech, Hard Beech, and Silver Beech, and two varieties: Black Beech and Mountain Beech. They may grow in combination, as pure forests (primarily in subalpine regions), or with other tree species, in lowland to montane to subalpine ecosystems. Silver Beech will hybridize with other Beeches. When in mixed forest, it is associated with trees including Miro and New Zealand Cedar in upper montane forests; with Rimu, Black Pine, Tōtara, and *Podocarpus dacrydioides* in lower montane forests; with Mountain Toatoa and Mountain Tōtara near timberline; and also with Tawa and Kauri. The strawberry fungus (*Cyttaria gunnii*) is found exclusively on the Silver Beech and has distinctive orange-yellow golf-ball-like fruiting bodies.

Maximum Age: Unknown.

Maximum Height and Girth: Up to 100 ft. (30 m) in height; up to 6.5 ft. (2 m) in circumference.

Animal Community: Kererū, the New Zealand pigeon, eats the flowers and leaves. Kākā birds peck at the bark looking for insect larvae. The leaves and trunks are chewed by caterpillars, moths, and many species of beetles. The tree supports other insects such as gall mites, weevils, whiteflies, scales, mealybugs, gall midges, and thrips.

Traditional Uses:
Tools and Objects: The bark was used to produce a black dye for coloring flax and cabbage tree leaves (Maori).

Modern Uses:
The bark has been used for tanning leather. The wood has been made into tubs, baskets, and wine casks. Around the turn of the 19th century, the wood was popular in France for making bedsteads and sideboards. It remains popular for artisan wood-turning.

Threats and Conservation: Not threatened.

Silver Beech Root Fairy (2015)

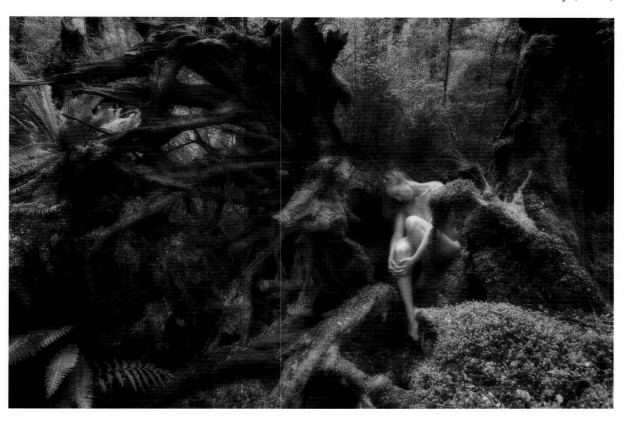

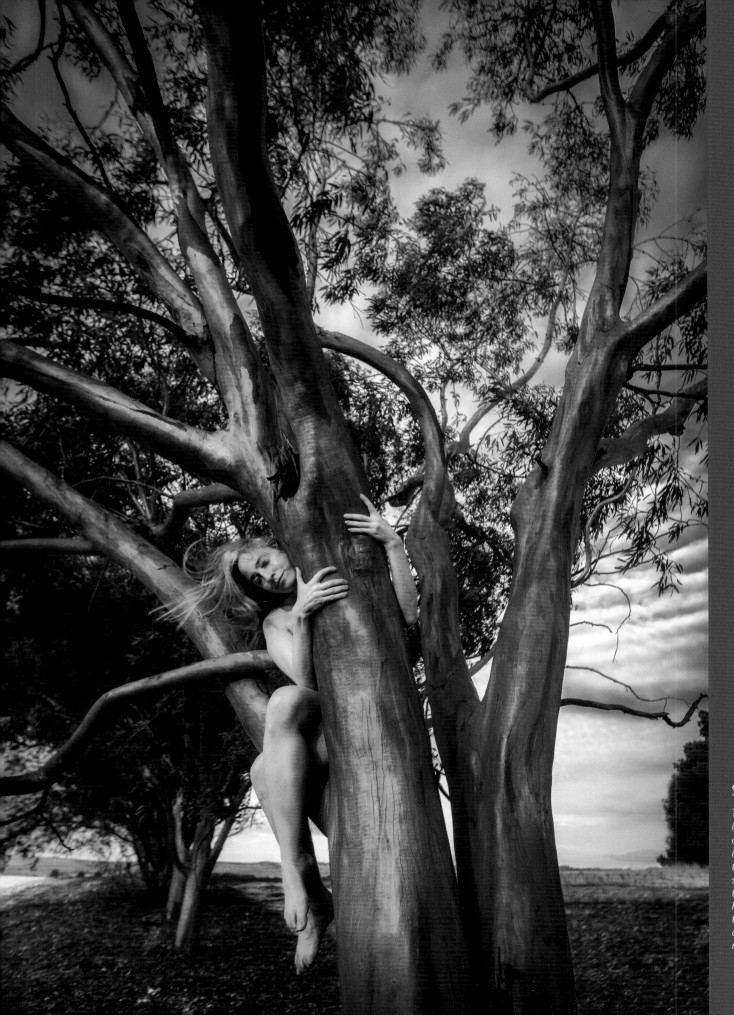

Silver Gimlet Girl (2015)

The most impressive feature of this individual tree is not its size, or even its shape. It is, of course, the beautiful copper color of its bark—which is not silver at all! I imagine that, at some point in its lifespan, some part of the tree shimmers in silver, as eucalypts have the trait of sometimes changing the colors of their bark and leaves throughout their lifespan. I had to get close to this tree's hard limbs with its smooth bark—like a muscular, bronze-tanned man in the hot Australian sun. This tree stands in a row planted with other Silver Gimlets, bordering a field near Yacka, South Australia, Australia.

Silver Gimlet

Eucalyptus campaspe

Other Names: Gimlet, Silver-Topped Gimlet

Distinctive Characteristics: This single- or multi-trunked, medium-sized evergreen eucalypt has striking shiny copper, bronze, to salmon-red-colored smooth bark that weathers to a greenish or greyish brown, shedding in thin strips. The heartwood is brown, tinted with pink, red, grey, mauve, or orange. The trunk tends to form spiral fluting. Silver refers to the silvery grey leaves.

Distribution: Native to limited areas in southern Western Australia in a 50 km radius around Coolgardie. Elevation: between about 1,300 and 1,640 ft. (400–500 m).

Ecosystem: Lives on low hilly areas in open woodland with other Eucalypts like Salmon Gum, Snap and Rattle, and Redwood.

Maximum Age: Estimated at 100 years.

Maximum Height and Girth: 33 ft. (10 m) in height; about 5 ft. (1.5 m) in circumference.

Animal Community: The nectar is consumed by various birds and bees.

Modern Uses: The wood has been used for mine props and firewood.

Threats and Conservation: Not threatened.

Silver Gimlet Leaves and Flower Buds

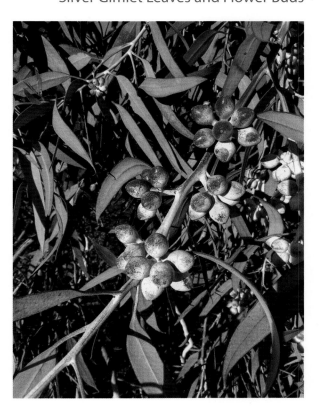

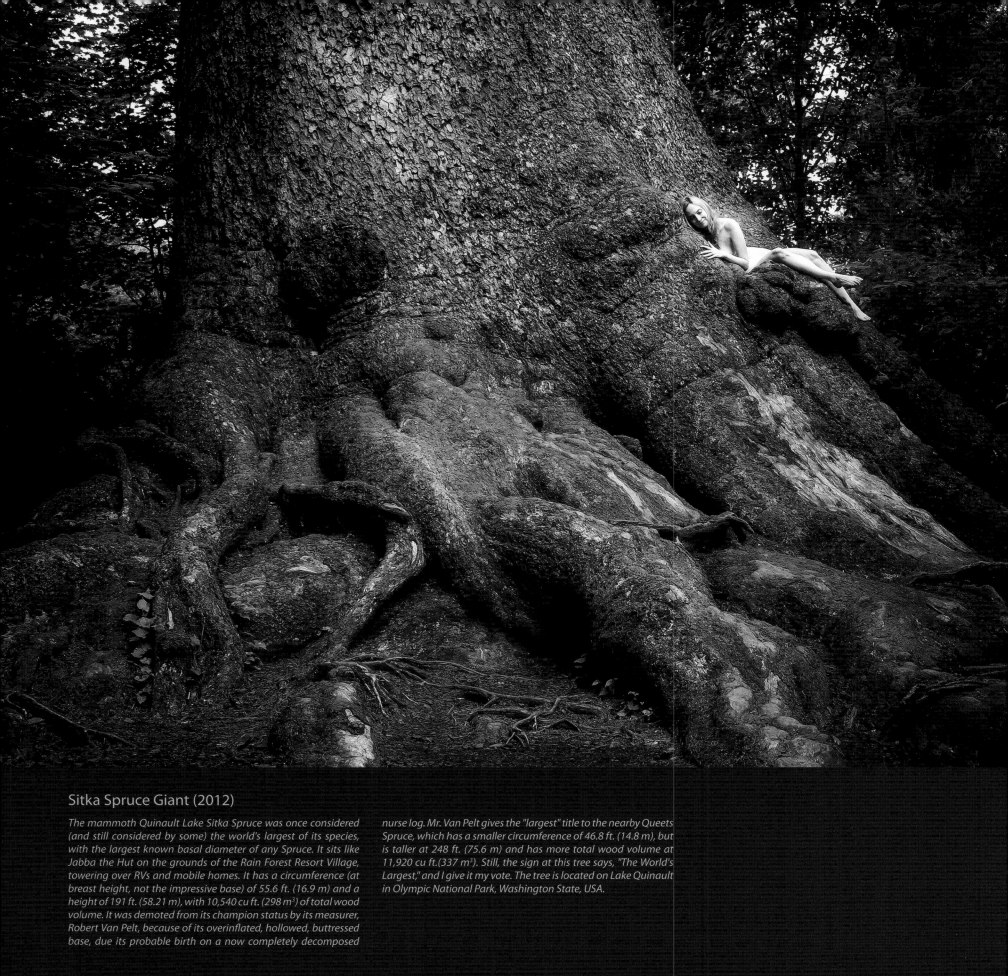

Sitka Spruce Giant (2012)

The mammoth Quinault Lake Sitka Spruce was once considered (and still considered by some) the world's largest of its species, with the largest known basal diameter of any Spruce. It sits like Jabba the Hut on the grounds of the Rain Forest Resort Village, towering over RVs and mobile homes. It has a circumference (at breast height, not the impressive base) of 55.6 ft. (16.9 m) and a height of 191 ft. (58.21 m), with 10,540 cu ft. (298 m³) of total wood volume. It was demoted from its champion status by its measurer, Robert Van Pelt, because of its overinflated, hollowed, buttressed base, due its probable birth on a now completely decomposed

nurse log. Mr. Van Pelt gives the "largest" title to the nearby Queets Spruce, which has a smaller circumference of 46.8 ft. (14.8 m), but is taller at 248 ft. (75.6 m) and has more total wood volume at 11,920 cu ft.(337 m³). Still, the sign at this tree says, "The World's Largest," and I give it my vote. The tree is located on Lake Quinault in Olympic National Park, Washington State, USA.

Sitka Spruce

Picea sitchensis

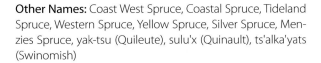

Other Names: Coast West Spruce, Coastal Spruce, Tideland Spruce, Western Spruce, Yellow Spruce, Silver Spruce, Menzies Spruce, yak-tsu (Quileute), sulu'x (Quinault), ts'alka'yats (Swinomish)

Distinctive Characteristics: Named for Sitka, Alaska, this tall evergreen conifer has a straight, columnar trunk, with a narrowly conical crown and horizontal branches. Old trees may have no branches for the lowest 100 ft. (30 m). Although it is one of the tallest tree species, Sitka Spruce can also grow stunted from wind, forming hedges along beaches. It is able to live directly on the coast because it has adapted to tolerate salt spray. It is, however, intolerant of shade. It generally propagates on nurse logs, resulting in inflated buttressed bases, which are sometimes partially hollow underneath. It can also reproduce from epicormic branches.

Distribution: Native to Alaska, British Columbia, Washington, Oregon, and Northern California, along coastal fog belts, inland to 124 mi. (200 km). Elevation: sea level to 2,296 ft. (700 m).

Ecosystem: Common along cool, moist headlands, alluvial floodplains, and within rainforest and bogs in pure stands or with Western Hemlock, Mountain Hemlock, Yellow Cedar, Redwood, Grand Fir, Douglas Fir, Shore Pine, Red Alder, White Spruce, Western White Pine, Port Orford Cedar, Black Cottonwood, and Lodgepole Pine.

Maximum Age: Approximately 800 years.

Maximum Height and Girth: 317 ft. (97 m) in height; 50 ft. (15.24 m) in circumference.

Animal Community: Deer, elk, porcupines, bears, and hares browse the new foliage, and squirrels feed on the cones. Bald eagles and peregrine falcons roost in the branches.

Traditional Uses:
Medicine: The sap provided an effective medicine for burns, boils and various skin infections, venereal diseases, internal swelling, heart trouble, and toothaches. The inner young bark was made into a tea or chewed to treat sore throats (Quinault). A decoction of the roots was used in the treatment of diarrhea (Kwakiutl). The bark, cambium, and sap were taken as a laxative (Nuxalk). Cones and bark were used for pain relief. Sapling bark and ripe cones were used

for steam beds and stomachaches (Bella Coola). The sap, cones, and bark were used in various ways to treat rheumatism (Bella Coola, Gitksan, Haisla, Hanaksiala, Oweekeno).

Food: The inner bark, fresh or dried into cakes, was eaten (Haida, Tlingit, Kitasoo, Tsimshian). It was also ground into powder and used as a thickener in soups or bread. The sap was chewed as a kind of gum (Makah, Haisla, Hanaksiala, Oweekeno, Southern Kwakiutl, Makah and Quinault). Newly grown needle tips are edible and may be used to flavor spruce beer.

Fiber: The roots, peeled, split, and dried, were used to make cordage, baskets, and water-tight hats and baskets (Makah, Haida, Tlinglit, Bella Coola, Hahwunkwut, Kwakiutl, Oweekno, Poliklah, Quiluete, Nitinaht Quinault, Quileute).

Tools and Objects: The softened, sticky pitch of the tree was warmed and used as glue and a protective, varnish-like waterproof coating on boats and harpoons (Nitinaht, Quinault, Quileute). The samplings were made into pole snares to hunt mammals (Quileute). The wood was used to make toys (Hoh).

Art and Ceremony: The boughs of sharp needles were used widely for spiritual protection and in ritual, for protection from death and illness (Bella Coola). They were used by shamans, hunters, and fishers during preparatory and purification rituals (Tsimshian); rubbed on skin for protection against evil (Thompson); used in winter dance ceremonies to make costumes and protection for the dancers for ceremonies to initiate children (Nitinaht). Similarly, used to hit and rub boys to increase strength and tolerance (Hanaksiala) and in the girls' puberty potlatch ceremony for protection (Hesquiat).

Shelter: The wood was used as building material (Hoh, Quileute, Hesquiat).

Modern Uses: Sitka Spruce was highly exploited during World Wars I and II to make wooden aircraft frames and propellers. The trees are still used to make boats and oars. It is also used to build ladders and scaffolding, as well as musical instruments such as pianos, harps, violins, and guitars. It is also manufactured into pulpwood lumber, plywood, pulpwood, and various paper products.

Threats and Conservation: Not threatened.

San Juan Spruce Resting (2013)

It was a usual, chilly, drizzly day on the south of Vancouver Island, and with the gracious assistance of The Ancient Forest Alliance—a local activist organization victorious in saving old growth—I met a tree who is the largest Sitka Spruce in Canada and considered the second largest in the world. The San Juan Spruce is 125.7 ft. (38.3 m) in circumference and 205 ft. (62.5 m) tall. Spruces tend to grow straight up, with one major trunk. As they age, they may form what are called "reiterations," wherein the trunk branches off a new major growth (sometimes with an elbow at the side, sometimes at the very top) that reaches vertically. What makes this tree so charismatic is the reiteration that begins 14 feet up from its base and continues upward to become the tallest part of the tree. The coat of soft moss and ferns that hug the tree invited me to do the same, making my rainy TreeGirl photo shoot incredibly pleasurable. This famous tree is accessible in Port Renfrew, Vancouver Island, British Columbia, Canada. On your way there, give a generous donation to the Ancient Forest Alliance for all their incredible work to preserve what giants and their fragile ecosystems are left on this over-logged island.

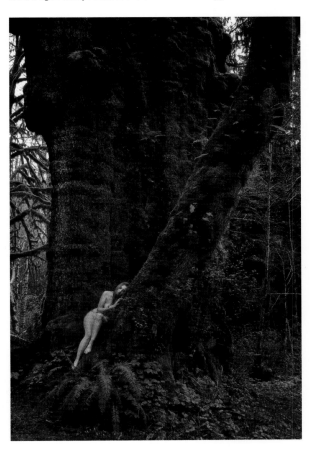

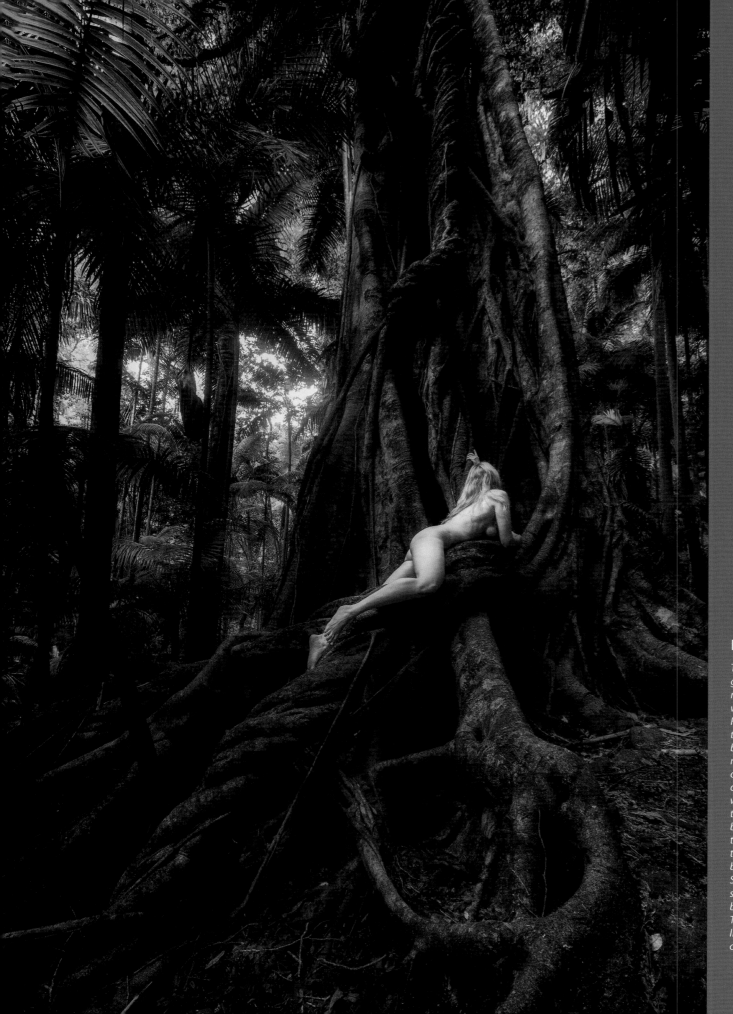

Fig Leaved Lover (2015)

The day I spent in Border Ranges National Park with my gracious arborist host, Robert Stavrou, was one of the most adventurous in TreeGirl history. The first photo-worthy tree we found was this magnificent Small Leaved Fig. What makes this tree so striking is the giant vine twisting around the tree; where the tree ends and the vine begins is a puzzle to unravel. It was one of those Tarzan moments when I felt completely and comfortably feral climbing the tree, as the sunlight broke through the forest canopy right next to me. Bliss. Immediately afterward, we found another photogenic tree: a Blue Quondong that produces bright blue fruits. I lay down between its buttressed roots, only to arise with about a hundred baby ticks crawling on my face. (Luckily for me, they were not the Paralysis Tick!) After that, a spontaneous rain shower brought our photography day quickly to a halt mid shot. So back to the truck we went, but not until I'd pulled a sneaky leech off my bloody big toe. Never before had I been so humbled at the potent life-force of the rainforest. This tree, ticks and leech, among many other organisms live in Border Ranges National Park, directly on the border of New South Wales and Queensland, Australia.

Small Leaved Fig

Ficus obliqua

Other Names: Small-Leafed Fig, Small Leaf Fig, Polynesian Banyan, Strangler Fig, Figwood, baka or baka ni viti (Fiji), burrawarra/baira (Dharawal)

Distinctive Characteristics: Like many Fig trees, the Small Leaved Fig develops a wide trunk with tall, ribbony, buttressed roots and forms a thick canopy of glossy, evergreen leaves. It can grow as a single-trunk tree, but more likely it grows as a parasitic 'strangler fig': born from a seed dispersed by a bird or bat atop a tree, it starts life as an epiphyte, absorbing nutrients and water from the air, dropping down multiple aerial roots from the host's branches that eventually take root in the earth. Over time, roots aggressively intertwine the trunk and fuse together, creating a latticework. Out-competing the host for space, nutrients and light, eventually the new tree becomes freestanding. The leaves and branches bleed a milky sap if cut. The tree has small fruits that turn from yellow-orange, to orange dotted with red, to dark-red figs.

Distribution: Native to eastern Australia, New Guinea, eastern Indonesia, and the southeastern Polynesian Islands. In Australia, it ranges from the Cape York Peninsula along the northeast coast of Queensland through New South Wales. Elevation: sea level to 3,300 ft. (1,000 m).

Ecosystem: Lives in subtropical rainforests, savanna woodland, sclerophyll forests, and gallery forests.

Maximum Age: Estimated at 500 years.

Maximum Height and Girth: 200 ft. (60 m) in height; 54 ft. (16.5 m) in circumference.

Animal Community: Many bird species consume the fruit and disperse the seeds, including double-eyed fig parrot, rainbow lorikeet, southern cassowary, brown cuckoo-dove, rose-crowned fruit dove, wompoo fruit dove, wonga pigeon, topknot pigeon, silvereye, pied currawong, blackfaced cuckoo-shrike, olive-backed oriole, Australasian figbird, green catbird, regent bowerbird, satin bowerbird, and Lewin's honeyeater. Polynesian species include the many-coloured fruit dove and crimson-crowned fruit dove. Bats such as the spectacled flying fox and grey-headed flying fox also eat the fruit. The leaves serve as a food source for the larvae of the butterfly species known as common crow and no-brand crow, as well as the geometer moth.

The tree is pollinated by two species of fig wasp: *Pleistodontes greenwoodi* and *P. xanthocephalus*.

Traditional Uses:

Medicine: Its white latex sap has been used to treat swollen joints and limbs (Fiji) and boils (Samoa, Fiji, Tonga). Liquid extracted from the root bark has been used to treat headaches and improve breast milk. A cold poultice of leaves was applied to treat venereal lesions (Fiji). A tea was made from the stem bark or leaves to treat respiratory ailments (Fiji). An infusion of the leaves was used to treat breast tumors (Tonga). It was mixed with other plants to treat convulsions (Fiji).

Threats and Conservation: Not threatened.

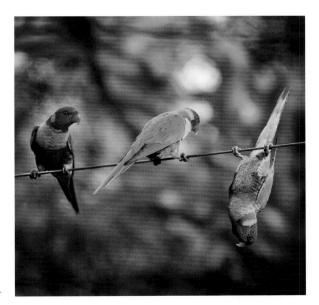

Rainbow Lorikeets (2015) ▸

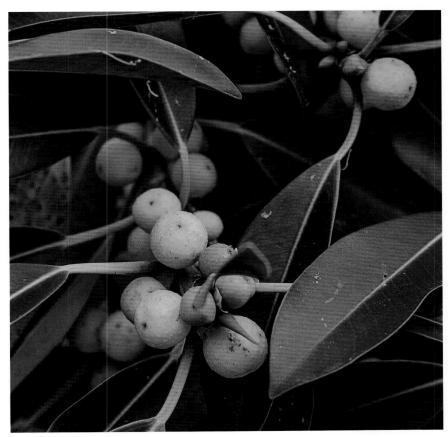

Small Leaved Fig Fruit ▸

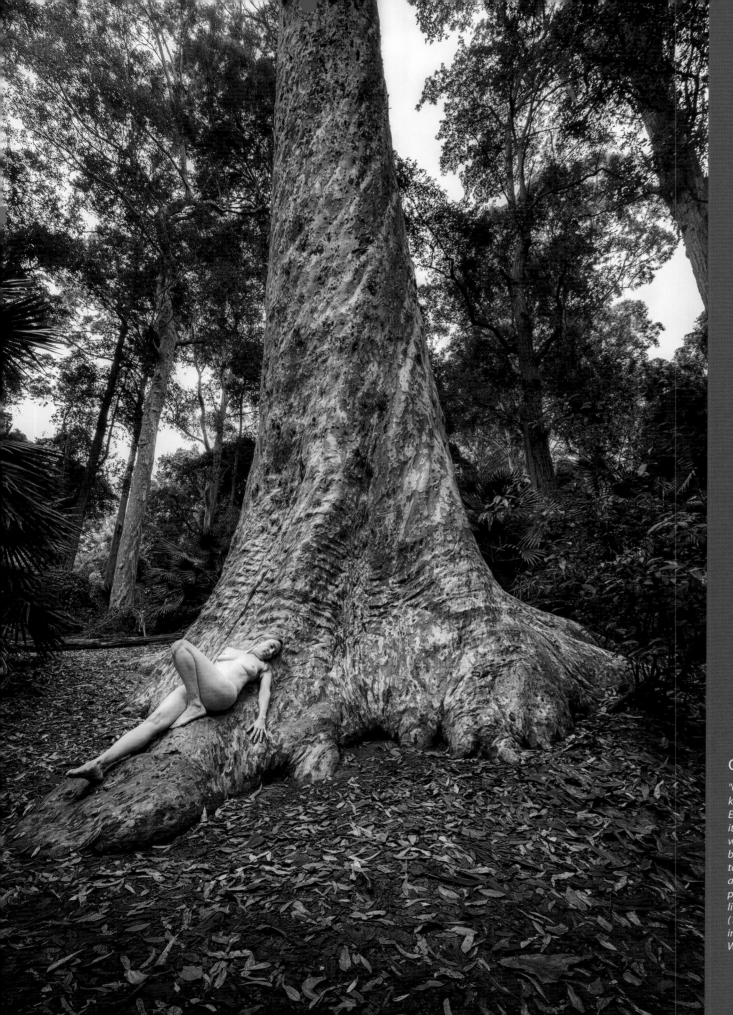

Old Blotchy's Girl (2015)

"Old Blotchy" is one of those secret trees that everyone knows about. Hidden down an overgrown trail, this Eucalypt stands tall in a small clearing. Right after I found it, so did three families with their gaggle of preschoolers who proceeded to hug and photograph the tree. These budding treegirls and treeboys had no idea what I was up to, but after they left, I introduced myself to Old Blotchy at the base of his flared roots and quickly shot some pictures before the next viewers arrived. He seemed to like all the attention that day. Old Blotchy is 35.30 ft. (10.76 m) in circumference and 193 ft. (59 m) tall, and lives in Murramarang National Park, near Turmeil, New South Wales, Australia.

Spotted Gum

Corymbia maculata

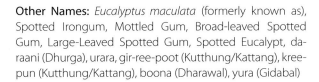

Other Names: *Eucalyptus maculata* (formerly known as), Spotted Irongum, Mottled Gum, Broad-leaved Spotted Gum, Large-Leaved Spotted Gum, Spotted Eucalypt, daraani (Dhurga), urara, gir-ree-poot (Kutthung/Kattang), kree-pun (Kutthung/Kattang), boona (Dharawal), yura (Gidabal)

Distinctive Characteristics: This tall, straight, evergreen tree has shedding patches of bark, mottled in coppery brown and numerous shades of grey; the underside is a smooth, creamy, yellow-white to green. The base of older trees is flared.

Distribution: Native to coastal areas of southeastern Queensland and New South Wales, Australia, where it is widely distributed, with an isolated population in east Gippsland, Victoria. Elevation: sea level to 2,100 ft. (650 m).

Ecosystem: Found in temperate and subtropical open forests and on coastal plains, mainly on slopes and ridges, in tall, open forests in pure stands or with Ironbark, Blackbutt, Tallowwood, Grey Gum, Sydney Blue Gum, Grey Box, White Mahogany, Brush Box, and Pink Bloodwood.

Maximum Age: Estimated at 500 years.

Maximum Height and Girth: 298 ft. (91 m) in height; 34 ft. (10.4 m) in circumference.

Animal Community: Koalas eat the leaves. Bees, bats, and birds such as rainbow lorikeets eat the nectar. Glider possums nest in hollows.

Traditional Uses:
Medicine: The gum, mixed with water and swallowed, was used as an anti-inflammatory. The leaves were crushed and soaked in water to be used as an antibacterial and antiseptic.

Tools and Objects: The wood and bark were used to make dishes, bowls, weapons, and tools.

Shelter: The bark was used to construct shelters.

Transportation: The bark was used to make canoes.

Modern Uses: An important commercial timber species, the strong, bendable wood has been used to make houses, ships, bridges, mining structures, railroad ties, carriages, fencing, flooring, furniture, tool handles, farm implements, poles, plywood, and charcoal. The flower nectar is valued for honey production.

Threats and Conservation: Not threatened.

Spotted Gum Bark

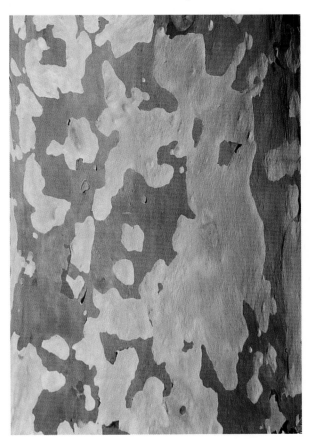

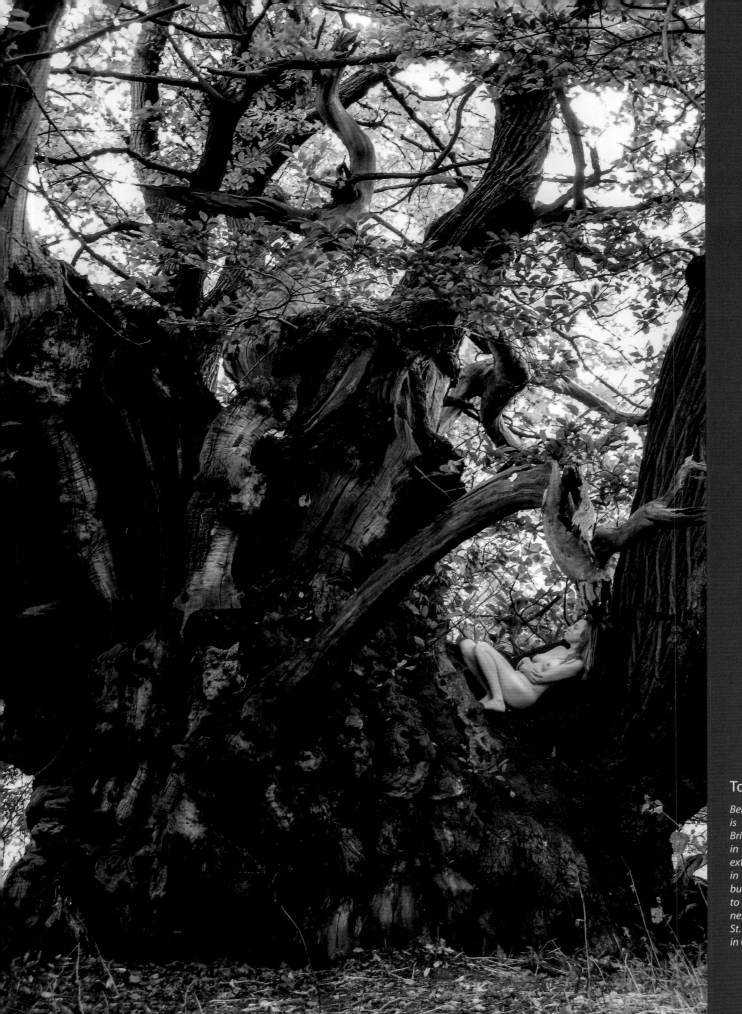

Tortworth Chestnut Nest (2009)

Believed to be 1,200 years old, the Tortworth Chestnut is the most famous and celebrated living chestnut in Britain. Its circumference is currently at 36 ft. (11 m), but in its grander days, before losing some major limbs and extensive canopy, it was much larger at 52 ft. (15.48 m) in circumference (measured in 1766). Now a craggy but noble remnant of its former self, the tree continues to hold a regal presence. Also a grand home for a small nest of bees, it lives a quiet, reclusive life in the back of St. Leonard's Churchyard in the small town of Tortworth, in Gloucestershire, England.

Sweet Chestnut

Castanea sativa

Other Names: Chestnut, European Chestnut, Portuguese Chestnut, Spanish Chestnut, Marron (French)

Distinctive Characteristics: Sweet Chestnuts are large, deciduous, broadly columnar trees that are well known for their longevity and large, edible nuts encased in prickly husks.

Distribution: Native to Southern Europe and Asia Minor; introduced to Britain by the Romans.

Ecosystem: Deciduous woodlands with English Oak, Sessile Oak, Downy Oak, Turkey Oak, Holm Oak, European Beech, Sycamore, and White Fir. Elevation: between about 650 and 4,000 ft. (200–1,200 m).

Maximum Age: Estimated at over 2,000 years.

Maximum Height and Girth: 150 ft. (46 m) in height; 64 ft. (19.5 m) in circumference.

The world record for the oldest known specimen (over 2,000 years) as well as the "Greatest Tree Girth Ever" is The Hundred Horse Chestnut (Castagno dei Cento Cavalli) on Mt. Etna, Italy. When measured in 1780, its center was hollowed and had split into a ring of separate trees, measuring at a supposed circumference of 190 ft. (58 m) What is left of the deteriorated ring of three tree fragments is 52 ft. (16 m) in circumference.

Animal Community: The nuts are an important food source for wildlife such as squirrels, wild boar, and jays, and are used as fodder for domesticated animals.

Traditional Uses:
Medicine: The leaves and bark have been used to treat inflammation, hemorrhaging from childbirth, hemorrhoids, rheumatism, fever, congestion, and nerve pain. Decoctions of the bark and fruit were used to treat chronic digestive disorders and improve bile flow (Russia). A shampoo was made from infusing leaves and fruit husks.

Food: Sweet Chestnuts are in the same family (Fagaceae) as other nut producers such as oaks, beeches, and chinquapins. Before the introduction of wheat, chestnuts (along with acorns) were the major source of carbohydrates for European and Mediterranean cultures, and they remained a staple food for certain regions where grains would not grow well. They fell out of popularity in the 1800s, when they gained a reputation as "poor people's food."

Tools and Objects: In Britain, Sweet Chestnuts were historically and are still used for coppicing (the repeated harvesting of wood from stumps that then regrow usefully straight from sprouts).

Modern Uses: In Europe today, the starchy, carbohydrate-rich nuts are traditionally eaten roasted at Christmastime and are also used in candies, cakes, and puddings. In addition, chestnuts can be used to make a type of flour and soup thickener. The traditional Corsican variety of polenta is still made with ground chestnuts. It is also made into a beer. Because of its high tannin levels, chestnut wood is rot-resistant and has traditionally been used to make outdoor objects such as posts, fencing, and stakes. It is also prized for making furniture, barrels, and roof beams.

Threats and Conservation: Not threatened.

Chestnut

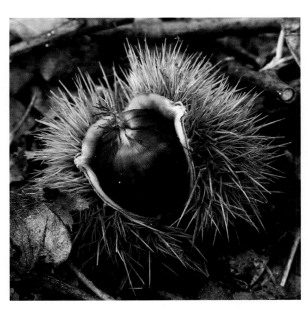

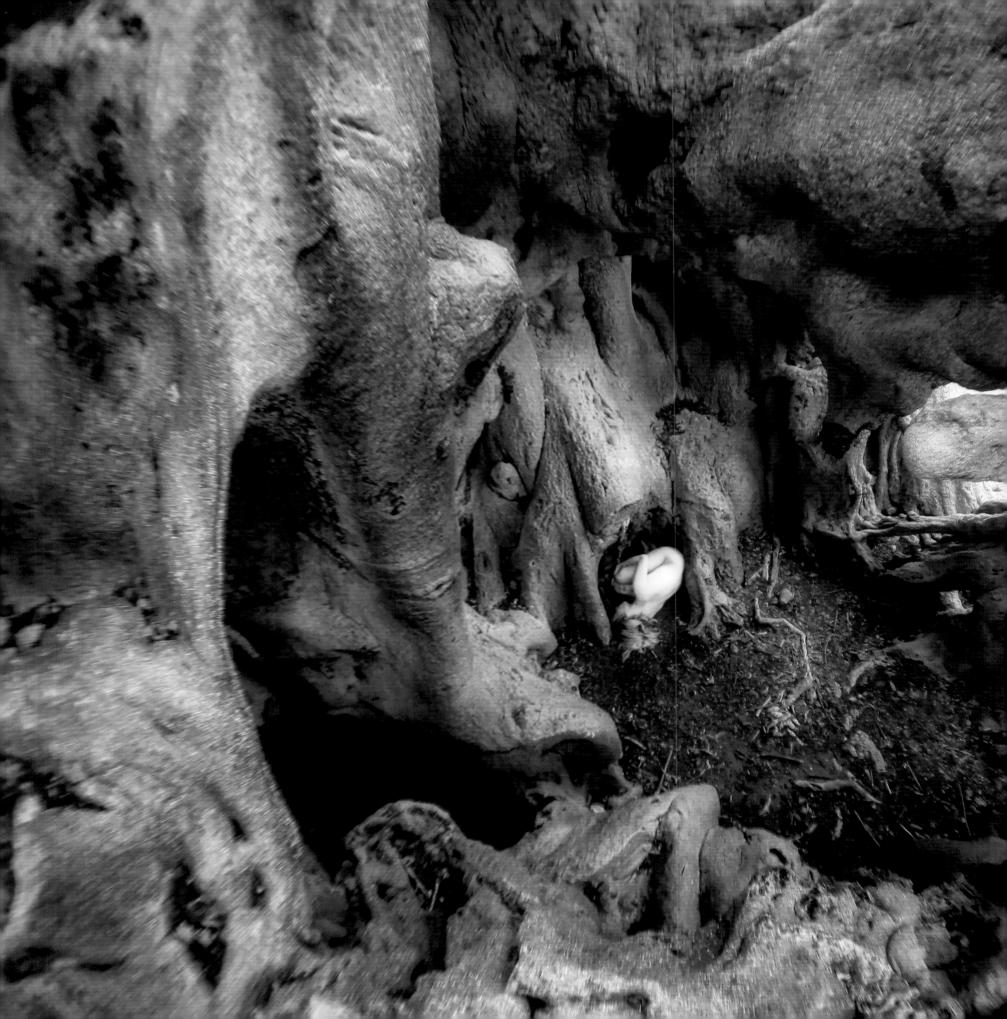

Trees are sanctuaries.
Whoever knows how to speak to them,
whoever knows how to listen to them,
can learn the truth.

~Hermann Hesse

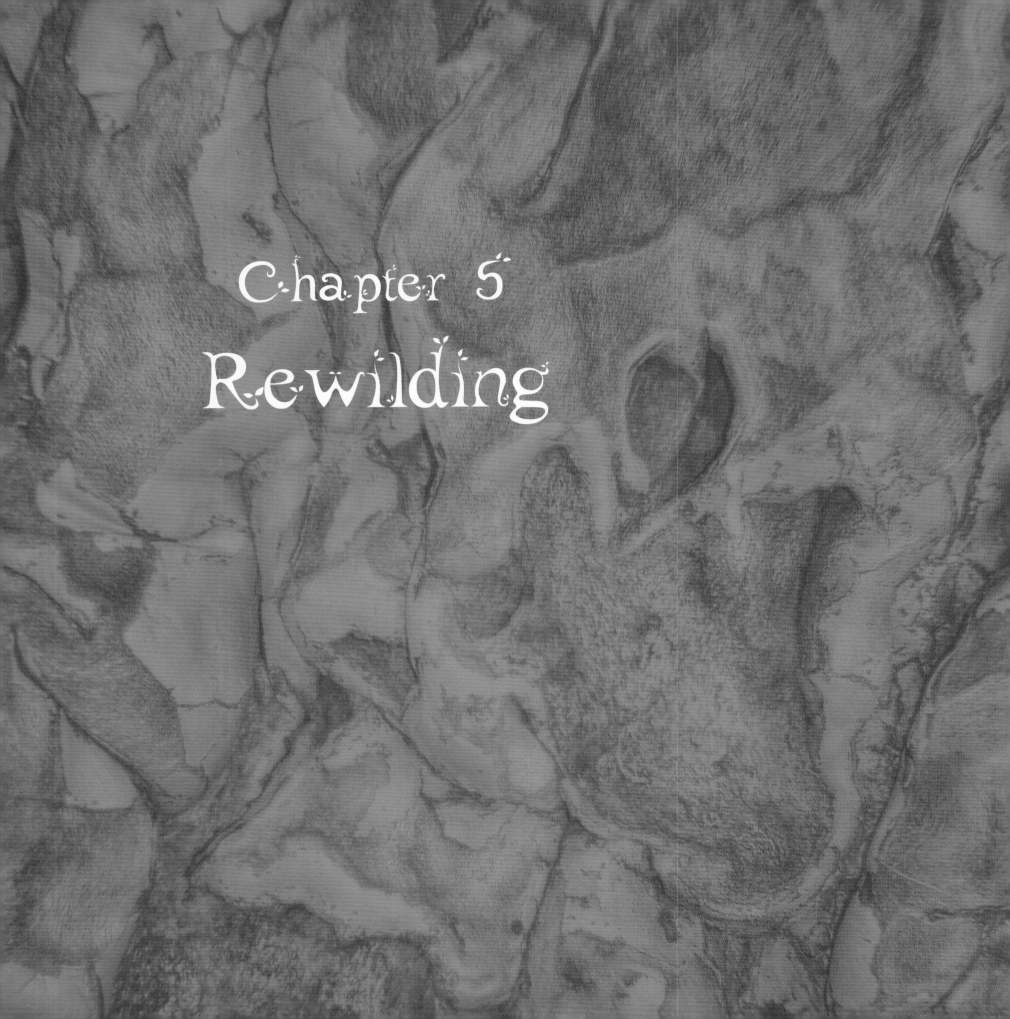

Chapter 5
Rewilding

Rewilding

is a return to a state of wildness. Much has been written about the need to rewild the planet as a conservation biology strategy, crucial to preserving and restoring functional ecosystems and to reversing the loss of biodiversity while reweaving the worldwide ecological web. This rewilding process includes creating wildlife corridors, reforestation, restoring wetlands, reintroducing extirpated megafauna (and microfauna) to their original habitats, and other forms of restoration ecology. But the truth is, we cannot rewild the planet without rewilding the human species as well. It is an illusion to think we can manipulate separate parts of Nature to regenerate and preserve them into a less damaged state without reviving human culture at the same time. We are going to have to willingly let ourselves out of our cages.

Rewilding is not as simple as just walking barefoot, abandoning our computers and cell phones, eating a tamed-down Paleo diet, wearing hand-tanned animal skins, or shitting in compost buckets (though it can include those). It must include questioning and unlearning the cultural and societal conditioning we joined when we were born into industrial civilization. It implies a genetic remembering—a re-cultivation and rebonding with our extended family of Nature. Rewilding is also a state of awareness—a paradigm even—of perceiving, understanding, and interacting with the world.

Today, the religion of Western science is both archaic in its attempt to categorize the natural world into linear boxes, and ever-expansive in changing what we think we know about how the entire material world—and all of reality—is constructed. Science keeps "discovering" how everything is interconnected—something that indigenous cultures have known all along. The more-than-human world of Nature is indeed a *natural resource*—not a reservoir of material to exploit, but a resource of information, wisdom, guidance, community and energy that can nurture us into being better co-inhabitants of Earth. All of this requires that we spend "quality" time, with a sincere *intention* to cultivate intimate relationships with the wild, just like we do with other members of our family, our companion animals, our friends and community, our gardens, and the place we call home. There may be a learning curve when engaging with a new person or place, or it may come quite naturally. Can intimacy with Nature be relearned?

Previous p. 134:
Fisheye Inside Holboom (2008)

African Baobab, (Adansonia digitata), *Namibia. See also p. 19.*

Yes. It's just a matter of accessing a language and behavior that lives innately in our brains, bodies, hearts, and souls.

The first step in rewilding yourself is to recognize the flood of distractions the modern world has created via the many companies, systems, agencies, institutions, and industrial complexes that want your money or that somehow benefit from you being distracted from life. Distractions such as the Internet, the media, shopping, entertainment, social obligations, and even work keep us busy, in a high level of stress and addictive patterns. Of course, there are unfortunate aspects of our culture that require us to be functioning members of our money-driven, over-bureaucratic society. Recognize that this culture is completely out of control, and that you are not crazy for fantasizing about quitting your job and moving to Hawaii to spend all your time snorkeling (or whatever a peaceful paradise would be for you). That may be your wild, but sane, inner *Homo sapiens* calling to you. But if you don't want or are unable to become completely feral, then look at where you do have choices. Unplugging from electronic technologies is an instant (although understandably challenging) option. How about changing careers to work outdoors or in support of the natural world? How about finding newness in discovering Nature rather than acquiring more and more new objects? How about redesigning your home to be more natural or creating an outdoor living space? [1] You can make small, simple changes to reclaim your attention, time, energy, money, and even your soul by choosing to be physically, sensorially present with the infinite and ever-changing living world in and around you.

Here are some questions to help you wade into this territory: What brings you curiosity and awe? What emotions and sensations arise when you are in contact with Nature: do you feel longing, appreciation, fascination, love, calmness, apprehension of the unknown? Where do you feel those emotions and sensations in your body? What in Nature fascinated you as a child? How about now?

Affinity is personal to your soul; you'll be drawn to whatever attracts your attention and what chimes in your heart. Perhaps you are drawn to birds, butterflies, flowers, fungi, stars, the ocean, or trees. Discover what your own bridge to Nature is, and *build it*. Use the resources you already have (experience, physical skills, knowledge, wonder, books, videos, or even a kayak…), and ask Nature for guidance on how to build your bridge. Listen to this guidance. Maybe your bridge involves learning birdwatching or bird calls. Maybe it's painting *en plein air* (outdoors), nature photography, or making ephemeral art out of natural objects and

leaving your creation in nature. Maybe it's fishing. You can also find others who have an affinity for Nature, then explore and build your bridge with them. Or take an imaginary friend along to cross your bridge to Nature with you. Just take a step.

Here are seven ways in which you can rewild yourself through Nature Connection.

1. MAKE CONTACT

Go outside of the built environment every day and make contact with the earth beneath your feet [2]—even if the temperature is cold; even if it's raining, snowing, or windy; even if it's dark; even if you only have one minute; even if you are tired and don't feel like it; even if you don't have a plan for what you will "'do'" (we have enough on our to-do lists already!). Going outside is more than just getting some fresh air (although that is a really great motivator, too). Mom sent us kids outside to play for reasons more than beyond maintaining her sanity; she knew it was good for us. Here are some ideas:

Every day, touch the earth with your bare hands, bare feet, bare anything, or bare everything. Our feet contain 7,000 nerve endings, 107 ligaments, 19 muscles, 26 bones, and 6 meridians [3]—and the $48-billion-dollar U.S. shoe industry [4] keeps us separated from the ground most of the time. There is a growing movement of barefoot walking and hiking, and even a Society of Barefoot Living, [5] endorsed by some podiatrists. [6] While I don't endorse being barefoot all the time (due to ouchy hazards and because, personally, my feet get cold easily), I do encourage a bit of exploration with our bare paws every once in a while. Take your shoes off and feel the temperature, the texture, and the sensations of the world beneath your feet.

Equally rewarding is to put your palms down on the earth. There are a total of 17,000 touch receptors and free nerve endings in the palm of the hand for passing on sensations of pressure, movement, temperature, and vibration. [7] Or find a place where you can lie down on your back, belly, or side, with some part of your body touching the earth. Don't be afraid to get earthy! Explore your comfort level with various degrees of skin contact. Every inch of human skin touching the skin of another organism—in TreeGirl's case, the bark of a tree—is another gateway for information and pleasure to be transmitted through sensory receptors. If you only have sixty seconds every day to go outside, simply making skin contact with the earth with your hands, feet, or entire body is very powerful in itself, and just might keep you looking and feeling young.

Every day, touch something new in Nature. Through touch, we interact, express emotion, and give and receive information, energy, and healing with others. Nature is no exception. As the arborist Alex Shigo emphatically said, "Touch Trees!" Explore touching tree bark. Have you ever been stunned by the beauty of a bark and felt moved to feel it? How about the chunkiness of the soft, fibrous bark of the Coast Redwood or the Giant Sequoia? Have you ever been tempted to peel some bark off a Eucalypt, Birch, or Sycamore? Have you ever petted the soft fur of a moss carpet on a tree trunk? Touch something new every day. Why? What will happen? I don't know. You don't know. Isn't that inviting?

Every day, smell something new, even if it doesn't appear to be fragrant. Everything has a scent. Be a dog for a day, be a hummingbird, be a bee (or be a human) and get your nose close to something alive. Smell deeply. Smell the air as you breathe it in.

Every day, listen to something new. It could be the birds, other animals, the rain, or the creaking of a tree in the wind. It could be a plant. Have you ever listened to a plant? Do they make sounds? Put your ear to the trunk or a hollow of a tree. Put your ear to the ground. What does the earth sound like? Do any of these have something to tell you?

Every day, look at something new. Or, look at the same thing every day in a different way. Look at something far away, or get up close to it. It takes effort and focused intention to really, truly see and observe something. It takes even more focused intention (and surrender) to allow something (with or without eyes) to *look at you*.

How about tasting? Most plants are not poisonous, and may even be edible, but are just not that tasty. Many flower petals are edible. Taste some bark. Perhaps just a kiss or a lick, without ingesting a plant, is all you need. Taste the rain, taste the snow. You can taste the air as well. You should probably avoid tasting a moving organism, such as an insect or slug, as that takes additional skills and knowledge.

Things to remember about making contact:
- Be safe! Take some risks and do some things you may never have done, but don't hurt yourself.
- Research (through books or the Internet) the plants and insects that are poisonous or can cause allergic reactions. Avoid plants that may have been treated with pesticides or that are close enough to roads to be contaminated with vehicle exhaust.

- Always be respectful of another organism. Avoid touching animals that do not want to be touched. Even if it doesn't move or seem to have feelings, honor its integrity. Ask permission. If it moves, put it back where you found it.
- Avoid drinking wild water. Unfortunately, nearly all wild water on the planet—even from underground springs—is contaminated in some way by agricultural pollutants, harmful microorganisms, and industrial toxins.
- Practice gratitude. Say thank you.

2. ENGAGE IN FOREST BATHING (SHINRIN-YOKU)

Forest bathing sounds like something TreeGirl would be into; being naked in the forest is a relaxing and cleansing experience. But there are no bars of soap or towels with this bath. In fact, forest bathing is more like *basking* in the atmosphere of the forest. The practice of forest bathing, as an ecotherapy, is a growing international practice of walking mindfully through and interacting with a forest (as opposed to exercise) as a source of physiological and psychological medicine.

The term *shinrin-yoku* (pronounced something like "sheen-drleen yo-koo," and meaning "bathing in the medicine of the forest") was coined in 1982 by the Forest Agency of the Japanese government. Their intention was to find non-extractive ways to use their forests, while benefiting their people by encouraging a healthy lifestyle and decreasing stress levels in one of the most fatally stressed countries in the world. Inspired by the Buddhist practice of walking meditation and by Shintoism,[8] the ancient indigenous religion of Japan, the intention of shinrin-yoku is to connect with Nature through all five senses by not simply walking, but engaging the whole body to achieve relaxation and health. Not surprisingly, researchers have found that certain trees in forests have special properties that are conducive to healing the mind and the body.

The science of forest therapy is big in Japan; the Japanese government has funded approximately $4 million worth of forest-bathing research since 2004, including laboratory tests and studies involving controlled groups who are hooked up to a host of monitoring devices prior to, during, and after their forest walks.[9] Through these scientific studies, forest bathing has been found to decrease levels of stress hormones such as cortisol and adrenaline in the blood, saliva, and urine; reduce sympathetic nervous system activity and increase parasympathetic nervous system activity (calming us down); lower blood pressure and pulse rate; be psychologically relaxing and reduce tensions; increase the number of natural-killer immune

cells and enhance their activity; improve mood; and boost overall immune function.[10] Some of these biological effects have been attributed to the natural aromatherapy of certain trees (such as pines) that release fragrant phytoncides,[11] organic chemicals that defend plants from harmful bacteria, fungi, and insects. These are naturally anti-microbial, so it makes sense that they would have that same immune-boosting effect on humans who inhale them.

The Japanese are serious about reducing stress. Forest bathing has now become a recognized, and even prescribed, relaxation and stress-management activity. Japan's forty-eight official Forest Therapy trails receive an estimated five million walks a year. The Japanese have a cultural love of their forests, which cover 67% of the country's landmass. (However, to preserve these forests, the country imports almost all of its wood from the U.S., some of which is old-growth timber. So how sustainable are they actually with that "not in my backyard" philosophy?) Japan intends to designate a total of 100 Forest Therapy sites within ten years. They have even trademarked the term "forest therapist."

Korea (which calls forest bathing *Sanlimyok*) has tagged along, with the construction of a multi-million-dollar Forest Therapy Center where people can stay for extended trips. At the Forest Therapy Center, there are even designated, gender-segregated forests where people can walk and lie on the ground naked.[12] In more conservative China, a "celestial forest bathing" facility created headlines in 2012 when some people mistakenly compared it to nude sunbathing or nudist beaches.[13] Spas and resorts in Europe and the U.S. are increasingly listing forest bathing on their menus of offerings as well. Indeed, returning to Nature has become a fashionable trend.

Forest therapy and nature connection research and practice are also being conducted in Korea, Scandinavia, Ireland, Germany, Spain, Finland, Taiwan, the U.K., Australia, and the U.S., and international alliances are being formed as the movement ripples outward. It is generally agreed that the term shinrin-yoku describes the Japanese-originated practice, while other nations use the terms "forest bathing" or "forest therapy," referring to shinrin-yoku as their inspiration. Sometimes all these terms are used interchangeably.

In forest bathing, a trained guide (or self-guiding signage) leads participants to particular stops along a designated trail that has been assessed for its quality of experience, phytoncide potential, and safety. Studies have shown that the greater the biodiversity of an area, the greater the healing.[14] (To me this is not surprising; the more life forms there are in a place, the more life-force energy is present there.) Specific sensory activities (termed *invitations* by some practitioners) slow the mind and body; participants are encouraged to give generous attention to curiosity and discovery, reaping the benefits of relaxation and the simple delights of sensory pleasure. All the senses are opened and engaged, as participants are guided through deeper ways of sensing, not only through the five major senses—seeing, smelling, touching, tasting, and hearing—but also by intuiting and sensing with their whole bodies.

A participant may be invited to "lie down on the ground and enjoy the colors in the sky" or "notice the warmth of sunlight on your skin" or "ask a question to a tree and listen for an answer," or any number of other experiences. An invitation may or may not have an emotional or spiritual impact, but participants often have a profound realization of their own grief over their disconnection from Nature. For example, the simple act of slowly and mindfully touching a trickle of water in a delicate, slow-moving creek may become a holy moment, even though someone may have touched water coming out of their faucet many times.

One very special practice is to brew a light tea at the end of the walk from plants that have been mindfully wildcrafted along the way. In northern California, where I live, nettle leaves, blackberries, wild rose petals, Douglas Fir tips, and wild mint are just some of the wild edibles that can be decocted, depending upon the season. Ceremonially drinking such a tea can bring surprise and simple pleasure. Do your research and find out what is seasonally available, edible, and best for making tea in your area. (If you cannot find any, perhaps there is a more ecologically diverse place where there are edibles.) After literally drinking in the forest, one cannot help but feel gratitude for the wild gifts of the Earth.

The leisurely approach of forest bathing can be considered part of the growing Slow Nature movement, a new aspect of the worldwide Slow movements (Slow Food, Slow Money, Slow Education, Slow Agriculture, Slow Art, etc.) that embrace consciously turning away from our fast-paced, desacralized, over-efficient, automated, hyper-computerized culture and returning to a more meaningful world.

Forest bathing may sound simple; yes, almost anyone can take a walk in the woods. But most of the time it's not easy to leave one's busy, plugged-in life to take pleasure in feeling one's body walking slowly, or actually being still, in Nature. It is sometimes just as challenging to simply be aware, really take notice, and be able to focus on something for more than eight seconds (the current average attention span of adult humans [15]). Many people could use some support in letting go of the goal to *get*—get somewhere, get exercise, get things done, get more stuff—instead of just *being*. A trained forest therapy guide has holistic expertise in ecopsychology, health, ecology, edible plants, facilitation, and outdoor leadership.

The intention is to heal not only our physical and mental ills, but to also heal our hearts' and spirits' disconnection from the Earth, cultivating long-term nature connection. These walks through the forest can be simple, yet profound, paths to expanding intimacy and rewilding our multi-sensory selves. For me, photographing myself and others in the nude with trees or in the landscape, is a natural extension of Forest Therapy, or Forest Ecotherapy. To

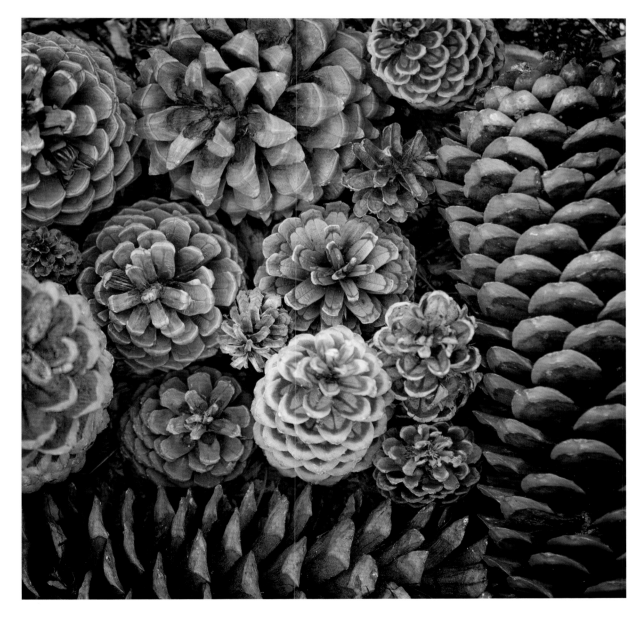

A Naturalist's Collection of Pine Cones

be witnessed by nature, without judgement, and then to be captured in an authentic state of being, brings other levels of healing. The possibilities with Forest Therapy are almost endless.

3. BECOME A NATURALIST

How many of us picked up leaves, seeds, acorns, or pine cones as children, twirling them between our fingers, stowing them in a pocket, or pressing them into a book only to find them later? Or perhaps you have your own small collection on display, along with rocks, seashells, dried flowers, animal bones, and bird feathers you can't seem to let go of. Chances are, you are a naturalist and you didn't even know it. A naturalist is one who identifies and studies organisms in their environment from the direct observational perspective of curiosity and discovery, and the discovery never ends. In fact, this kind of study leads to all the science "-ologies" you can think of. Curiosity helps to establish any new relationship. Perhaps you have stories and memories of your explorations in Nature. That is a relationship based on a connection. Do you connect best with insects, with leaves, or with something else? Or with all of life? What is your affinity?

It is easy to extend your affinity by taking a simple walk in the woods or by finding a place to sit and silently observe and listen. Bring some field guides to reference, little bags for respectfully collecting, and a nature journal for sketching, painting, and recording. While there are plenty of new naturalist and citizen-scientist websites and apps to encourage nature connection, to me they seem to support techno-addiction, so I recommend keeping the smartphone turned off.

Some of the most important figures in the history of preservation and conservation—people who changed the course of humanity's relationship with nature—were, in essence, naturalists: Carl Linnaeus, Charles Darwin, John Muir, Henry David Thoreau, Rachel Carson, Teddy Roosevelt, Aldo Leopold, Jane Goodall, David Brower, and Jacques Cousteau, to name a few. These pioneers felt a deep bond with Nature from birth, a profound affinity and reverence that led them to dedicate their lives to protecting Nature and educating others about it. And it all started with a simple curiosity.

4. (RE)LEARN ANCESTRAL TECHNOLOGY AND TRADITIONAL ECOLOGICAL KNOWLEDGE

Perhaps you're working on your collection of acorns and seeds, and you wonder if you can eat or make something out of them. The next logical path to rewilding is to learn the skills and knowledge of our ancestors from tens of thousands of years ago. Today we call that ethnobotany, ancestral technology, or traditional ecological knowledge (TEK). These ways include identifying, foraging, and processing plants and trees for use in food and medicine; as dyes and fiber for weaving baskets, clothing, ropes, nets, and mats; to make musical instruments and ceremonial objects; or to construct shelter and watercraft. For example, the Western Redcedar tree (*Thuja plicata*) of the U.S. Pacific Northwest has been identified as the North American plant with the most documented ethnobotanical uses—368![16]

Ancestral technology includes not only traditional knowledge and skills of working with plants, but also bird language, animal language, and animal tracking; hunting and fishing; animal processing and hide tanning; making fire; reading the weather and landscape; weaving cordage and baskets from fibers; making tools of wood, stone, and bone; harvesting water; managing and tending landscapes; mapping and wayfinding;[17] practicing cultural arts such as painting with natural pigments; and even shamanic perception, communication, healing, and ceremony. All of these skills, varying slightly depending on the culture and bioregion, were used by peoples all over the world for thousands of years after generations of experimentation, refinement, and practice, practice, practice. These skills will become valuable again in the coming generations, as the mass production of goods (which we take for granted) diminishes due to resource depletion.

There are growing numbers of resources available for acquiring these skills: books, museums, websites, workshops, convergences, and people. I suggest that you start locally

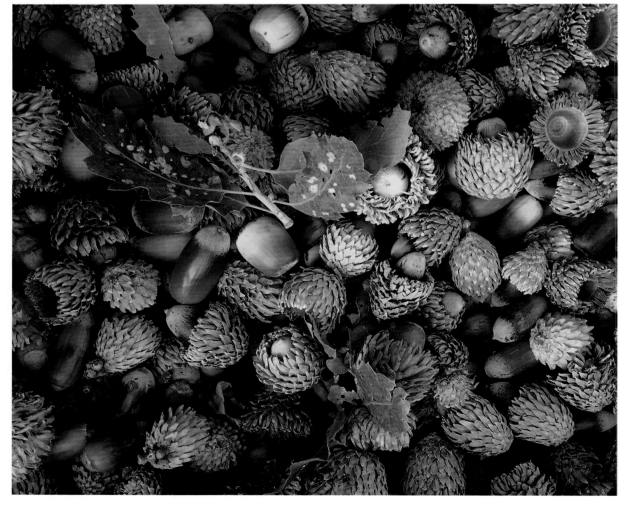

Valonia Oak Acorns

and bioregionally, learning skills people used historically in your area or from your own cultural heritage, knowing that they are *traditional*, not dead or "primitive." There are probably still elders in your area who are teaching the younger generations many of these living skills. Find a mentor. Explore what you are drawn to. Practice your new skills by teaching others, and practice respect and humility. Enjoy the pleasure of knowing that you are using your hands, mouth, senses, brain, and psyche in ways your pre-industrial ancestors did.

5. (RE)LEARN INTERSPECIES COMMUNICATION

Perhaps you've heard of the scientific research studies of communication between humans and primates, horses, dolphins, or whales. We commonly talk to our companion dogs and kitties, and we hear them as well. The truth may be, our ancient ancestors knew this skill of listening, interpreting, and dialoguing with other species very well; this ability is hardwired into our brains and our spirits. Before the advent of theistic religions, animistic and earth-based cultures all over the world lived (and still do) in a paradigm in which *everything* is alive and has a spirit, and thus could communicate. It makes the world a lot more interesting.

Much of re-accessing this ability to communicate with other species—in my case, mostly trees—comes through the risky realm of *trust*. As modern humans, we are taught through the education system, the media, our society, and even our parents to doubt ourselves, our intuition, our inner knowing, our knowing beyond the rational-analytical, and the trustworthiness of others. Those "others" include species we appear to not look or be like. We are taught that we humans are the most evolved species on the planet, and so we leave the others behind with our big smarty-pants heads and opposable thumbs. We as *Homo sapiens sapiens*, however, are a very young species (195,000 years old) compared to the rest of the four-billion-year-old planet. While every life form on Earth has its own language, way of being, character, vitality, and presence, can we trust that we are more like other species than not? Many genetic strains have evolved together. Did you know that humans share at least 50% of our genes with trees?[18] Why wouldn't we be able to communicate with them? You are both living beings. The point is: if a tree could talk to you, what would the tree say? If you did talk to a tree, and you know it understood you, what would you say? It's all about the intention.

We know that animal species communicate via songs, sounds, tracks, scat, signs, movement, and behavior. If we listen and look closely enough, we can hear not only their mating calls, warning cries, and social chatter, but also their fears and grief, protests of outrage, wisdom, advice, and perhaps empathy and caring for others and sometimes even us. We can extend this listening to beyond just critters with eyes, ears, claws, and vocal chords. Scientific research continues to indicate that plants, trees, and fungi relate, communicate, and respond with others through chemicals, vibrational frequencies, and growth patterns.[19] But perhaps we don't need scientific validation to believe that life communicates with life, and that we are part of that dialogue. We inherently have the ability to hear the song of the Earth, singing in concert together. This concert can also be heard, or seen, as a story. Every landscape, every forest, every place tells a story, or multiple stories. Everyone in the community of beings that makes up that place has a part in that story. It is a story you can learn to read with your eyes, hear with you ears or feel with your intuition.

So the skill of interspecies communication really starts with openness and trust, and continues with respect, intuition, and patience. I invite you to do an experiment. Set the intention of communicating with, conversing with, or 'hearing' a tree. After you have set your intention, trust that perhaps there is a tree who has some wisdom for you. Once you have intuited and trusted who it is, then introduce yourself to the tree. Approach this tree with as much respect as you would another human you were about to engage with; treat the tree as a "who" instead of an "it." You may try to introduce, or show yourself, as who you most deeply are on a soul level. You don't have to do this out loud, but speak from your heart. "Listen" for a response. The tree's communication may come to you in any number of ways: you may hear words "in your head," feel something in your body, see images, or have an unexplainable knowing. Continue by asking this tree a question, or *being* a question, and "listening" for a response. Take it another step and ask this tree to tell you, or show you, its life story. These conversations can be very short, or may be endless. (My experience with trees is that the older or bigger a tree is, the easier it is to communicate with, but this is not always so.) Whatever you experience, trust it. And practice, practice, practice.

6. CREATE SLOW PHOTOGRAPHY

The aim of any Slow Movement is to slow down an experience or process to how things were before the modern world sped up and made things so efficient or complicated that we forgot how they work or what role we play in them. The point of Slow Photography is to enjoy the creative process, not just the product—to really be present to what you are photographing, paying more attention to your subject than to capturing the moment in time. For some photographers, Slow Photography is about taking very long exposures. For others, it is about bringing back old-school manual cameras, cumbersome and slow-exposure large-format cameras, or even the elementary pinhole camera. All of these techniques require time. Look where photography is today, compared to even twenty years ago; nearly every adolescent and adult in the developing world has a camera on them at all times (via their cell phone) and can quickly and effortlessly take hundreds of photographs with the pressure of one finger. What happens as a result is that photography can become an almost unconscious experience, with the camera replacing our own sense of vision.[20]

And, in this current age of instant gratification and infinite choices, we are able to view, download, cut and paste, save, or even purchase virtually any image we want online at any time (many of which are duplicative). This ability to capture and manipulate an endless stream of images leaves us visually numbed by image redundancy. While some small portion of the photographs taken today may become timeless art, the majority of digital files produced on a daily basis are just plain deletable junk taking up computer memory, turning modern photography into post-modern meaninglessness. Ultimately, everything is ephemeral.

Why not challenge ourselves to do something *different*? It is all about one's intention. In 2012 and 2013, two photographers took on a project of photographing the same tree for 365 days in a row. Ciaran Burke documented a willow tree in Ireland for his onetree365 project,[21] and Mark Hirsch documented an oak tree in Iowa using his iPhone camera for his ThatTree project. Both photographers published their works on their own websites and in print books.[22] Even though these were digital camera processes, the photographers had a vision that challenged them into slowness every day. They proved that there are endless ways of seeing a tree (or any subject matter).[23]

In my own photographic process as TreeGirl, my "subject matter" is often more like a dance partner. The "fourth wall," as it is often referred to in theater, is removed as I step into and engage with the scene. My *slow* process is like another state of consciousness. I begin by researching noteworthy or unusual trees, then travelling to find them. Once there, I sometimes spend hours, even days, with just one tree—but sometimes only minutes, if time and circumstance allow no more. I have to slow myself down in order to be present with my surroundings, because I am not simply behind the camera, witnessing and observing a scene; I am part of the composition, engaging with another conscious being. Every sensory experience is heightened, whether I am listening and waiting for other people on the trail, hiding from people (or becoming invisible), or experiencing the ecosystem I've entered: the wind, the smells, the sounds, the hazards, the changes in light and temperature, the colors, and the community of creatures that inhabit the home I'm intruding upon. I explore and assess the tree and its surroundings for their aesthetic qualities, energy, and safety, and determine where I might fit in. I take all that in, then I communicate my best intentions to the tree and ask permission to collaborate with it.

My creative process is very physical. It sometimes involves long, arduous hiking, sometimes in inclement weather. It includes physically setting up the camera, undressing, being physically vulnerable, walking barefoot, touching and climbing the tree, getting scratched up, getting earth stuck on my body, encountering smaller (sometimes hazardous) creatures, getting comfortable or being uncomfortable, dismounting from the tree, and quickly getting dressed. Years after an encounter, I can remember vividly the texture and scent of the bark, the weight and mass of the trunks of different species, like the body and skin of different lovers. Whether or not I get a good shot out of the encounter, the experience is burned into my body somatically.

After the shooting, there are more slow processes: photo editing, more research, writing about the natural history of the species, remembering, reflecting, and sharing my experiences. Slow Photography, like all manifestations of the Slow Movement, gives us hope for rewilding the art of living. The value is in the experience, and not the end result. Some day, none of my digital images will exist. This book, made of trees, will decompose like fibers of wood on the forest floor or burn in someone's fire. One day, purely because of resource depletion, we will not have digital cameras to capture a moment artistically in time. But humanity will always be creative; it will always have art. What is yours?

7. CLIMB TREES

Tree climbing is what ecopsychologist Peter Kahn[24] calls a "core human interaction pattern." Starting with our sight and leading with curiosity and urge, it is a natural engagement of our hands, arms, legs, and feet. A tree with low-hanging, sturdy branches is an invitation to exploration and escape from the all-too-familiar terrestrial world. There is nothing quite like wrapping our bare fingers (and perhaps bare toes) around a giant branch, finding balance as we access our inner primate or squirrel. Perhaps we are

drawn to a better view or perspective, or maybe to fulfill a desire to be other than human—that we had wings, or claws, or a tail instead. Our bodies long to follow our eyes and hearts upward in awe, while our spirits long to ascend to the magical upper realms. Maybe some of us can't help but climb because we simply wish we *were* trees.

Tree climbing—the kind that requires safety ropes, harnesses, helmets, and support—is a popular recreational and therapeutic activity for adults and children, including the handicapped. Imagine seeing what a squirrel or bird can see 100 feet up in a tree. Learning tree-climbing techniques can lead not only to new perspectives, confidence, and enjoyment, but it can also prepare one for a career in

conservation, research, tree care, animal rescue, or tree-sitting activism. Professor Steve Sillett is a botanist and research scientist who has discovered and measured some of the world's tallest trees. He is best known as the first person to ascend and explore the hidden world of the old-growth canopies of California's Coast Redwood trees, discovering that entire unique ecosystems reside up there.[25] What other discoveries are yet to be made?

However you choose to climb a tree, make sure you do it safely! Utilize your intuition, your best judgment, your humility, and a buddy to help you up and down. Climbing in the nude is a whole other skill with its own hazards and pleasures!

[1] www.comehometonature.com, accessed 4/29/2016.

[2] Of course, if you are physically unable to get outside because of a physical condition, have someone bring nature to you.

[3] Elizabeth Marazita, LAc, Professor of Foot Reflexology & Qi Gong, The University of Washington, Nursing School & Bastyr University, CAM program. www.medfinds.com/therapy_definitions.html?definition=reflexology, accessed 9/5/2015.

[4] Statistic Brain Research Institute. National Shoe Retailers Association, U.S. Census Bureau, April 13, 2015. www.statisticbrain.com/footwear-industry-statistics, accessed 9/5/2015.

[5] www.barefooters.org, accessed 9/5/2015.

[6] L. Daniel Howell, *The Barefoot Book: 50 Great Reasons to Kick Off Your Shoes* (Turner Publishing Company, Kindle Edition, 2010).

[7] www.ncbi.nlm.nih.gov/books/NBK279362, accessed 9/5/2015.

[8] *Shinto* refers to the "spirit or divine essence" that manifests in various living forms, such as rocks, rivers, and trees.

[9] Florence Williams, "Take Two Hours of Pine Cone Forest and Call Me in the Morning," *Outside Magazine*, Dec. 2012, www.outsideonline.com/1870381/take-two-hours-pine-forest-and-call-me-morning, accessed 4/28/2016.

[10] Michiko Imai, "An Introduction to the Forest Therapy Society of Japan, Forest Therapy, and Forest Therapist," *Forest Medicine* (Nova Science Publishers, 2013).

[11] Experimental studies have shown that plant oil vapor can enhance the production of the brain's own calming chemical gamma-aminobutyric acid (GABA), while also boosting mood-regulating serotonin function. Eva M. Selhub and Alan C. Logan, *Your Brain on Nature: The Science of Nature's Influence on Your Health, Happiness, and Vitality* (HarperCollins Canada, Kindle Edition, 2013), p. 78.

[12] www.koreatimes.co.kr/www/news/nation/2011/07/117_91135.html, accessed 4/14/16.

[13] www.dtinews.vn/en/news/017004/22280/celestial-forest-bath--trial-nude-program-in-guangdong-park-sparks-debate.html, accessed 4/14/16.

[14] Richard A. Fuller, Katherine N. Irvine, Patrick Devine-Wright, Philip H. Warren, and Kevin J. Gaston, "Psychological benefits of greenspace increased with Biodiversity," *Biological Letters* 2007 3, doi: 10.1098/rsbl.2007.0149, published 22 August 2007.

[15] http://time.com/3858309/attention-spans-goldfish, accessed 4/28/16

[16] Daniel Moerman, *Native American Ethnobotany* (Portland: Timber Press, 1998).

[17] Orientation techniques for unmarked routes by land or sea, such as Polynesian voyagers' long-distance, non-instrument navigation and mapping using the stars, sun, wind, birds and ocean swells.

[18] Olavi Huikari, *The Miracle of Trees* (New York: Walker Publishing, Co, Inc., 2012).

[19] Peter Wohlleben, *The Hidden Life of Trees: What They Feel, How They Communicate—Discoveries from a Secret World* (Vancouver: Greystone Books, 2016), and Daniel Chamowitz, *What a Plant Knows: A Field Guide to the Senses* (New York: Scientific American/Farrar, Straus and Giroux, 2013).

[20] www.slate.com/articles/life/obsessions/2011/01/the_slowphotography_movement.html, accessed 4/29/16.

[21] www.onetree365.com, accessed 8/18/14.

[22] www.thattree.net, accessed 8/18/14.

[23] See the following books: Nancy Ross Hugo and Robert Llewellyn, *Seeing Trees: Discover the Extraordinary Secrets of Everyday Trees* (Portland: Timber Press, 2011); Françoise Reynaud, *The Tree in Photographs*. Los Angeles: J.Paul Getty Museum, 2011); and Cedric Pollet, *Bark: An Intimate Look at the World's Trees*. London: Francis Lincoln, 2010.

[24] "A Nature Language," in Peter Kahn, Jr., and Patricia H. Hasbach, Eds. *Ecopsychology: Science, Totems, and the Technological Species* (Cambridge: The MIT Press, 2012).

[25] Institute for Redwood Ecology. http://www.humboldt.edu/redwoods, accessed 8/14/2014.

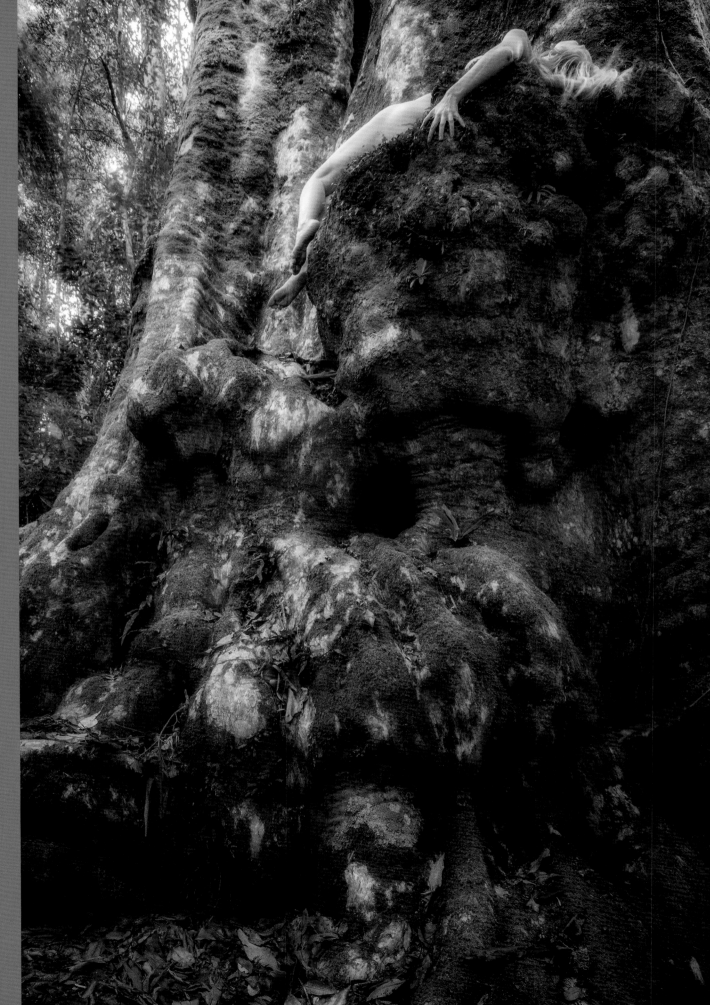

Brush Box Ecstatic (2015)

See also p. 33.

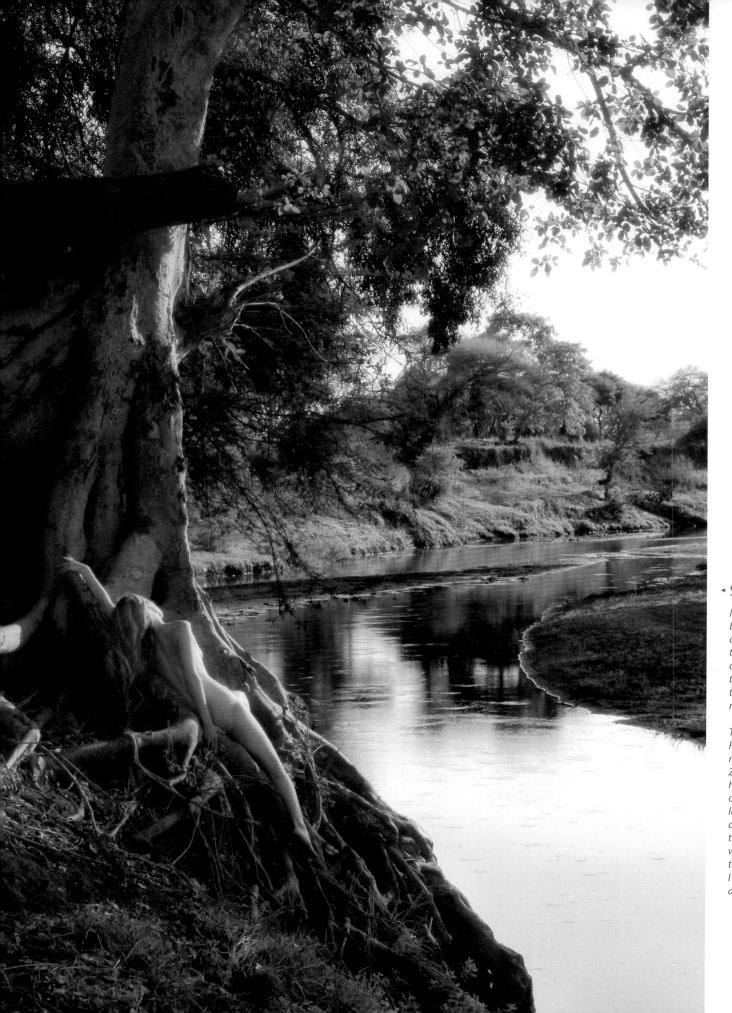

◄ Sycomore River Fig (2008)

Nearing sunset on a long day of shooting, my safari guide, Brett Greenaway, brought me to this modest Sycomore Fig anchored to the edge of a river, its roots like a web clinging to the bank. Its grandeur was not in its girth, but in its creative buttressing. The beautiful amber light, striking the mottled bark and reflecting the lushness surrounding the placid water of the Limpopo River, was an invitation to nestle in the still arms of my riparian companion.

This tree rests gracefully on the banks of the Limpopo River, in northeastern South Africa, near Pafuri, at the northern tip of Kruger National Park, directly bordering Zimbabwe and Mozambique. The area is a biodiversity hotspot for over 400 species of birds and 100 species of mammals, such as water buffalo, elephant, nyala, leopard, lion, impala, hippo, crocodile, and waterbuck, among others that find refuge from poachers by crossing the river to the safety of the South African side. While we were there, my guide took a walkie talkie call alerting him to look out for some nearby poachers. Knowing I was safe, I found refuge with the Sycomore Fig while relaxing in the art of camouflage.

Sycomore Fig Big (2008) ▶

Sycomore Fig

Ficus sycomorus

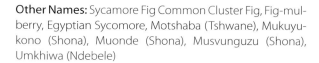

Other Names: Sycamore Fig Common Cluster Fig, Fig-mulberry, Egyptian Sycomore, Motshaba (Tshwane), Mukuyu-kono (Shona), Muonde (Shona), Musvunguzu (Shona), Umkhiwa (Ndebele)

Distinctive Characteristics: The Sycomore Fig is a common deciduous to semi-deciduous tree with large, dark-green leaves, and clustering figs. It exudes milky latex. It has distinctive buttressing, fluting trunk, and patchy green-yellow to orange-brown bark that exfoliates in papery strips, revealing a yellow inner bark underneath. Like all figs, Sycomore Figs depend on one particular species of symbiotic wasp for pollination.

Distribution: Native and common throughout Africa south of the Sahara, excluding west Africa, and dipping into northern South Africa, as well as the Southern Arabian peninsula, Lebanon, Cyprus, restricted areas in Madagascar.

Ecosystem: Found in rich soils in riparian areas and in mixed woodlands with Natal Mahogany, Fever Trees, Nyala Trees and Sausage Trees.

Maximum Age: Approximately 800 years.

Maximum Height and Girth: 66 ft. (20 m) in height; 65 ft. (20 m) in circumference.

Animal Community: The small figs are an important food source for many mammals including elephants, Wahlberg's epauletted fruit bats, thick-tailed bush babies, and birds including African green pigeon and purple crested lourie.

Traditional Uses:

Medicine: Used for lung ailments, sore throats, inflammation, swollen glands, and diarrhea.

Food: The fruits are edible.

Tools: The soft wood was used as a friction base for fire-starting.

Art and Ceremony: Wood was used to make drums. In ancient Egypt, Sycomore Fig was a prominent tree, used for mummy caskets, and associated with the afterlife, and the rebirth of the god Osiris.

Threats and Conservation: Not threatened.

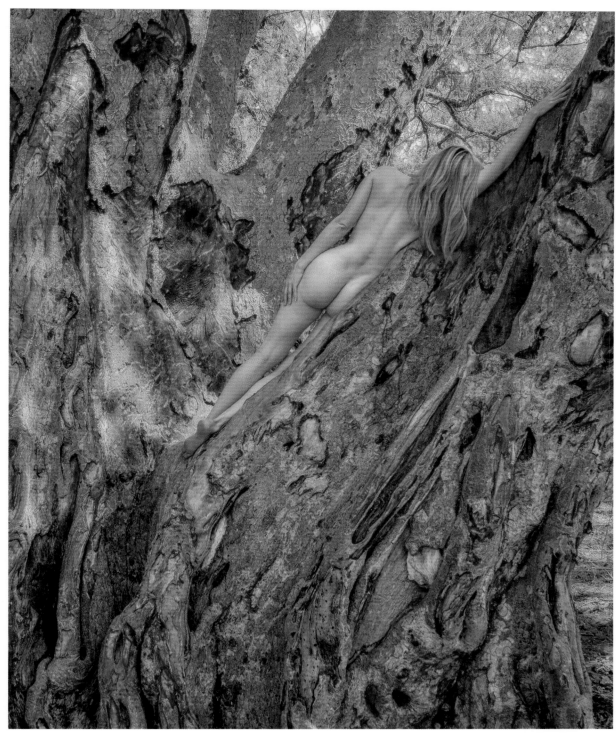

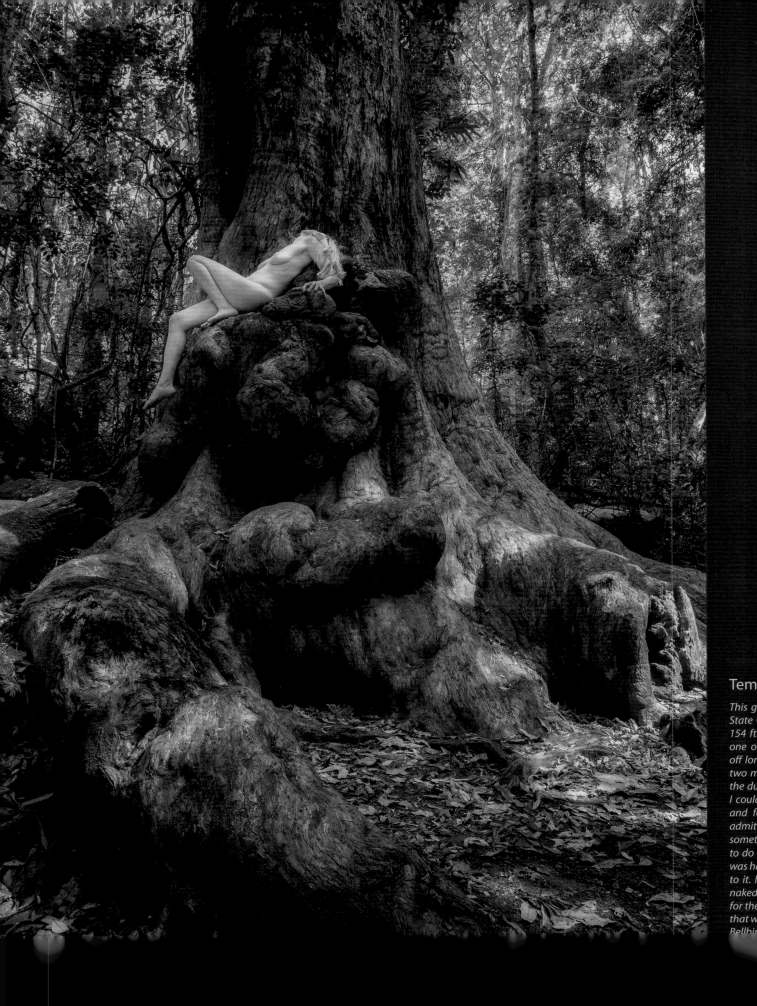

Tempestuous Tallowwood (2015)

This giant Tallowwood is known as Big Foot, and is the State Champion at 27 ft. (8.25 m) in circumference and 154 ft. (47 m) tall. The tree sits right on the trail, where one of its giant, octopus-like, buttressed roots was cut off long ago to make way for trekkers. I only had about two minutes to get this shot on the busy trail. I climbed the dusty, duff-covered trunk to a big, burly ledge where I could sit. The bark was quite soft and fibrous, looking and feeling unlike other Eucalypts I had met. I must admit that this tree wasn't completely crazy about me; sometimes it's just not love at first sight. It did allow me to do my TreeGirl thing, after I asked it nicely. Perhaps it was having a bad day, or didn't like the trail being so close to it. It was definitely a wild creature, unaccustomed to naked treegirls climbing on it, but seemed to tolerate me for the few moments I shared its space. Still, I am grateful that we met, and we parted cordially. This tree lives on the Bellbird Circuit in Lamington National Park, Queensland.

Tallowwood

Eucalyptus microcorys

Other Names: Australian Tallowwood, Tee

Distinctive Characteristics: This evergreen tree has soft, spongy, red-brown bark that is rough and fibrous, with surface pores and horizontal cracks. The base can form buttressed roots and burls. The special characteristic and name-giving feature is the greasy feel of the wood.

Distribution: Native to Australia, along the north coast of New South Wales near Newcastle to southeastern Queensland near Fraser Island, westward into higher altitudes of the Great Dividing Range. Elevation: sea level to 2,460 ft. (750 m).

Ecosystem: Tallowwood occurs mainly in tall, open forests in rainforest fringes on slopes, broad ridges, and sheltered valleys, sometimes forming over-stories, with Sydney Blue Gum, Blackbutt, Flooded Gum, White Mahogany, Silver-top Stringybark, Pink Bloodwood, White-Topped Box, Turpentine, and Brush Box. In drier forests, it occurs with Red Bloodwood, White Stringybark, and Grey Box.

Maximum Age: Estimated at over 300 years.

Maximum Height and Girth: 230 ft. (70 m) in height; 32 ft. (10 m) in circumference.

Animal Community: The nectar is consumed by bees, birds such as honeyeaters and lorikeets, and bats such as flying foxes. Koalas eat the leaves.

Traditional Uses:
Fiber: The leaves were used to make dye.

Modern Uses: A favored hardwood timber, the durable, rot-resistant wood has been used in heavy bridge and mining construction, and to make carriages, railroad ties, poles, sills, wheel spokes, tool handles, outdoor furniture, landscaping, fencing, joinery, boats, plywood, flooring, and decking.

Threats and Conservation: Not threatened.

Koala (2015)

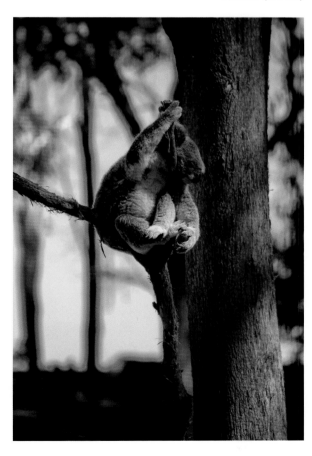

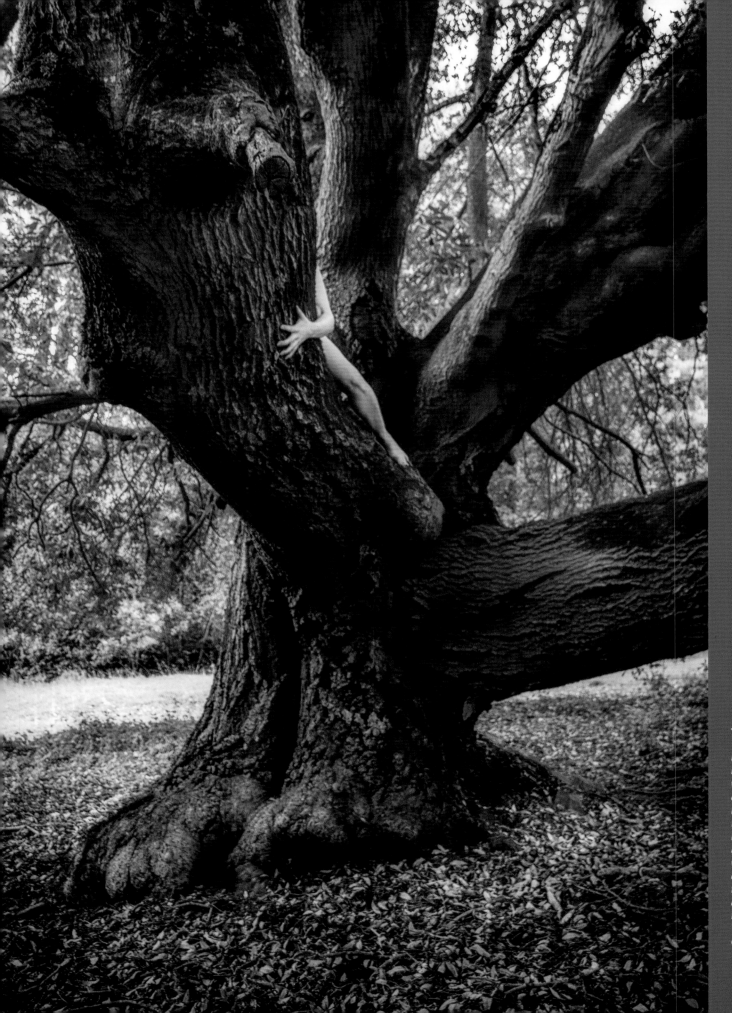

Tanoak Lizard (2012)

In 2012, I was on a personal retreat at Mt. Madonna Center in the Santa Cruz mountains, working on this book. There was plenty of yoga, ayurvedic massage, meditation, and delicious vegetarian food to distract me from writing, but I chose to take a hike and find the big Tanoak tree I was told about. Sure enough, it was tucked away near some residences. I remember it being very cold that morning, but the thrill of intertwining with a new species quickly warmed my body. I knew this was a once-in-a-lifetime chance, as Tanoaks all over California are under threat from sudden oak death. I also gathered as many acorns as I could carry from other Tanoaks nearby to make acorn food when I returned home. This is one of two good-sized Tanoaks who live on permanent retreat at Mt. Madonna Center, Watsonville, California, USA.

Tanoak

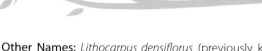

Notholithocarpus densiflorus

Other Names: *Lithocarpus densiflorus* (previously known as), Tan oak, Tanbark-oak, Sovereign Oak, cíšqhale ("beautiful tree", Kashaya Pomo), xunyêep (Kurok)

Distinctive Characteristics: The evergreen Tanoak is the only species in this genus. However, they are often confused with true oaks (*Quercus* sp.) because of their similar name, and that they have acorns. Compared to other acorns, Tanoak's are distinctively light-colored, unshiny and woody, with a cap that is covered in short, bristle-like spines. While the two both belong to the Beech Family, Tanoak's flowers are more like those of its relatives, Chinquapins and Chestnuts. Its evergreen leaves make it easy to distinguish from a true Oak; they are long and elliptical, with chevron-patterned veins and serrated edges. Most often the leaves are 2–3 in. (5–7 cm) long, but under certain conditions they can reach up to 9 in (23 cm) long.

Distribution: Native to southwest Oregon through the California Coast Range to near Santa Barbara, with inland populations throughout the Siskiyou Mountains, and from the southern tip of the Cascade Range along the western slope of the Sierra Nevada to Yosemite National Park. Elevation: sea level to 8,000 ft. (2,438 m).

Ecosystem: Found in moist mixed evergreen forest communities, including Coast Live Oak, Coast Redwood, Madrone, Canyon Live Oak, Douglas Fir, Oregon White Oak, California Bay Laurel, Monterey Pine. At higher elevations in the Sierra Nevada, it is found with Red Fir, Knobcone Pine, and Port Orford Cedar. Typically associated with manzanita, hazelnut, Pacific rhododendron, huckleberry, and salal.

Maximum Age: 400 years.

Maximum Height and Girth: 150 ft. (45.7 m) in height; 28 ft. (8.5 m) in circumference. Can also grow as a shrub or a dwarf tree.

Animal Community: Tanoak trees provide habitat and food for many animals, including black bears, northern flying squirrels, California ground squirrels, Allen's chipmunks, dusky-footed wood rats, mice, mule deer, chipmunks, squirrels, raccoons, and foxes. Many birds seek food and shelter in these trees, such as woodpeckers, Steller's jays, northern flickers, nuthatches, wild turkey, and varied thrush chickadees.

Traditional Uses:

Medicine: Acorns were sucked on to treat coughs (Kashaya Pomo). An infusion of the bark was used to wash face sores and as a toothache remedy (Costanoan, Ohlone).

Food: The acorns were historically a favorite staple food of California tribal groups, including the Pomo, Yurok, Coast Miwok, Miwok, Wappo, Ohlone, Shasta, Tolowa, , Tavitam, Yuki, Tsnungwe, Hoopa (Hupa), Lassik, Esselen, Hahwunkwut, Karuk, Shasta, Mendocino Indians, Costanoan, Chumash, and Salinan. Acorn nutmeats were ground and leached before cooking, then roasted whole or prepared into porridge, paste, bread, or soup. The flour was stored for later use. Many edible mushrooms were collected on living and decaying trees.

Tools: The tannin-rich bark was used for dye (Costanoan and Tolowa). Tanoak saplings were used to make heavy-duty baskets.

Art and Ceremony: Strung acorns were made into a musical instrument (Kashaya Pomo). Dreams of Tanoak were considered a sign of good luck (Sinkyone). Acorn festivals and dances have been an important cultural practice.

Modern Uses: The bark was once the main commercial Western source of tannin for leather before modern chemicals replaced them (the *tan* in *tanoak* is for tannin), so the tree was over-exploited. Tanoak wood has been used for making railroad ties, cabinets, decking, plywood, flooring, veneer, paneling, tool handles, baseball bats, paper pulp, biofuel, and firewood. Acorn food continues to be important for contemporary California Indians culturally, socially, and spiritually.

Threats and Conservation: Tanoak has been considered a weed, or trash tree, in forestry and has been the target of irradication by herbicides. It is also one of the species most severely devastated by sudden oak death (SOD), caused by the pathogen *Phytophthora ramorum*, indicated by black bleeding sores and brown leaf tips. Dying trees are left standing with completely dried foliage. It is believed by some that the death of trees from SOD is due to the modern lack of sacred cultural reliance on the acorn as a food source; the trees no longer feel needed. From an ecosystem management viewpoint, researchers have discovered that these trees have become more susceptible to pathogen epidemics due to the severe lack of wildfires and controlled burns—which were traditionally managed by native peoples—as well as the acidification of forests due to industrial pollution. Partnerships between tribes, universities, the US Forest Service, the National Park Service, and independent organizations have formed to manage Tanoak woodlands. Some trees are being treated via preventive measures, including re-mineralization of the soil and fire mimicry, while conservation easements are being created to preserve forested areas.

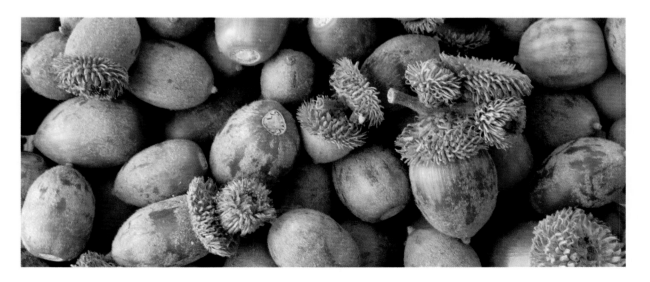

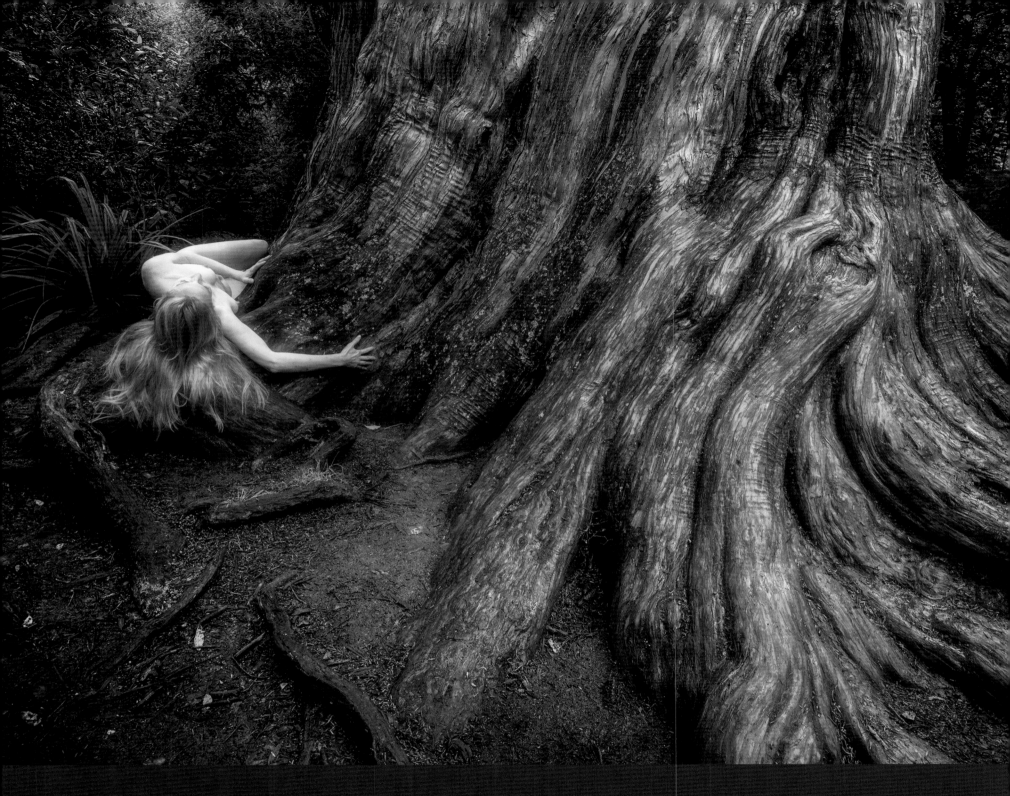

Totara Twisted (2015)

This ancient Totara tree, estimated at 1,000 years old, lives in a remnant of a large Podocarpus forest, reduced in area by logging. Visiting this tree, one gets the sense that it is lonely for others of its species. The day I visited was a chilly autumn day. I wore every layer of clothing I had brought on my epic TreeGirl expedition.

I had just arrived at the tree and was setting up my camera when an Asian couple rudely cut in front of me. They were on a different photo mission: their official pre-wedding wedding photo. They hid in the bush to change into their wedding attire, and after about ten minutes came back out to pose with the tree. The bride didn't seem very comfortable with the tree, the dress, or the man she was about to marry. Of course, I would be overjoyed to get my wedding photo taken in front of a millennium-old tree, and I wouldn't even need the man! They finally left after about forty-five minutes, and I wished them luck. Now it was my turn. I checked out the "back side" of the tree and found it more pleasing to my artist's eye. I, too, hid in the woodsy dressing room and hesitantly took off my five layers of clothing, returning to lie down on the very cold, wet base of the tree. Not surprisingly, even with the cold, it was an incredibly enjoyable experience photographing and videotaping myself, twisting my body around the gnarly base of the trunk, which looked like a giant rope mop being perpetually wrung. The tree is 28.37 ft. (8.65 m) in circumference and 85.30 ft. (26 m) tall, and on the short Big Tree Track in Peel Forest Park, north of Geraldine, North Island, New Zealand.

Tōtara

Podocarpus totara

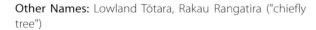

Other Names: Lowland Tōtara, Rakau Rangatira ("chiefly tree")

Distinctive Characteristics: The trunk on these evergreen trees is thick, gnarly, and slightly buttressed at the base. The bark is reddish to brown to silver, thick, furrowed, and long-plated. The seeds grow on top of an edible red fruit.

Distribution: Native to and widely distributed throughout the North Island and northeastern South Island of New Zealand. Elevation: up to 2,000 ft. (600 m).

Ecosystem: Lowland, montane, and lower subalpine forests with Mataī and Kahikatea trees.

Maximum Age: Over 1,000 years.

Maximum Height and Girth: At least 167 ft. (51 m) in height; at least 35 ft. (10 m) in circumference.

Animal Community: Birds including the Tūī, kereru, yellow-crowned parakeets, weka, and bellbirds eat the seeds and fruit. Insects include six native moths, prickly stick insect, and longhorn beetle, as well as insects who are Tōtara specialists: gall mite, gall midge, thrip, mirid, and aphid. Invasive possums can heavily browse the tree.

Traditional Uses:

Medicine: The smoke from burning wood and the boiled bark were both used medicinally for hemorrhoids and venereal disease. An infusion of the leaves was taken for upset stomach. The inner bark was boiled to treat fevers (Maori).

Food: The red fruit was a favorite food (Maori).

Tools and Objects: Wood was used for carving bowls; branch wood was used for axe handles and beaters weapons, musical instruments, and toys; sheets of inner bark were used as water containers; wood was used as splints to support fractured bones; sticks were used for producing fire by friction (Maori).

Art and Ceremony: The wood was prized for large ceremonial carvings (Maori).

Shelter: Sheets of inner bark were used as roofing (Maori).

Transportation: It was the primary wood used to make war canoes, called waka, due to its length, relatively light weight, straightness and rot resistance. A single log would be hollowed out, some capable of carrying 100 warriors, and took at least a year to make using stone adzes (Maori).

Modern Uses: Bridge and warf building, fence posts, floor pilings, railroad ties. The chemical compound in the tree, *Totarol*, is antibacterial and anti-microbial, and is currently being used in natural medicines and cosmetics. The bark has been used to make home made dye for wool.

Threats and Conservation: Not threatened.

Totara Too (2015)

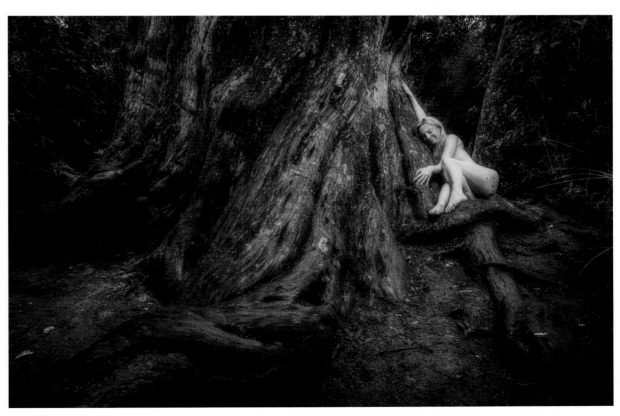

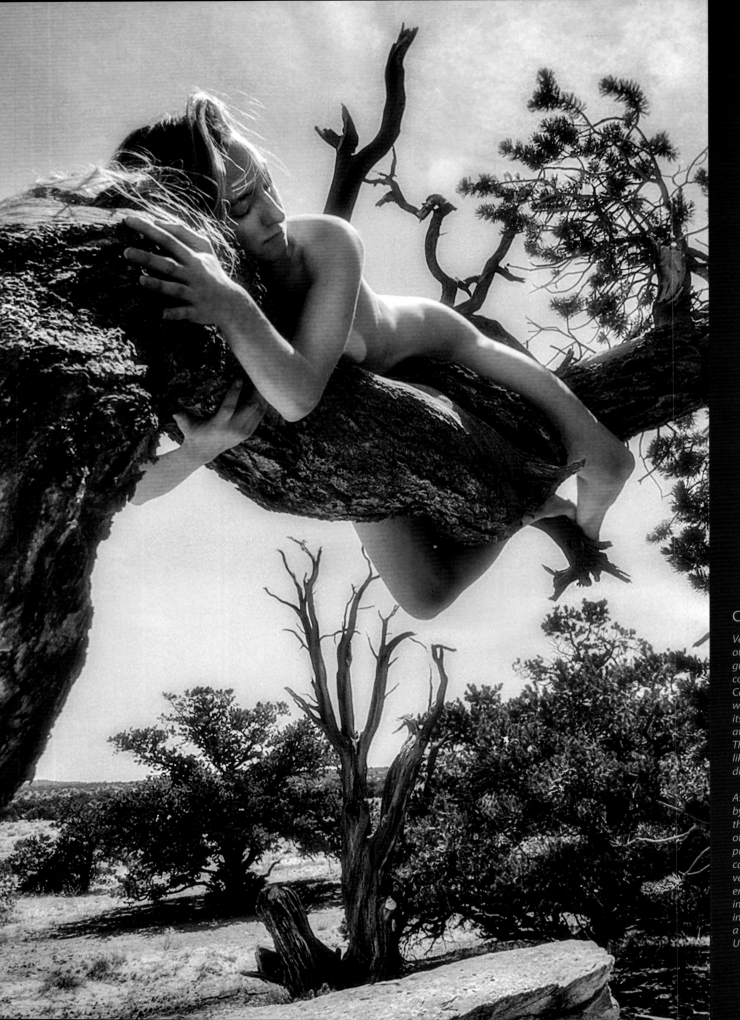

Caterpillar Pinyon (2003)

Venturing into the desert always provides an experience of raw beauty, awe, and humility. This time, my wilderness guide-friend Caiyloirch and I were on a mission of co-creating environmental art photography near Canyonlands National Park. This Two-Needle Pinyon Pine was perfectly poised for a mountain lion to take a nap on its contorted trunk. We didn't see any mammals around at this hot time of day, so I tested the trunk for a photo. Those desert trees are hard and rough—tough survivors, like all the organisms of a harsh ecosystem. As long as I didn't move too much, I felt very relaxed on top of the tree.

As our journey continued that evening, we were greeted by hundreds of caterpillars emerging in the dark from the sandy earth in some sort of orchestrated spring rite of fertility. Then, in the middle of the night, the ancient pictographs of well-defined shamanic figures on the canyon wall near our campsite awoke me and asked me very directly "who" I was. Our adventure the next day ended with a flat tire and a flat spare tire, followed by an incredibly long, boring tow-truck ride transitioning back into the civilized world. This Two-Needle Pinyon Pine lives a wild life outside of Canyonlands National Park, Utah, USA.

Two-Needle Pinyon Pine

Pinus edulis

Other Names: Colorado pinyon, Mesa pine, Piñon pine

Distinctive Characteristics: The Two-Needle Pinyon is a common desert conifer of the southwest U.S. In mature trees, the dark-gray brown bark with reddish patches is furrowed with rough, scaly ridges. Short bright-green or blue-green needles are bunched in twos, to distinguish it from its Single-Leaf cousin. Mature cones are small, at 1¾ in. (4.5 cm), and lack prickles.

Distribution: Native to the Mojave Desert, Colorado Plateau, and the Great Basin Desert; widespread and abundant in Utah, Colorado, New Mexico, and Arizona, with sparse populations in Nevada, southwestern California, Wyoming, Texas, and Mexico. Elevation: 4,600–9,800 ft. (1,402–2,987 m).

Ecosystem: Arid, rocky flats or slopes, canyons, and some riparian zones. Common in desert shrublands and juniper-pinyon woodlands, along with Single-Leaf Pinyon Pines.

Maximum Age: Recorded at 973 years.

Maximum Height and Girth: 69 ft. (21 m) in height; 9.4 ft. (2.8 m) in circumference.

Animal Community: This pine is an important habitat and food source to a variety of desert-dwellers, including virtually any animal capable of cracking and digesting the nuts. Among these consumers are mice, chipmunks, squirrels, deer, bears, bighorn sheep, rodents, and birds. Jays and nutcrackers have been responsible for much of the redistribution and regeneration of the trees, caching them in the ground miles away from their original homes. *Dianthidium* bees use the sticky resin to help construct nests.

Traditional Uses:
Medicine: Smoke from the burning needles was inhaled for colds (Mescalero Apaches). The burning sap was inhaled for colds (Ramah Navajo).

Food: Pine nuts were a staple food (Piute, Washoe, Great Basin Shoshone).

Tools and Objects: The sap was used for waterproofing baskets and clay vessels, and later as glue for jewelry. The wood was also burned to cure animal hide leather for bags and tepee coverings (Mountain Ute).

Art and Ceremony: The sap was placed on hot coals to purify themselves after a funeral (Hopi, Tewa), the sap was also burned for other ceremonial purposes (Navajo).

Modern Uses: For modern industrial civilization, the wood has been exploited for charcoal, railroad crossties, lumber, fence posts, and even pulpwood. It is harvested commercially and non-commercially for pine nuts.

Threats and Conservation: There have been many threats to the survival of the Pinyon Pine. In the 1800s, hundreds of thousands of acres were deforested to supply charcoal for mining, smelting, and the railroad industry. During the last and current century, millions of hectares of productive Pinyon Pine woodlands have been destroyed due to conversion of land for livestock grazing on both private and public lands. Entire woodlands have been ripped down using giant chains pulled by heavy machinery, seen by many as an act of major ecological and cultural vandalism. Most recently, because of global climate change, widespread drought is making the trees susceptible to bark beetle infestations, resulting in a massive die-off of up to 80% of Pinyon-Juniper woodlands in certain areas.

Pinyon Pine Cones

A cliff chipmunk (Tamias dorsalis) enjoys the tasty pine cones.

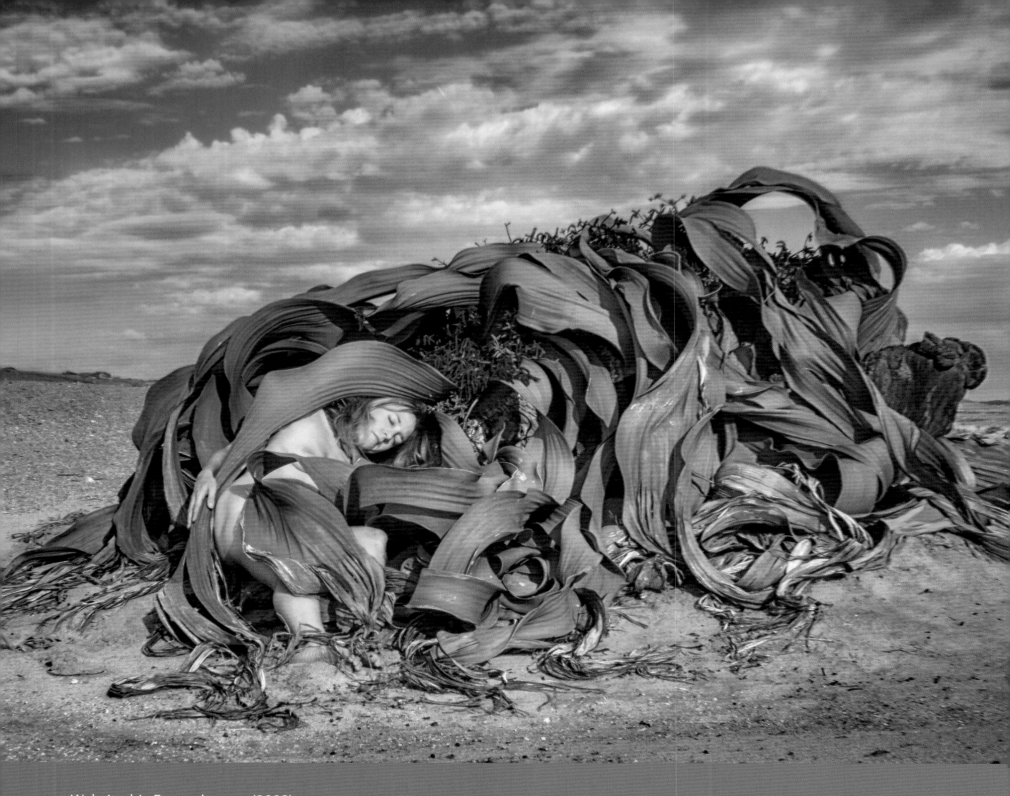

Welwitschia Entanglement (2008)

I had come to Namibia to photograph the largest Baobab in Africa, but I took a detour to the other side of the country to see the famous Welwitschia—a botanical wonder. If you are a plant or tree nerd, this is a prize. You can hardly call it a tree, but it has cones and a trunk. Often labeled a dinosaur stuck in time, it is one of those renegades that refuses to fit into the human scientific categorization of Mother Nature. While the other botany pilgrims stopped to photograph the designated Welwitschia behind the fence, my guide and I proceeded a few yards farther into the barren desert to meet one of the largest Welwitschias in the world,

unencumbered. How to cuddle up to this one was a challenge. But I managed to gently lift up part of one of the stiff, heavy, twelve-foot-long, wind-shredded leaves and experience what it is like to be a botanical fugitive—or perhaps, because it is a Near Threatened species, a refugee in one of the world's oldest deserts. The plant pictured may be 900 years old and lives on Welwitschia Drive in the coastal dunes of Namib Desert, east of Swakopmund in Namibia.

Welwitschia

Welwitschia mirabilis

Other Names: tumboa, n'tumbo (Angolan), tweeblaar-kanniedood (Afrikaans), !kharos (Nama, Damara), nyanka (Damara), khurub (Nama), onyanga (Herero), onion of the desert

Distinctive Characteristics: Classified in a Family and Order all its own, the ancient Welwitschia is an odd, slow-growing species that seems to be evolutionarily stuck in prehistory, blurring the categorization between plant and tree. Named for the Austrian botanical explorer, Friedrich Welwitsch, it is uniquely adapted to limited areas in the harshest, most barren deserts in southern Africa, where almost no other flora can survive. The slow-growing Welwitschia is a dioecious, cone-bearing plant that has a subterranean woody trunk without annual growth rings. The majority of the trunk lives below the sandy ground, with a taproot system that can extend up to 30 ft. (9 m) deep. The crown barely surfaces mere inches above the ground, up to 2 ft. (0.6 m) tall, in curvaceous meanders like the jaws of a giant open clamshell. This crown, which sports either small male or female cones on flowery stalks, is sometimes barely visible, as it is buried beneath a pile of very unusual leaves. The Welwitschia grows two—and only two (unique in the plant kingdom)—very wide, leathery, ribbed, ribbony evergreen leaves, which can reach over 12 ft. (4 m) long, splitting, shredding, and fraying into a heaping, tangled mass, thick, hard and heavy, beaten by the harsh desert wind. It is through these leaves that minute pores absorb the majority of the tree's much-needed water from fog, in a region that barely receives 2 in. (5 cm) of rain a year. This hearty creature sits alone, looking something like an octopus that might slither across the sand and swallow you up whilst you're not looking.

Distribution: Endemic to the Namib Desert in extremely limited patches in northwestern Namibia and southern Angola, within 30 miles inland, along the fog belt.

Ecosystem: Arid and semi-arid desert and mopane savanna, in spread-out colonies, usually solitary.

Maximum Age: Estimated to 1,500 years.

Maximum Height and Girth: 6.5 ft. (2 m) in height; 27 ft. (8.2 m) in circumference as a pile of leaves.

Animal Community: During drought, antelope, springbok, oryx, the endangered Hartmann's zebra, and rhino chew the leaves for their juice and spit out the tough fibers. It also acts as a shelter for birds, small reptiles and insects.

Traditional Uses: The core, especially of the female plant, was eaten raw or baked in hot ashes.

Threats and Conservation: Due to limited habitat, injury from off road vehicles, and a possible fungal pathogen that reduces female cone viability, it has been listed as Near Threatened by the IUCN Redlist. It is protected in some parks, and is being monitored.

Male Cones of the Welwitschia (2008)

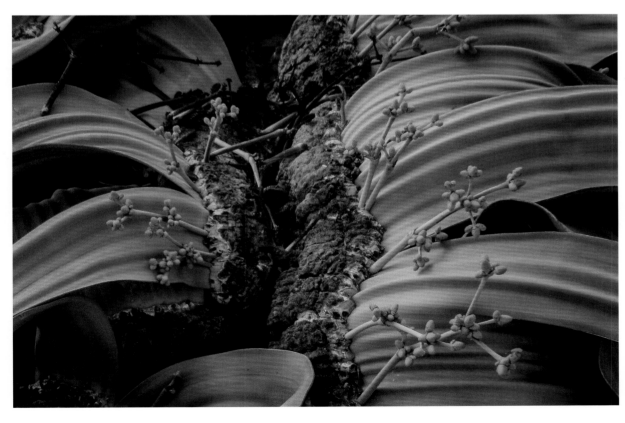

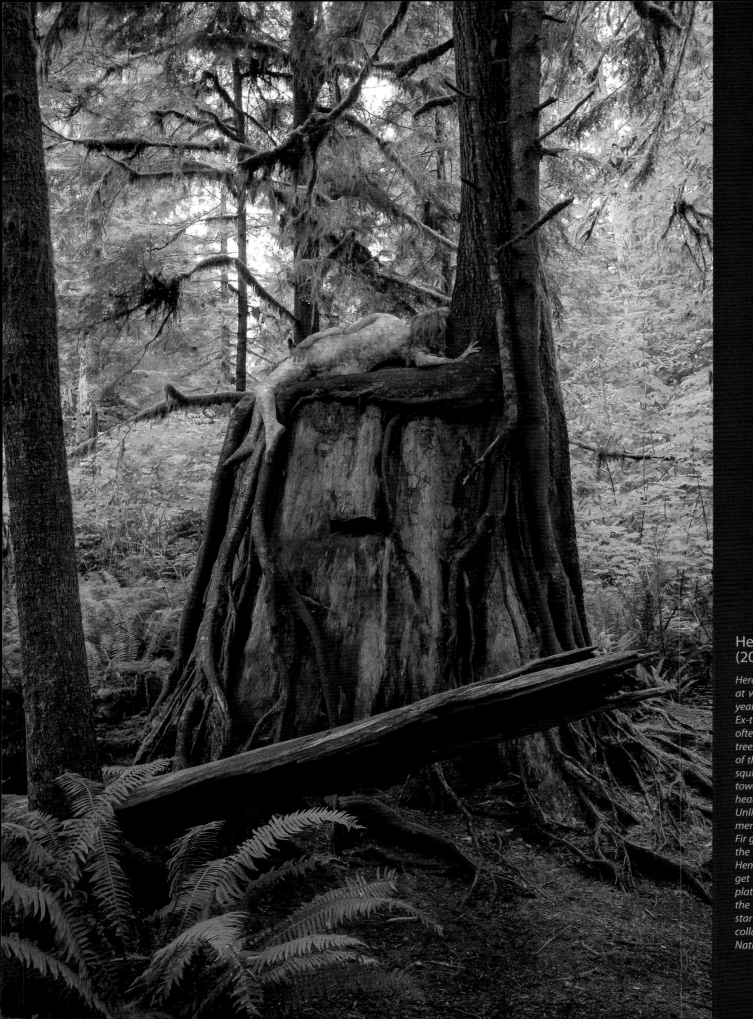

Hemlock Green Woman on Nurselog (2012)

Here is a picture-perfect example of Nature's collaboration at work in the rainforest. In this spot, about a hundred years ago, a Douglas Fir was cut down, leaving a stump. Ex-trees like this one, whether logged or fallen naturally, often become nurseries, or nurse logs, for other plants and trees. In this case, a Western Hemlock seed landed on top of the wet, decomposing stump and lucked out. Looking squid-like in its growth form, the trunk headed upward toward the light in the dark rainforest, while the roots headed down the stump and into the earth for nutrients. Unlike a strangler fig, the Hemlock is not parasitic; it is merely taking advantage of a good thing, and the dead Fir gets to continue on in another form. The Fir will nurse the Hemlock until it completely disappears, leaving a Hemlock on stilts. I could not help myself, and had to get in on the action. The top of the stump was a perfect platform for this TreeGirl to become one with the tree and the entire forest. If I stayed there, I probably would have started growing roots myself. You can find this ecological collaboration on a trail in the Quinault region of Olympic National Park, Washington State, USA.

Western Hemlock

Tsuga heterophylla

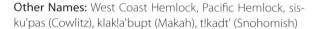

Other Names: West Coast Hemlock, Pacific Hemlock, sis-ku'pas (Cowlitz), klak!a'bupt (Makah), t!kadt' (Snohomish)

Distinctive Characteristics: This tall, evergreen conifer with dense short-needled foliage forms a conical shape with a distinctive, strongly drooping leader and branch tips. Older trees become cylindrical and may have no branches in the lowest 100–130 ft. (30–40 m). Their dangling cones are small, about 1 in. (1.5–2.5 cm). In the rainforest, the opportunistic Hemlock reproduces by seed, commonly establishing itself on decaying logs, producing an aggressive root system that is adept at scavenging nitrogen from dead wood. The root system is also covered in mycorrhizae-forming fungi that aid in its reproduction. Nurse logs such as the Douglas Fir decompose underneath the cascading root systems of Western Hemlock as they establish themselves, resulting in a root system that is stilted above the ground.

Distribution: Native along the western U.S. in Alaska, Washington, Oregon and California in coastal regions and separately inland in Montana and Idaho. In Canada it is found in British Columbia and Alberta. Elevation: sea level to 5,900 ft. (1,800 m).

Ecosystem: A shade-tolerant understory tree in low- to mid-elevation, common in temperate coastal rainforests, found mostly with Douglas Fir, Sitka Spruce, Western Redcedar, and Coast Redwood. It also grows alongside Pacific Silver Fir, Paper Birch, Mountain Hemlock, Engelmann Spruce, Western Larch, Grand Fir, Western White Pine, Lodgepole Pine, Red Alder, Black Cottonwood, Port Orford Cedar, Noble Fir, Big Leaf Maple, Alaska-cedar, Incense Cedar, Tanoak, Sugar Pine, California Bay Laurel, Subalpine Fir, Ponderosa Pine, and Pacific Yew.

Maximum Age: Recorded at 1,238 years.

Maximum Height and Girth: 271.7 ft. (83 m) in height; 28.5 ft. (8.6 m) in circumference.

Animal Community: Old-growth Western Hemlock stands provide hiding, habitat, and thermal cover for many wildlife species, including small mammals such as the northern flying squirrel, deer mice and red tree vole, as well as large mammals. It is also an important browse species for Roosevelt elk and black-tailed deer in coastal Oregon, Washington, and British Columbia. Rabbits and snowshoe hare

clip off the main stems of western hemlock seedlings, as do mountain beavers. Many bird species also call Western Hemlock forests their home. The northern spotted owl in Washington and Oregon and barred owls in British Columbia depend on Hemlock-dominated forests. Nesting bird species such as the yellow-bellied sapsucker and northern three-toed woodpecker use cavities in trees for nesting.

Traditional Uses:

Medicine: The needles, inner and outer bark, and sap were all used medicinally in various ways: As an antirheumatic (Bella Coola, Hesquiat), tuberculosis remedy (Chehalis, Shuswap, Klallam, Hesquiat), antihemorrhagic (Klallam, Nuu-chah-nulth, Cowlitz), heart medicine (Bella Coola), eye medicine (Cowlitz), gastrointestinal aid (Gitksan), appetite stimulator (Klallam), cold and flu remedy (Quileute, Thompson) antidiarrheal (Kwakiutl), burn dressing (Bella Coola, Kwakiutl), birthing aid (Kwakiutl), dermatological aid (Cowlitz, Kwakiutl, Nitinaht, Quileute, Bella Coola), disinfectant, and antiparasite (Makah). New needle growth was eaten or made into a bitter tea, rich in vitamin C. An infusion of the inner bark or twigs was also used in the treatment of kidney or bladder problems, as well as a gargle for mouth and throat problems. The sap was mixed with deer tallow as a salve to prevent sunburn (Hesquiat, Makah). A moxa of the twigs was burned to eliminate warts (Kwakiutl).

Food: The inner bark was prepared and pressed into cakes for preservation, eaten as a staple food (Gitksan and Haisla) and as an emergency food (Haida, Bella Coola, Haisla, Hanaksiala, Kitasoo, Nitinaht, Tsimshian, Oweekeno, Coast Salish, Wet'suwet'en, Nuxalk). Small branch tips were used as flavoring when cooking meat (Cowlitz).

Tools and Objects: Many tribes used the boughs to collect herring eggs from the bottom of waterways during the spring (Tlingit, Bellabella, Haisla, Hanaksiala, Hesquiat, Kitasoo, Klallam, Nitinaht, Makah). The bark was used as a tanning agent and dye (S'Klallam, Lummi, Makah, Quileute, Snohomish, and Chehalis), and a red dye was made from the bark to color basket materials and mountain goat wool (Coast Salish). Various tribes dyed fishnets brown to make them invisible to fish (Chehalis, Hesquiat, Nitinaht, and Bella Coola). The easily carved wood was made into implements such as spoons, roasting spits, combs, spear shafts, children's bows and elderberry picking hooks. Large feast bowls were carved from the wood of bent trunks (Haida).

Art and Ceremony: Boughs used with other plants for ritual purification and protection (Haisla, Kwakiutl, Oweekeno, Nitinaht, Hesquiat and Hanaksiala). Wood knots were used to make jewelry for women of high social rank (Hanaksiala). The wood was used to carve ritual masks (Southern Kwakiutl).

Shelter: Boughs were used as a temporary building material for hunters, fishers, shamans and their initiates, and for girls following their first menstruation (Kwakwaka'wakw, Oweekeno, Quinault, Haisla, Hanaksiala).

Modern Uses: In the early days of modern logging, Hemlocks on both sides of the U.S. were regarded as trash trees and often discarded in favor of other species. As availability of these other species declined from exploitation, both Western and Eastern Hemlock soon gained importance as commercial resources. Today both are harvested for lumber for general construction, railroad ties, mine timbers, and marine pilings. The wood is also used to make boxes and crates, kitchen cabinets, flooring, veneer for plywood, paper pulp products, as well as in the manufacture of rayon, and cellophane.

Threats and Conservation: Not threatened.

Western Hemlock Needles and Cones

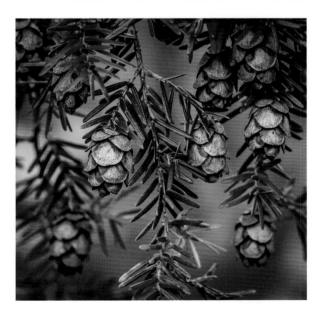

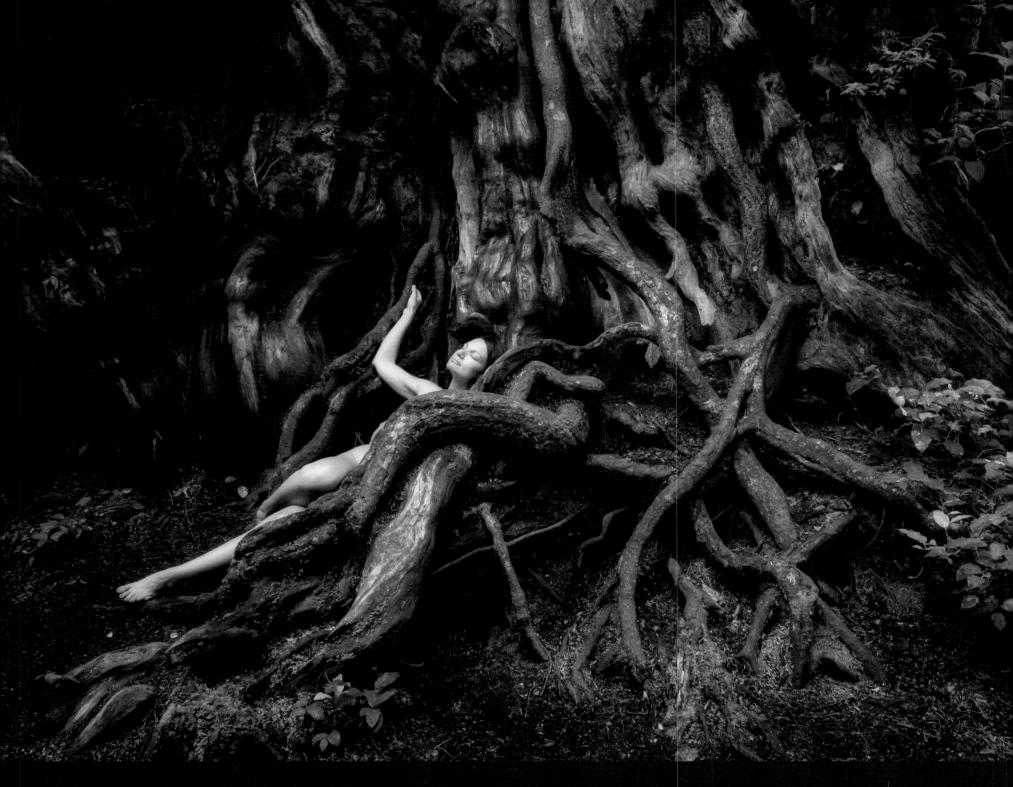

Becoming Redcedar Roots (2012)

The Kalaloch Redcedar is the ultimate gnarly tree: cavernous, mossy, covered with burls, roots, and worn-off bark, exposing what looks like muscles and bones, holding a giant together. She is a classic rainforest nurse tree, growing other flora species such as salal, huckleberry, and two young Hemlock trees with roots cascading down from above. Strange stilts, looking like branches reaching for the forest floor, give evidence of other, long-gone nurse logs. Her crown is comprised of twelve staghorn reiterations (branch-like growths out of the main trunk that grow upward), giving the unique appearance of a candelabra. An ecosystem

unto herself, she provides a home for many forest critters and looks like a castle fit for a village of fairies or gnomes.

The day we visited her, it was gently raining, but that did not keep my friend and me from nestling in her many twisted nooks and crannies. From 1955 to 1977, she was the national champion Western Redcedar, with a recorded circumference of 61.57 ft. (18.74 m), a height of 123 ft. (37.4 m), and a total wood volume of 12,370 cu ft. (350 m³). She is probably at least one millennium in age. Because of her deterioration of wood

volume, and the subsequent discovery of larger red cedars, the champion is now the nearby taller Quinault Lake Redcedar (also known as the "Quinault Big Cedar"). In 2015, a major portion of the Kalaloch Redcedar's trunk broke and collapsed to the forest floor, continuing the cycle of regeneration. Who knows what the tree will look like in another 100–1,000 years? This tree can be easily found on the west side of the rainy Olympic National Park, Washington State, USA.

Western Redcedar

Thuja plicata

Other Names: Western Red Cedar, Pacific Redcedar, Shingle Cedar, Giant Cedar, Giant Redcedar, Canoe Cedar, Inland Redcedar, Giant Arborvitae, Western Arborvitae, Western Flat Cedar, lata'wa, p'alans (Chehalis), xatea'tcl (S'Klallam), x'pai'epl (Lummi), t'sa'p is (Quileute), t'ci'tum (Quinault), xaxapi'ats (Skagit), xelpai'its (Snohomish), q!wê'le (Skokomish)

Note: The compounding of the words in "Redcedar" and "Red-cedar" are used joined instead of separately to distinguish it from a true cedar such as "Deodar Cedar" (Cedrus deodara) which has short clustered needles instead of scale-like sprays.

Distinctive Characteristics: An evergreen conifer with a conical crown, arching branches, and buttressed trunk. It can also develop independent vertical stems, called "reiterations," which resemble secondary trunks that form multiple tops on very old trees. As tops die back new ones will sprout. Redcedars are long-living trees and are prone to becoming gnarled and hollow, with several sparsely foliated or even spiked tops (dead tops sticking up above the live part of the tree). The fibrous bark is reddish-brown to gray. Leaf sprays are scaled, flat, and pleasantly aromatic when crushed. The nooks of the deeply fissured bark can collect windborne soil and moisture, becoming seedling nurse logs from which many plants and trees can grow. They have decay-resistant organic compounds that contribute to their longevity. "Arborvitae," the common name for the genus *Thuja*, is Latin for "tree of life," and some native peoples referred to it as "long life maker."

Distribution: Native to the western U.S. in Alaska, Washington, Oregon, and California in coastal regions and separately inland in Montana and Idaho; in Canada it lives in coastal British Columbia and Alberta. Elevation: sea level to 7,513 ft. (2,290 m).

Ecosystem: Found in wet, temperate, mixed conifer rainforests, riparian zones, bogs, and mountainsides, but can also live in dry environments with Douglas Fir, Sitka Spruce, Big Leaf Maple, and Western Hemlock.

Maximum Age: At least 1,460 years.

Maximum Height and Girth: 277 ft. (84 m) in height; 78.4 ft. (23.9 m) in circumference.

Animal Community: Cavities in old-growth provide animals such as skunks, bears, and raccoons spaces for dens, hiding, and thermal cover, while bird species such as hairy woodpeckers, yellow-bellied sapsuckers, tree swallows, Vaux's swifts, and chestnut-backed chickadees nest within its branches. In the northern Rocky Mountains, the leaves are eaten in winter by elk and deer.

Traditional Uses: Historically, the Western Redcedar has been a crucial cultural resource for the indigenous peoples of the Pacific Northwest Coast. This tree has the *greatest* recorded number of uses—totaling 368—of *any* plant used by Native Americans, including 188 fiber uses, 52 medicinal uses, six food uses, one use as a dye, and 121 other uses! Some northwest coast tribes even refer to themselves as "people of the Redcedar" because of their extensive dependence on this dominant tree species for basic materials. Legends tell of the origin of the Redcedar as a gift from the Great Spirit to provide for the people all of their needs.

Medicine: The tree has strong immune-stimulant, antifungal, and antibacterial properties. It contains oil extracts and leaf infusions contain thujaplicins, a group of chemicals that may prevent mature trees from rotting, used in applications to treat fungal infections, warts, hemorrhoids, venereal lesions, and skin blisters. A tea was made out of the bark and twigs to treat kidney and urinary tract infections, and an infusion of leaf sprays was used to relieve diarrhea, colds, and respiratory infections (Cowlitz, Makah, Bella Coola), sore throats, and tuberculosis (Clallam). A tea of the seeds and ends of limbs were used to break a fever. The buds were chewed for sore lungs or toothaches (Cowlitz). A poultice was used for skin ailments (Bella Coola, Haisla, Kwakiutl). The inner bark of a small tree was chewed or boiled and the liquid drunk to help bring about menstruation (Chehalis). It was also used also as a painkiller (Bella Coola).

Food: Although not a staple food source, the moist, inner bark was eaten fresh or dried for storage (Montana Indian, Coast Salish). The sap was used as chewing gum (Southern Kwakiutl).

Fiber: Strips of its strong, pliable bark up to 27 ft. (8 m) long were peeled off the tree. The roots, inner bark, and limbs were used to make baskets, as well as fish traps, ropes, and cordage (Bella Coola, Clallam, Quinault, Coast Salish, Suswap, Squaixin, Thompson, Wet'suwet'en, Haisla, Hesquiat, Nitinaht, Oweekeno, Quileute, Montana Indian, Kwakiutl, Kutenai, Hanaksiala, Chehalis, Hoh, Gitksan, Flathead, Okanagan-Colville). Bark was crafted into full dresses and ceremonial capes (Chehalis, Clallam, Haisla and Hanaksiala, Hesquiat, Kwakiutl, Hoh, Nitinaht, Okanagan, Coast Salish, Oweekeno). Beautiful, watertight woven hats were meticulously made from the fine, strong roots.

Tools: Redcedar wood was used to make hunting tools such as arrow and harpoon shafts, spear poles, fish clubs, cooking utensils (Haisla, Hanaksiala, Hoh, Hesquiat), drying racks for food, drills for fire starting, while wads of shredded bark were used as tinder. Bark was plated to make dishes, platters and line cooking pits (Cowlitz, Hoh). The wood was also used to make bentwood boxes (Tsimshian, Makah, Nitinaht, Oweekeno, Quileute), benches, combs, and spindles for spinning wool (Quileute). The soft bark was shredded fine to make padding for cradles (Chehalis), towels (Chehalis, Kwakiutl), mats (Bella Coola, Haisla, Kwakiutl, Coast Salish), blankets (Nitinaht), diapers (Hesquiat, Kwakiutl) and sanitary pads (Chehalis), and paintbrushes (Southern Kwakiutl). The sap was used as glue in many applications.

Art and Ceremony: Often common utilitarian objects in Pacific Northwest cultures were so lavishly decorated with designs that they could be viewed as art objects in themselves. The wood was made into drums (Okanagan-Colville, Pauite), rattles (Haisla, Quiluete, Hanaksiala), spirit whistles (Haisla, Hanaksiala), masks (Bella Coola, Nitinaht), ceremonial headbands, capes (Bella Coola), and jewelry (Shuswap). The inner bark was worn around the neck and legs by shamans (Haisla, Hanaksiala). The wood was also used to make both cradles and coffins (Haisla, Hanaksiala). In fact, there is a strong association between cedar and death. Men would chew cedar branch tips to avoid nausea when burying a corpse (Lummi). Branches were used as a broom to sweep off the walls of a house after the removal of a corpse (Lummi), and smoking Redcedar branches were waved through the house to scare away the ghost after death (Skagit, Lummi). Boys used the branches ceremonially, rubbing themselves with them before a guardian spirit quest (Lummi). The wood was made into shamanic soul catchers in ritual healing (Oweekeno). Both Redcedar and Sitka Spruce were said to cause

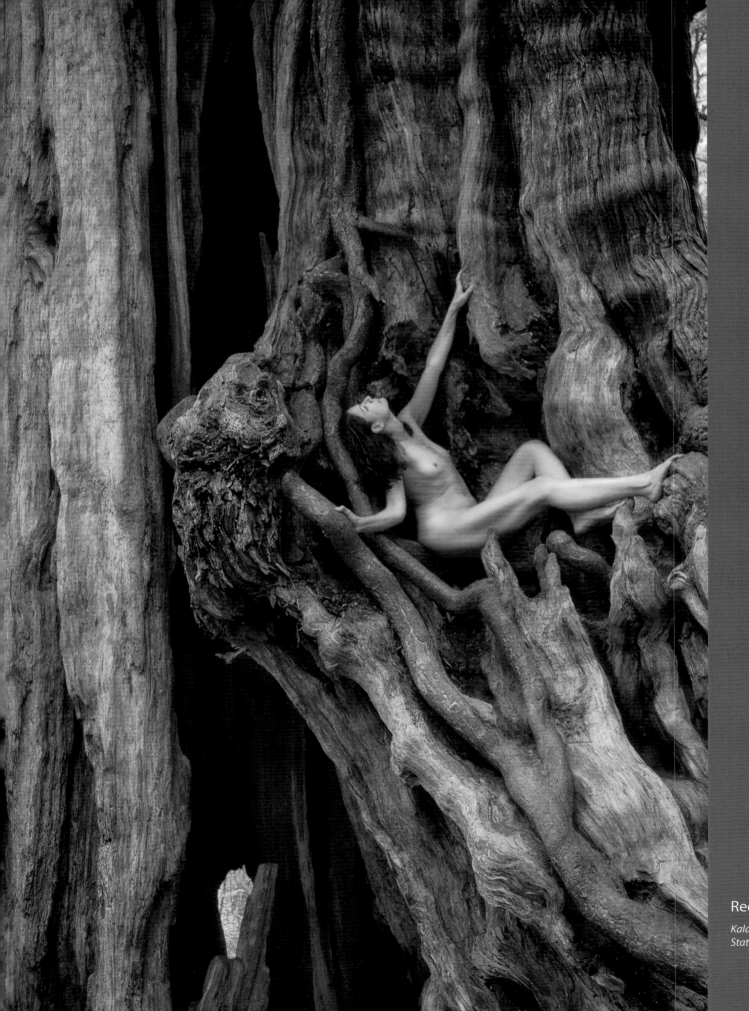

Redcedar Spirit Tree III (2012)

Kalaloch Redcedar, Olympic National Park, Washington State, USA.

vivid dreams for those who slept under them (Thompson). Whalers placed branches under their beds to make themselves ready for the hunt. Totem poles, those monumental sculptures that recount cultural legends, notable events, clan lineages—or were simply elaborate works of art—were primarily carved from Redcedar (Haisla, Makah, Gitxsan, Kwakiutl, Hanaksiala, Tsimshian, Tlingit, Kwakwaka'wakw, Haida, Tsimshian).

Shelter: Logs and planks were used for construction of house posts, siding, and roofs (Clallam, Haisla, Hanaksiala, Hesquiat, Hoh, Kwakiutl, Montana Indian, Okanagan-Colville).

Transportation: The grand size of the Redcedar, along with its rot resistance and soft, carve-able wood, served for its extensive use for paddles and dugout canoes, up to 100 ft. (30 m) long (Clallam, Hesquiat, Haisla, Hanaksiala, Kutenai, Kwakiutl, Makah, Montana Indian, Nitinaht, Okweekeno, Okanagan-Colville, Quileute, Coast Salish, Thompson, Tsimshian).

Modern Uses: Redcedar is an important timber wood valued for its lightweight, tight grain, durability, decay resistance, as well as insulative and aromatic properties. It is used for constructing sailboats and kayaks, as well as shingles, posts, utility poles, decking, and siding. It is also widely used for making beehives. The aromatic oil is used today for production of essential oils, perfumes, insecticides, soaps, deodorants, medicinal preparations, and shoe polishes.

Threats and Conservation: Not threatened.

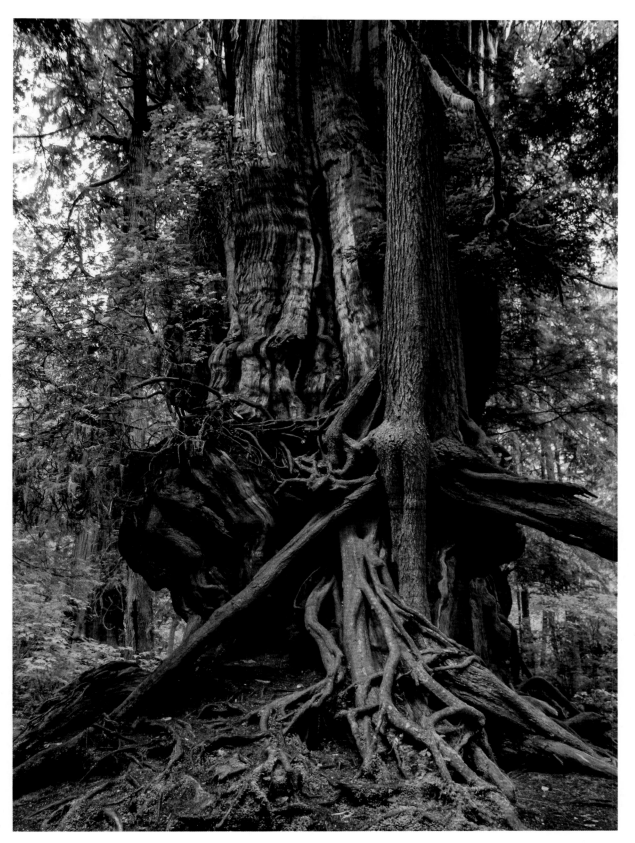

Kalaloch Redcedar (2012) ▸

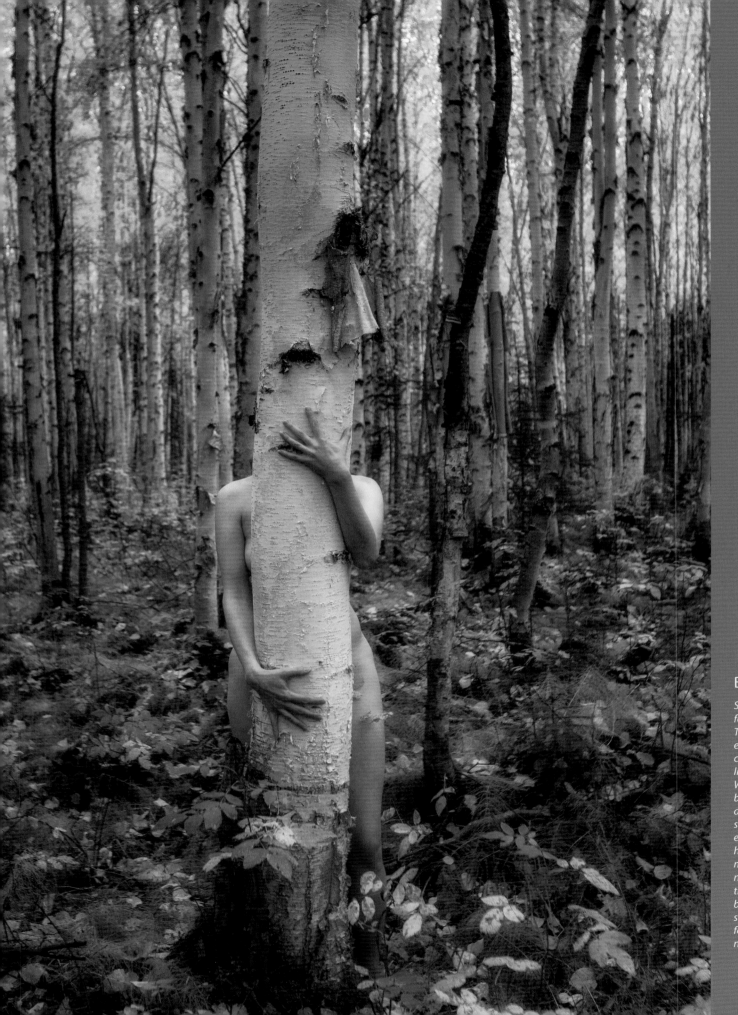

Birch Forest Embrace (2013)

Some trees species tempt you with their glorious beauty for days before you can stop and touch them for yourself. This was the case when I was on a ten-day autumnal equinox photography trip in Alaska, seeking a rare up-close glimpse at wild polar bears and the elusive northern lights. All along the hundreds of miles we travelled, the White Birch trees were on fire in the distance, or just beyond reach outside our whizzing vehicles. On the last day of the trip, I conspired with our guide, Laurent, to stop our vehicles on the road for a short break. While everyone else stood mesmerized, watching a pair of hunters disembowel and decapitate a full-grown male moose they had just shot, my guide and I snuck off for no more than five minutes to secretly get the Birch shot that had been calling to me all week. If only I could have been left behind, instead of getting back in the car, I am sure I would have disappeared into the glow of the forest forever. This grove of trees lives somewhere off the road near Fairbanks, Alaska, USA.

White Birch

Betula papyrifera

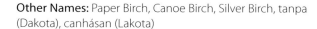

Other Names: Paper Birch, Canoe Birch, Silver Birch, tanpa (Dakota), canhásan (Lakota)

Distinctive Characteristics: Paper birch is a medium-sized, single- or multiple-stemmed, deciduous tree with heart-shaped leaves and a tall, thin canopy, most notably distinguished by its bark. Mature bark is variable in color, from bright white to silver to yellowish to reddish-brown. The bark is smooth, with noticeable black marks, scars, and variable lengths of dark, horizontal lenticels (pores providing a pathway for the direct exchange of gases between the internal tissues and atmosphere) that sometimes resemble eyes. The bark flakes and peels easily in fine, paper-like horizontal strips, revealing a pale pink to copper-orange underside. It scars easily, and removing bark by cutting or peeling it from a live tree will kill it. Sapling bark is dark reddish-brown with pale lenticels. The bark is highly weather-resistant; the wood is quick to burn and slow to rot.

Distribution: Native to all Canadian provinces; Alaska, Washington, Idaho, Montana, North Dakota, Nebraska, Minnesota, Maine, New Hampshire, Connecticut, Vermont, Massachusetts, Pennsylvania, New York, Michigan, Wisconsin, Iowa, and isolated patches in Illinois, Indiana, Colorado, Oregon, North Carolina, and New Mexico. Elevation: sea level to 10,000 ft. (3,000 m).

Ecosystem: White Birch has a wide range; it is a common species in nearly forty forest types, from boreal to montane and subalpine forests, to rugged mountain slopes, bordering bogs and swamps, and in the northern great plains. It is found in coniferous, deciduous, and mixed woodland communities with Jack Pine, Maple, Beech, Douglas Fir, Ponderosa Pine, Western White Pine, Balsam Fir, White Spruce, Black Spruce, Quaking Aspen, Western Hemlock, Sitka Spruce, Poplar, and Larch, but also in pure stands.

Maximum Age: About 150 years.

Maximum Height and Girth: Up to 130 ft. (40 m) in height; up to 18 ft. (5.5 m) in circumference.

Animal Community: Due to its abundance, Birch bark is a winter staple food for moose. Porcupines and beaver also eat the inner bark. White-tailed deer eat the leaves in the fall. Snowshoe hares and rabbits browse Paper Birch seedlings and twigs. Birch seeds are an important food source for voles and shrews, as well as for many winter birds, including American goldfinch, pine siskin, northern junco, blue jay, and the chickadees, redpolls, and sparrows. Ruffed grouse eat the buds and catkins. In spring, Birch flowers attract many insects, which in turn attract large numbers of migrating warblers. Birches can also be important nesting sites for red-tailed hawks and vireos, as well as cavity nesting birds, such as chickadees, nuthatches, swallows, and woodpeckers. Small strands of the bark are the key materials used by vireos in hanging nests, while many other birds and red squirrels use it to line their dens and nests. Yellow-bellied sapsuckers create holes to both drink the sap and attract ants. Hummingbirds drink the sap as well.

Traditional Uses: Ethnobotanically, White Birch is one of the most proliferate tree species that was utilized. The waterproof, lightweight, durable, flexible bark—which could be cut and bent to make virtually any shape—had a wide variety of uses. The bark could be sewn together with fiber and sealed with tree pitch.

Medicine:
Sap: The sap was boiled with spruce pitch to make an ointment for wasp stings or tapped from trees in early spring and taken for colds (Thompson) and coughs.
Bark: Bark powder used to treat diaper rash and other skin rashes (Algonquin, Quebec). An infusion of inner bark was used as an enema (Chippewa). The dried inner bark was ground and added to pitch and grease as an ointment for persistent scabs and rashes. A poultice of outer bark was used to bandage a burn (Woodlands Cree). The bark was also used ritualistically for contraception. A decoction of the bark was taken for internal blood diseases. The inner bark was used as a pain reliever (Shuswap), and a decoction of the inner bark was used to treat dysentery (Menominee).
Branches and wood: A decoction of wood was taken to cause sweating, for back pain, to ensure an adequate supply of milk for breastfeeding, and as a gynecological aid. A decoction of the branch tips was used as a tonic and for teething sickness. The wood was mixed with other materials and used to treat gonorrhea. Burned bark ashes were used to "shrivel the womb" (Iroquois).
Roots: The roots were used as a flavoring for medicines (Ojibwa). The root bark was cooked with maple sugar into syrup to treat stomach cramps.

Food: The sap was drunk as a tea, chewed as a gum, to sweeten medicines, or boiled down and made into syrup (Algonquin). The raw sap, sometimes mixed with fish grease, was used as food (Tanana). The inner bark could be eaten as food in an emergency (Montagnais). Soft, rotten wood was burned to smoke-cure meat and fish (Woodlands Cree).

Fiber: Thick bark was made into casts or slings (Tanana).

Tools and Objects: The bark was used for making baskets, storage containers, trays, cooking pots, dishes, mats, household utensils, baby carriers, racks for curing hides, torches, moose and bird calls, fishing gear, spears, bows, arrows, and children's toys. The wood was burned as fuel, and the bark was shredded and used as tinder to start campfires.

Art and Ceremony: Shamans contacted the plant spirit of the Birch to heal sick people (Koyukon). The wood was used to make rattles and drums. The bark was crafted into decorative fans. Stencils were made out of the bark to decorate containers. The bark was also used as paper, and drawn on with charcoal.

Shelter: Both the bark and the logs were used to construct wigwams, tipis, and sod-roofed houses.

Transportation: The wood and bark of Birches were used to construct snowshoes, sleds, and canoes up to 24 ft. (7.3 m) long and carrying 50 paddlers (Abenaki, Ojibwe, Huron, Kickapp, Pennacook, , Fox, Sauk, Lenape, Algonquin, Passamaquoddy, Micmac, Maliseet, Wabanaki, Cree, Iroquois, Têtes de Boule, Beothuk, Dogrib, Penobscot).

Modern Uses: The sweetener xylitol is made commercially from Birch sap. The wood is used for making furniture, flooring, spools and spindles, popsicle sticks, veneer, plywood, pulpwood, oriented strand board, and woodstove fuel. The sap is boiled down to produce birch syrup.

Threats and Conservation: White Birch is stressed by climate change in its southern range. Although it is not officially listed as threatened, in some U.S. states it is severely imperiled by the bronze birch borer (*Agilius anxius*).

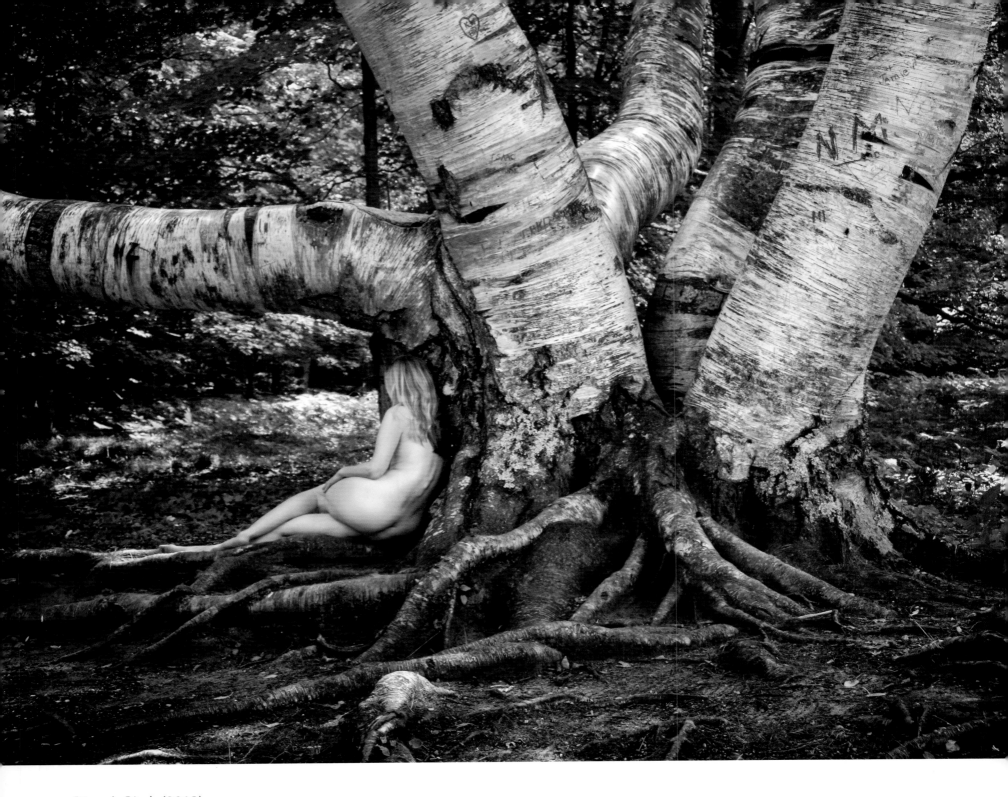

Diane's Birch (2013)

Some trees live simple lives as the human world passes by them, but also hold precious memories in their roots, bark, and limbs. I had heard about this giant White Birch tree in Michigan for years from my treegirl girlfriend. Finally, she coaxed me to visit her and her beloved tree while she was recovering from cancer. We made our epic day-trip to Beaver Island, healing our dear friendship as we shared an intimate discovery of the tree and its enchanted branches. Throughout the extended photo shoot of each of us, my trickster camera decided when to shoot and when not (or perhaps

the tree did). Pieces of this tree's enchanted white bark grace my altar to this day. This book is dedicated to the memory of Diane Osborne, my best and most beloved TreeGirl assistant, who left our tree-filled material world six months later for an even more magical place. You can make a pilgrimage by ferry to this tree on Beaver Island, Michigan, USA, and make your own memories with this magnificent tree.

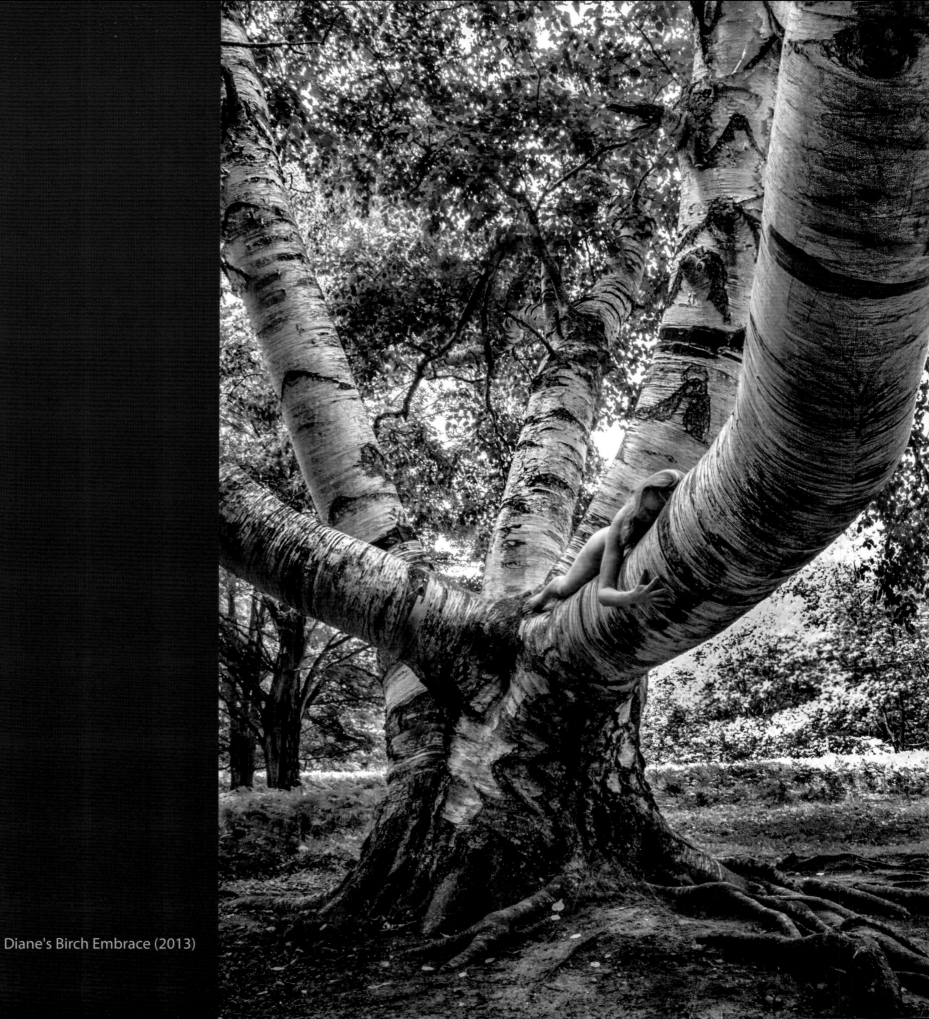

Diane's Birch Embrace (2013)

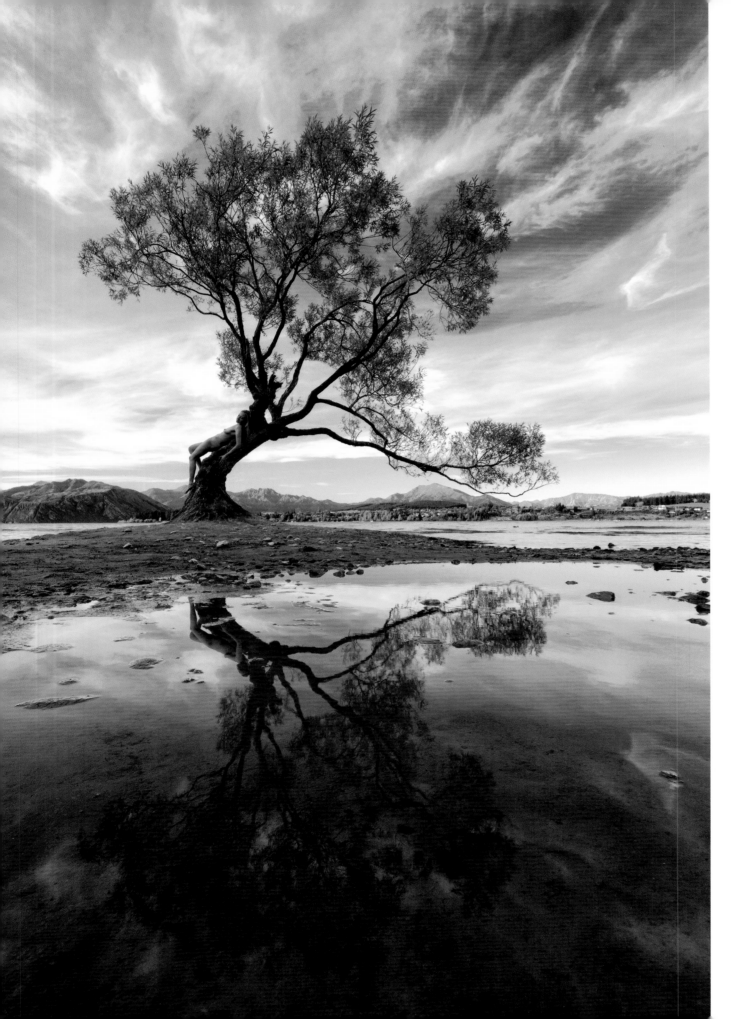

Wanaka Water Willow (2015)

This lone, twisted White Willow tree, on Lake Wanaka on the North Island of New Zealand, has been photographed thousands of times against a brilliant sunset backdrop by both professional and non-professional photographers. I'm guessing I'm the only one who has done it in the nude, but when I saw the tree, it seemed obvious to me what was needed to make a good shot. I was on the first day of a nature photography course. I couldn't wait to intertwine with the tree and knew just where I would fit in as soon as I saw it, but I had to wait a few hours until everyone else was done shooting in order for me to get in the tree. I was completely sleep-deprived after a couple of intense days of international travel, so I passed the time by sleeping face-down on the warm rocky shore, like a big bean bag.

Finally, right after dusk, everyone was done shooting and it was my turn. This was a rare case of TreeGirl photographing herself witnessed by anyone other than the rare safari guide or friend. However, the soon to be fun-loving group was open-minded enough to allow me to do my TreeGirl thing quickly, while the remaining light and privacy allowed, in the course of about one minute.

White Willow

Salix alba

Other Names: Willow, Saille

Distinguishing Characteristics: White Willow can grow as a shrub or a tree, with single or multiple trunks. Deciduous, the green leaves are distinctively lanceolate, 2–4 in. (5–10 cm) long and .25–.6 in. (5–1.5 cm) wide, with undersides covered in very fine, white, silky hairs that shimmer in the wind. Flowers are long catkins that resemble caterpillars. (But this is not the species that produces 'pussy willows'). Young branches are slender and straight; some varieties are colorful in winter and prized by landscapers. Older trees often have a twisted and leaning crown, with bark is that rough and deeply fissured. Willows respond well to cutting by pollarding and coppicing, re-growing with fast, straight, flexible shoots. These shoots, if cut, regenerate easily by simply sticking them in the moist earth. There are many subspecies, cultivars, and varieties.

Distribution: Native to Great Britain, Europe, and western and central Asia. Elevation: below sea level to 945 ft. (3,100 m).

Ecosystem: Grows in wetland areas and along watercourses.

Maximum Age: 100 years (longer if pollarded).

Maximum Height and Girth: 100 ft. (30 m) in height; 12.5 ft. (3.8 m) in circumference.

Animal Community: Caterpillars of puss moths, willow ermines, eyed hawk-moths, and red underwings feed off the new foliage. The catkins provide an important source of early nectar and pollen for bees. The branches are utilized by many species of birds for food, shelter, nesting, and roosting.

Traditional Uses:

Medicine: Infusions of the bark were used to treat headache, fever, pain, rheumatism, diarrhea, inflammation, and dandruff. The leaves were infused into a tea as a sedative.

Food: The inner bark was eaten.

Tools and Objects: Flexible shoots were important materials for making baskets, wicker furniture, hats, fencing, livestock pens, fishing traps, coracles (small Welsh boats), and wattle (long sticks braided and covered in mud for building). Gunpowder was also made from the wood.

Art and Ceremony: Willow, associated with water, the moon, fertility, and magic, was identified with the Greek goddesses Persephone, Helice, Selene, Hecate, and Artemis; the Roman goddess Luna; the Sumerian goddess Belili and god Bel; and the Celtic goddess Brigit. Green George, a man comprised of willow branches, was a Romanian character in a seasonal fertility ritual. Branches were used in Celtic funerary ceremonies. Willow is the fourth tree in the Celtic Ogham tree alphabet. Its branches were used for dowsing and to make magic wands.

Modern Uses: Willow was the original source of salicylic acid, used in the pharmaceutical production of aspirin. The lightweight wood is used to make cricket bats (var. *caerulea*), flooring, and boats. It is also being used artistically to create living sculptures, and in landscaping and restoration for erosion control.

Threats and Conservation: Not threatened.

Ihlara Valley Willow (2015)

These handsome White Willows lead a serene life in the dramatic Ihlara Valley gorge in Turkey, formed thousands of years ago from volcanic rock carved out by the Melendiz Suyu river, which is now a tranquil stream. The stream, a haven amidst the surrounding desert, flows for 8.5 miles (14 km), studded with Willows, Poplars, wildflowers and small waterfalls, providing habitat for dozens of species. Due to the valley's plentiful water supply and hidden location, this became the settlement of the first Christians escaping from Roman soldiers in the 4th c.

Along the Valley, there are an estimated 10,000 archeological locations carved into the canyon walls of volcanic tuff, including an estimated 60–104 dwellings, Byzantine churches, and one of the world's largest rock-cut monasteries, some with remnant wall paintings depicting religious scenes. An easy half-day hike winds along the flat canyon floor. While most day-tourists make a short trip to explore the historical sites, I was thrilled to find this haven of lush, wild nature, and made friends with the trees along the trail. This White Willow tree, probably 100 years old, sits peacefully on the banks of the stream with other large willows, drinking in the precious water of the Anatolian desert. You can find it in the Ihlara Valley about an hour's drive from the underground, carved city and tourist destination of Cappadocia, Turkey.

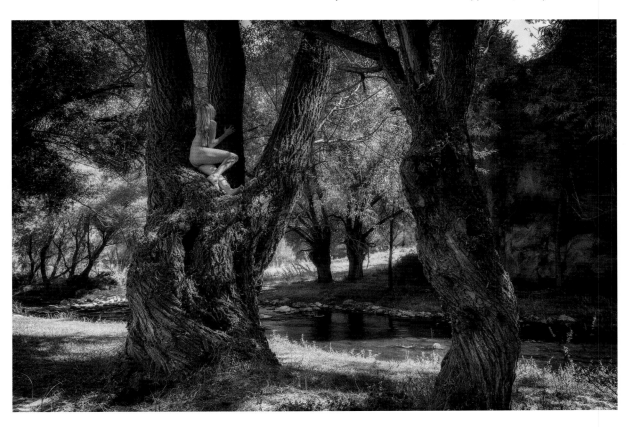

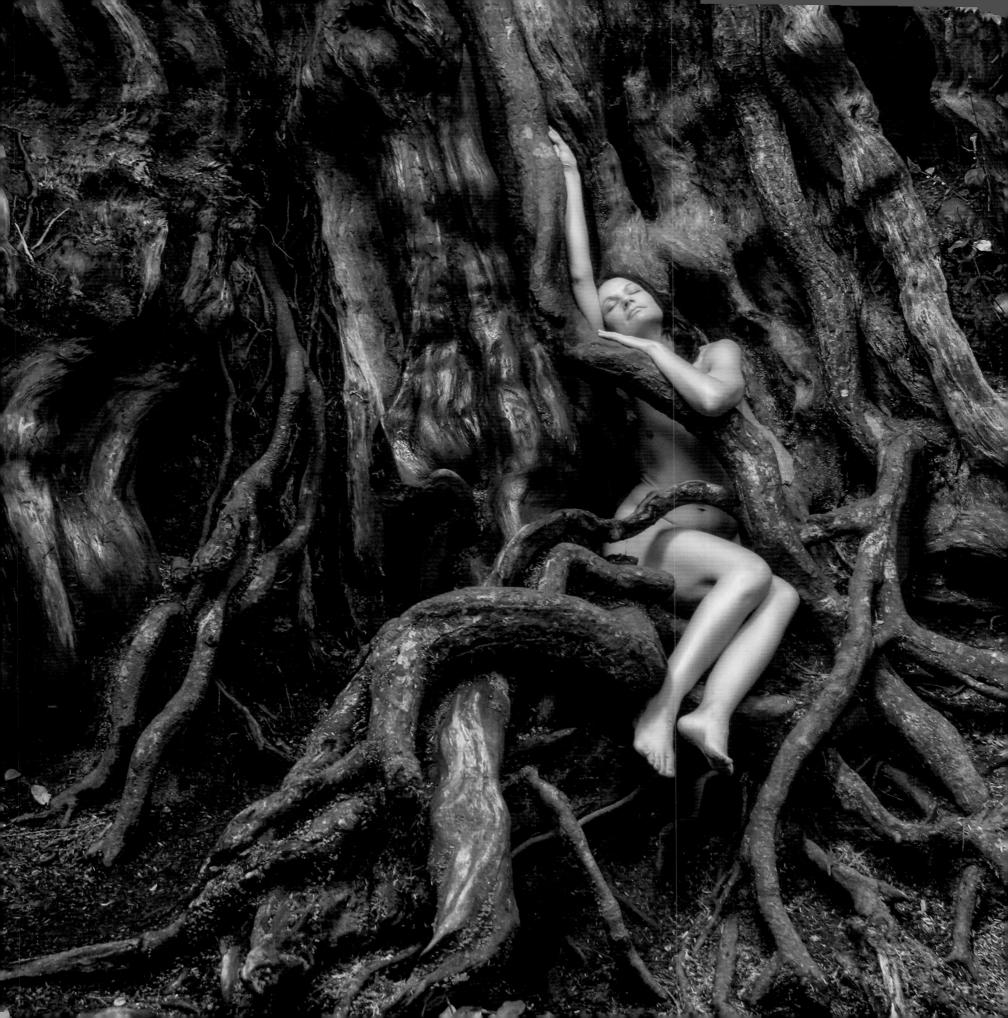

Afterword

It is not only possible, but essential, for us as humans to regain our intimacy with wild Nature. More than that, there is a yearning, sometimes secret, in almost every one of us to feel a sense of belonging, purpose, connection, and intimacy with others and with something greater than ourselves. Imagine a world in which all humans recognize everything that is wild as an extension of ourselves. How would that be different? There are an infinite number of pathways to engage the senses, heart, intellect, and spirit in interaction with the Wild. I hope this book has inspired you. A whole wild world is accessible, waiting for you, and trees are only the beginning.

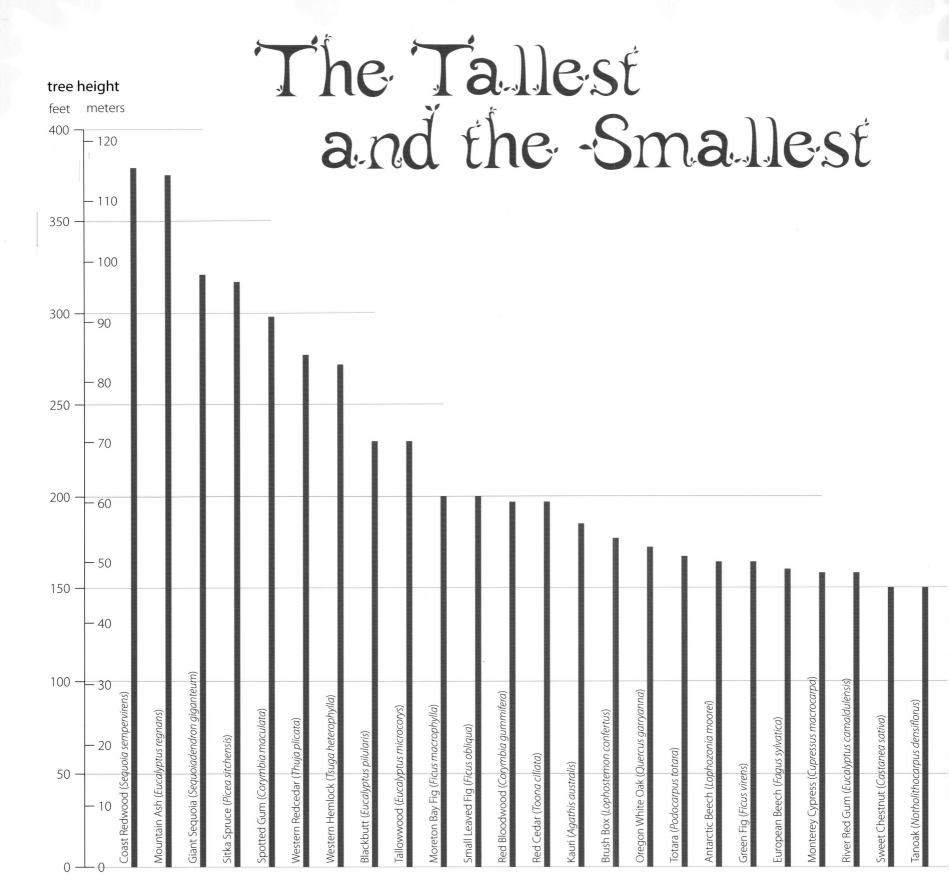

The Tallest and the Smallest

tree height

feet   meters

tree diameter

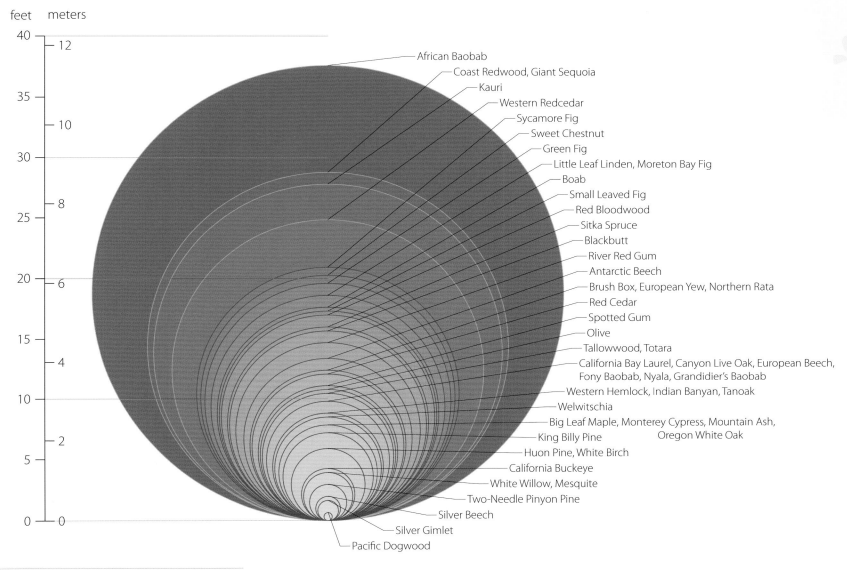

feet | meters

40 — 12
35 —
30 — 10
25 — 8
20 — 6
15 — 4
10 —
5 — 2
0 — 0

African Baobab
Coast Redwood, Giant Sequoia
Kauri
Western Redcedar
Sycamore Fig
Sweet Chestnut
Green Fig
Little Leaf Linden, Moreton Bay Fig
Boab
Small Leaved Fig
Red Bloodwood
Sitka Spruce
Blackbutt
River Red Gum
Antarctic Beech
Brush Box, European Yew, Northern Rata
Red Cedar
Spotted Gum
Olive
Tallowwood, Totara
California Bay Laurel, Canyon Live Oak, European Beech,
Fony Baobab, Nyala, Grandidier's Baobab
Western Hemlock, Indian Banyan, Tanoak
Welwitschia
Big Leaf Maple, Monterey Cypress, Mountain Ash,
Oregon White Oak
King Billy Pine
Huon Pine, White Birch
California Buckeye
White Willow, Mesquite
Two-Needle Pinyon Pine
Silver Beech
Silver Gimlet
Pacific Dogwood

Northern Rata (*Metrosideros robusta*)
King Billy Pine (*Athrotaxis selaginoides*)
Little Leaf Linden (*Tilia cordata*)
White Birch (*Betula papyrifera*)
Huon Pine (*Lagarostrobos franklinii*)
Big Leaf Maple (*Acer macrophyllum*)
Canyon Live Oak (*Quercus chrysolepis*)
California Bay Laurel (*Umbellularia californica*)
Silver Beech (*Lophozonia menziesii*)
White Willow (*Salix alba*)
Grandidier's Baobab (*Adansonia grandidieri*)
Indian Banyan (*Ficus benghalensis*)
Nyala (*Xanthocercis zambesica*)
European Yew (*Taxus baccata*)
Fony Baobab (*Adansonia rubrostipa*)
Mangrove (*Rhizophora sp.*)
Two-Needle Pinyon Pine (*Pinus edulis*)
Sycamore Fig (*Ficus sycomorus*)
Pacific Dogwood (*Cornus nuttallii*)
African Baobab (*Adansonia digitata*)
Boab (*Adansonia gregorii*)
Olive (*Olea europaea*)
California Buckeye (*Aesculus californica*)
Mesquite (*Prosopis juliflora*)
Silver Gimlet (*Eucalyptus campaspe*)
Welwitschia (*Welwitschia mirabilis*)

Scientific Names Index

Indigenous Names Index

Index

Bibliography

Abram, David. *The Spell of the Sensuous: Perception and Language in a More-Than-Human World*. New York: Vintage Books, 1996.

Agrawal, Arun, Ben Cashore, Rebecca Hardin, Gill Shepherd, Catherine Benson, and Daniel Miller. "Economic Contributions of Forests." Paper presented at the United Nations Forum on Forests, Tenth Session, Istanbul, Turkey, 20 March, 2013. Accessed September 22, 2014. http://www.un.org/esa/forests/pdf/session_documents/unff10/EcoContrForests.pdf.

Allen, Richard and Kimbal Baker. *Australia's Remarkable Trees*. Carlton, VIC, Australia: The Miegunyah Press, 2009.

Altman, Nathaniel. *Sacred Trees: Spirituality, Wisdom & Well-Being*. New York: Sterling Publishing, 2000.

Altonen, Brian L. "Co-Evolution: Man and Plants." Accessed August 12, 2014. https://brianaltonenmph.com/natural-sciences/4-projects/plantae-the-evolution-of-plant-chemicals/co-evolution-man-and-plants/.

Always Ayurveda: A Sister Concern of Planet Ayurveda. "*Ficus benghalensis*." Accessed March 20, 2016. http://www.alwaysayurveda.com/ficus-benghalensis/.

American Forests. "Oregon White Oak (*Quercus garryana*)." Accessed June 20, 2016. https://www.americanforests.org/bigtree/oregon-white-oak-quercus-garryana-4/.

Amorok, Tina. "The Eco-Trauma and Eco-Recovery of Being." *Shift: At the Frontiers of Consciousness*, no. 15 (June–August, 2007). Accessed February 12, 2012. http://library.noetic.org/library/publication-articles/seeing-where-we-are

Anderson, Eric. *Plants of Central Queensland: Identification and Uses of Native and Introduced Species*. Clayton South, VIC, Australia: CSIRO Publishing, 2016.

Anderson, Kat. *Tending the Wild: Native American Knowledge and the Management of California's Natural Resources*. Berkeley: University of California Press, 2013.

Atlas of Living Australia. "*Eucalyptus microcorys*: Tallowwood." Accessed March 27, 2016. http://bie.ala.org.au/species/urn:lsid:biodiversity.org.au:apni.taxon:252952#tab_names.

Atlas of Living Australia. "*Eucalyptus regnans*: Mountain Ash." Accessed February 28, 2016. http://bie.ala.org.au/species/urn:lsid:biodiversity.org.au:apni.taxon:306354#tab_names.

Australia Walkabout Wildlife Park. "Vegetation Communities." Accessed March 4, 2016. http://www.walkaboutpark.com.au/index.php/walkabout-s-vegetation.

Australian Native Plant Society (Australia). "*Corymbia maculata*." Accessed March 26, 2016. http://anpsa.org.au/c-mac.html.

Bailey, Frederick M. *Contributions to the Queensland Flora*. Whitefish, MT: Kessinger Publishing, LLC, 2010. Originally printed in 1898.

Bainbridge, David. "The Rise of Agriculture: a New Perspective." *Ambio* 14, no. 3 (1985): 148–151.

Bernstein, Jerome S. *Living in the Borderland: The Evolution of Consciousness and the Challenge of Healing Trauma*. New York: Routledge, Taylor & Francis Group, 2005.

Bibi, Yasmin, Sobia Nisa, Fayyaz M. Chaudhary, and Muhammad Zia. "*Toona ciliata* – Antibacterial Activity of Some Selected Medicinal Plants of Pakistan." *BioMed Central Complementary and Alternative Medicine*. Accessed January 20, 2016 http://bmccomplementalternmed.biomedcentral.com/articles/10.1186/1472-6882-11-52.

Big River Internet. "Healing Secrets of Aboriginal Bush Medicine." Accessed February 28, 2016. http://www.bri.net.au/medicine.html.

BioNET-EAFRINET. "*Prosopis juliflora* (Prosopis or Mesquite)." Accessed October 12, 2015. http://keys.lucidcentral.org/keys/v3/eafrinet/weeds/key/weeds/Media/Html/Prosopis_juliflora_(Prosopis_or_Mesquite).htm.

Boland, D. J., and M. W. McDonald. *Forest Trees of Australia*. Collingwood, VIC, Australia: CSIRO Publishing, 2006.

Bowcutt, Frederica. *The Tanoak Tree: An Environmental History of a Pacific Coast Hardwood*. Seattle: University of Washington Press, 2015. Kindle edition.

Breeden, Stanley. *Australian World Heritage Tropical Rainforest: A Journey of Discovery*. Archerfield, QLD, Australia: Steve Parish Publishing, 1999.

Bressette, Dana K. "Oregon White Oak, *Quercus garryana*." Native Plants PNW: An Encyclopedia of the Cultural and Natural History of the Northwest Native Plants. Accessed June 30, 2016. http://nativeplantspnw.com/oregon-white-oak-quercus-garryana/.

Breyer, Melissa. "25 Shocking Fashion Industry Statistics." *TreeHugger* (September 11, 2012). Accessed September 5, 2015. http://www.treehugger.com/sustainable-fashion/25-shocking-fashion-industry-statistics.html.

Brooker, Ian and David Kleinig. *Field Guide to Eucalypts. Vol. 1, South-Eastern Australia*. Hawthorn, VIC, Australia: Bloomings Books, 1999.

Brown, Peter M. *Rocky Mountain Tree Ring Research*. Accessed October 12, 2015. http://www.rmtrr.org/.

Brunner, Michael. *Bedeutende Linden: 400 Baumriesen Deutschlands*. Stuttgart, Germany: Haupt Verlag, 2007.

Burke, Ciaran. "Photograph One Tree, Everyday, For a Year!" *OneTree365*. Accessed August 18, 2014. https://onetree365.com/.

Caldecott, Moyra. *Myths of the Sacred Tree: Including Myths from Africa, Native America, China, Sumeria, Russia, Greece, India, Scandinavia, Europe, Egypt, South America, [and] Arabia*. Rochester, VT: Destiny Books, 1993.

Calflora Database. "*Cornus nuttallii*." Accessed January 25, 2015. http://www.calflora.org/cgi-bin/species_query.cgi?where-calrecnum=2387.

California Oak Mortality Task Force. "Native Plants Associated with Sudden Oak Death (SOD) and Their Use by California Indians – Fact Sheet N. 19: California Bay Laurel." Accessed May 28, 2012. http://www.suddenoakdeath.org/pdf/Tribal/Tribal_uses_for_Bay_Laurel.pdf.

Cambie, R. C. and J. Ash. *Fijian Medicinal Plants*. East Melbourne, VIC, Australia: CSIRO Publishing, 1994.

Campbell, Joseph and Bill Moyers. *The Power of Myth*. New York: Doubleday, 1988.

Carr, Nicolas. *The Shallows: What the Internet Is Doing to Our Brains*. New York: W.W. Norton & Company, 2010.

Chamovitz, Daniel. *What a Plant Knows: A Field Guide to the Senses*. New York: Scientific American/Farrar, Straus and Giroux, 2012.

Clarke, Philip. *Australian Plants as Aboriginal Tools*. Dural Delivery Center, NSW, Australia: Rosenberg Publishing, 2012. Kindle edition.

Craig, Jeffrey M., Alan C. Logan, and Susan L. Prescott. "Natural Environments, Nature Relatedness and the Ecological Theater: Connecting Satellites and Sequencing to Shinrin-Yoku." *Journal of Physiological Anthropology* (2016). doi 10.1186.s40101-016-0083-9.

Crowe, Andrew. *Which Native Tree? Native Trees of New Zealand – A Simple Guide to Their Identification, Ecology and Uses*. Auckland, New Zealand: Penguin, 2009.

Diaz, Henry F., Martin Beniston, and Raymond S. Bradley, eds. *Climatic Change at High Elevation Sites*. Dordrecht, Netherlands: Springer Netherlands, 1997.

DiZerega, Gus. *Faultlines: The Sixties, the Culture War, and the Return of the Divine Feminine*. Wheaton, IL: Quest Books, Theosophical Publishing House, 2013.

Dolman, Brock, Brian Swimme, and Paul Stamets. *From Kingdom to Kindom: Acting as if We Have Relatives*. Santa Fe, NM: Bioneers/Collective Heritage Institute, 2011. CD-ROM.

Dremann, Sue. "California Trees Get the Axe." *Palo Alto Online*. Accessed August 13, 2013. http://www.paloaltoonline.com/news/2009/09/15/california-avenue-trees-get-the-axe.

Dunn, P. and K. Ewing. Ecology and Conservation of the South Puget Sound Landscape. Seattle: The Nature Conservancy, 1997. Accessed June 30, 2016. http://cascadiaprairieoak.org/EcologyandConservationBook.html.

Earle, Christopher J. "*Athrotaxis selaginoides*: King Billy Pine." The Gymnosperm Database. Accessed March 5, 2016. http://www.conifers.org/cu/Athrotaxis_selaginoides.php.

Earle, Christopher J. "*Cupressus macrocarpa*: Monterey Cypress." The Gymnosperm Database. Accessed January 24, 2016. http://www.conifers.org/cu/Cupressus_macrocarpa.php.

Earle, Christopher J. "*Lagarostrobos franklinii*: Huon Pine." The Gymnosperm Database. Accessed January 20, 2016. http://www.conifers.org/po/Lagarostrobos.php.

Earle, Christopher J. "*Podocarpus totara*: Totara." The Gymnosperm Database. Accessed January 23, 2016. http://www.conifers.org/po/Podocarpus_totara.php.

Eckenwalder, James E. *Conifers of the World: The Complete Reference*. Portland: Timber Press, 2009.

Encyclopedia of Life. "*Adansonia rubrostipa*: Perrier's Baobab." Accessed April 3, 2016. http://eol.org/pages/5406370/details.

Encyclopedia of Life. "*Olea europaea*: Olive." Accessed March 24, 2016. http://eol.org/pages/579181/overview.

Essentially New Zealand. "The Totara Tree." Accessed February 5, 2016. http://www.essentiallynz.com/our-products/totara-tree/.

Eurobodalla Regional Botanic Gardens. "Aboriginal Heritage Walk: Traditional Aboriginal Uses of Eurobodalla Plants." Accessed March 26, 2016. www.erbg.org.au/features/aboriginalheritagewalk.pdf.

Evans, Julian. *The Forests Handbook. Vol. 1, An Overview of Forest Science*. Oxford Malden, MA: Blackwell Science, 2001.

Farjon, Aljos and Rudolf Schmid. "*Sequoia sempervirens*." The IUCN Red List of Threatened Species. Accessed November 26, 2015. http://dx.doi.org/10.2305/IUCN.UK.2013-1.RLTS.T34051A2841558.en.

Feedipedia. "Mesquite (*Prosopis juliflora*)." Accessed October 12, 2015. http://www.feedipedia.org/node/554.

Feininger, Andreas. *Trees*. New York: Viking Press, 1978.

Fern, Ken. "*Corymbia gummifera*." *Useful Tropical Plants*. Accessed March 4, 2016. http://tropical.theferns.info/viewtropical.php?id=Corymbia+gummifera.

Flint, Wendell D. *To Find the Biggest Tree*. Three Rivers, CA: Sequoia Natural History Association, 2002.

Floyd, A. G. *Rainforest Trees of Mainland South-Eastern Australia*. Lismore, NSW, Australia: Terania Rainforest Publishing, 2008.

Foreman, Dave. "Five Feathers for the Cannot Club." In *The Rediscovery of the Wild*, edited by Peter H. Kahn Jr. and Patricia Hasbach, 181–206. Cambridge: The MIT Press, 2013.

Forest Products Commission. "Western Australia Blackbutt." Accessed January 18, 2016. http://www.fpc.wa.gov.au/node/991.

Forestry Corporation of New South Wales. "Burrawan State Forest." Accessed February 27, 2016. http://www.forestrycorporation.com.au/visit/forests/burrawan-state-forest.

Fuller, Richard A., Katherine N. Irvine, Patrick Devine-Wright, Philip H. Warren, and Kevin J. Gaston. "Psychological Benefits of Greenspace Increase with Biodiversity." *Biological Letters* 3, no. 4 (2007): 390–394.

Funston, Malcolm, Peter Borchert, and Braam Van Wyk. *Bushveld Trees: Lifeblood of the Transvaal Lowveld*. Cape Town, South Africa: Fernwood Press, 2005.

Gibbons, Philip and David Lindenmayer. *Tree Hollows and Wildlife Conservation in Australia*. Collingwood, VIC, Australia: CSIRO Publishing, 2002.

Global Times. "Celestial Forest Bath: Trial Nude Program in Guangdong Park Sparks Debate." *Dan Tri International News*. Accessed April 14, 2016. http://www.dtinews.vn/en/news/017004/22280/celestial-forest-bath--trial-nude-program-in-guangdong-park-sparks-debate.html.

Goodrich, Jennie, Claudia Lawson, and Vana P. Lawson. *Kashaya Pomo Plants*. Berkeley: Heyday Books, 1996.

Gucker, Corey L. "*Cornus nuttallii*." *USDA Forest Service, Fire Effects Information System (FEIS)*. Accessed January 26, 2016. http://www.fs.fed.us/database/feis/.

Gucker, Corey L. "*Quercus garryana*." *U.S. Department of Agriculture, Forest Service, Rocky Mountain Research Station, Fire Sciences Laboratory*. Accessed June 20, 2016. http://www.fs.fed.us/database/feis/plants/tree/quegar/all.html.

Gumbaynggirr Language Database. New South Wales National Parks and Wildlife Service. Accessed March 26, 2016. http://www.environment.nsw.gov.au/resources/education/2010GumbaynggirrLanguageDatabase.pdf.

Gunther, Erna. *Ethnobotany of Western Washington: The Knowledge and Use of Indigenous Plants by Native Americans*. Seattle: University of Washington Press, 2003.

Greater Taree City Council's Environmental and Strategic Planning Section. "Indigenous Plant Names of Greater Taree." 2nd ed., 2008. Accessed March 27, 2016. https://www.hastingslandcare.org.au/images/media_transfer/docs/resources/Indigenous_Plants_of_Greater_Taree_Final_2008_Version2.pdf.

Hageneder, Fred. *The Meaning of Trees: Botany, History, Healing, Lore*. San Francisco: Chronicle Books, 2005.

Hageneder, Fred. Yew: *A History*. Stroud, England: Sutton Publishing Ltd., 2007.

Hankey, Andrew. "*Adansonia digitata*." *South African National Biodiversity Institute, Plantzafrica*. Accessed April 2, 2016. http://www.plantzafrica.com/plantab/adansondigit.htm.

Hanna, Neil and Diana Menefy. *Kāpia: New Zealand Kauri Gum*. Auckland, New Zealand: Jadepress, 2008.

Harte, Jeremy. *The Green Man (The Pitkin Guide)*. Andover, England: Pitkin Publishing, 2008.

Heinberg, Richard. Powerdown: *Options and Actions for a Post-Carbon World*. Gabriola, BC, Canada: New Society Publishers, 2004.

Heinrich, Ingo. "Dendroclimatology of *Toona ciliata* in Australia." *TRACE - Tree Rings in Archaeology, Climatology and Ecology* 3, Proceedings of the DENDROSYMPOSIUM 2004, April 22–24, 2004, Birmensdorf, Switzerland (2005). Accessed January 20, 2013. http://treering.de/sites/default/files/TRACE_pdf/Volume_3/Heinrich_TraceVol_3.pdf.

Hennessey, Tom. "Entirely by Hand… From the Ground Up." *Penobscot River Restoration Trust*. Accessed February 3, 2016. http://www.penobscotriver.org/content/4060/birch-bark-canoe.

Herbison-Evans, Don and Christine Ashe. Coffs Harbour Butterfly House. "*Corymbia maculata*." Accessed March 26, 2016. http://lepidoptera.butterflyhouse.com.au/plants/myrt/corymbia-maculata.html.

Hicks, Clive. *The Green Man: A Field Guide*. Helhoughton, England: Compass Books, 2000.

Hirsh, Mark. *That Tree*. Accessed August 18, 2014. http://www.thattree.net/.

Holland, Jennifer S. *Unlikely Friendships: 47 Remarkable Stories from the Animal Kingdom*. New York: Workman Publishing Company, 2011.

Holmes, Russ. "Plant of the Week: Pacific Dogwood (*Cornus nuttallii*)." *United States Department of Agriculture, Forest Service*. Accessed January 25, 2016. http://www.fs.fed.us/wildflowers/plant-of-the-week/cornus_nuttallii.shtml.

Hope, Cathy and Steve Parish. *Amazing Facts About Australian Native Plants*. Archerfield, QLD, Australia: Steve Parish Publishing, 2008.

Hornsby Shire Council. "*Eucalyptus pilularis* – Blackbutt." Accessed January 18, 2016. http://www.hornsby.nsw.gov.au/media/documents/environment-and-waste/bushland-and-biodiversity/native-tree-database-fact-sheets/Fact-sheet-Eucalyptus-pilularis-Blackbutt.pdf.

Hornsby Shire Council. "*Corymbia gummifera* – Red Bloodwood." Accessed February 27, 2016. http://www.hornsby.nsw.gov.au/media/documents/environment-and-waste/bushland-and-biodiversity/native-tree-database-fact-sheets/Fact-sheet-Corymbia-gummifera-Red-Bloodwood.pdf.

Howell, Daniel. "Foot Anatomy 101-Biofeedback." *The Barefoot Professor*. Accessed September 5, 2015. http://barefootprof.blogspot.com/2011/04/foot-anatomy-101-biofeedback.html.

Howell, L. Daniel, *The Barefoot Book: 50 Great Reasons to Kick Off Your Shoes*. Alameda, CA: Hunter House, 2010. Kindle edition.

Hugo, Nancy R. and Robert J. Llewellyn. *Seeing Trees: Discover the Extraordinary Secrets of Everyday Trees*. Portland: Timber Press, 2011.

Huikari, Olavi. *The Miracle of Trees*. New York: Walker & Company, 2012.

Hunter Region Botanic Gardens. "Gundabooka Aboriginal Plants Trail." Accessed March 4, 2016. http://www.huntergardens.org.au/about/gundabooka.cfm.

Hyde, Mark, Bart Wursten, Petra Ballings, and Meg Coates Palgrave. "*Ficus sycomorus* L. subsp. *sycomorus*." *Flora of Zimbabwe*. Accessed February 11, 2016. http://www.zimbabweflora.co.zw/speciesdata/species.php?species_id=164690.

Hyde, Mark, Bart Wursten, Petra Ballings, and Meg Coates Palgrave. "*Xanthocercis zambesiaca*." Flora of Zimbabwe. Accessed February 10, 2016. http://www.zimbabweflora.co.zw/speciesdata/species.php?species_id=131230.

Hyo-sik, Lee. "Nude Forest Bath to Open in Jangheung." *Korea Times*. Accessed April 14, 2016. http://www.koreatimes.co.kr/www/news/nation/2011/07/117_91135.html.

Imai, Michiko. "An Introduction to the Forest Therapy Society of Japan, Forest Therapy, and Forest Therapist." *Forest Medicine*, edited by Qing Li, 233–242. New York: Nova Science Publishers, 2013.

Immel, Diana L. "Tanoak: *Lithocarpus densiflorus* Plant Guide." USDA, NRCS National Plant Data Center. Accessed February 5, 2012. http://plants.usda.gov/plantguide/pdf/cs_lide3.pdf.

Institute for Quality and Efficiency in Health Care. "How Does the Hand Work?" *National Center for Biotechnology Information*. Accessed January 15, 2016. http://www.ncbi.nlm.nih.gov/books/NBK279362/.

Jacobson, Arthur L. "How Long Do Trees Live?" Accessed February 3, 2016. http://www.arthurleej.com/a-oldtrees.html.

Janssen, Peter, and Mike Hollman. *Trees of New Zealand: Stories of Beauty and Character*. Auckland: Hodder Moa, 2011.

Johnson, Derek, Linda Kershaw, Andy MacKinnon, and Jim Pojar. *Plants of the Western Forest: Alaska to Minnesota Boreal and Aspen Parkland*. Auburn, CA: Lone Pine Publishing and the Canadian Forest Service, 1995.

Jones, Christine A. "The Medicinal Properties and Bush Foods of Eucalypts." *Australian Plants Online*. Accessed February 27, 2016. http://anpsa.org.au/APOL9/mar98-2.html.

Jordan, Linlee. "The Moreton Bay Fig Tree." *Hpathy: Homeopathy for Everyone*. Accessed March 25, 2016. http://hpathy.com/homeopathy-papers/the-moreton-bay-fig-tree/.

Kahn, Peter H. Jr. and Patricia H. Hasbach, eds. *The Rediscovery of the Wild*. Cambridge: The MIT Press, 2013.

Kahn, Peter H. Jr. and Patricia H. Hasbach, eds. *Ecopsychology: Science, Totems, and the Technological Species*. Cambridge: The MIT Press, 2012. Kindle Edition.

Kals, Elisabeth, Daniel Schumacher, and Leo Montada. "Emotional Affinity Toward Nature as a Motivational Basis to Protect Nature." *Environment and Behavior* 31, no. 2 (March, 1999): 178–202. doi: 10.1177/00139169921972056.

Kaplan, Rachel and Stephen Kaplan. *The Experience of Nature: A Psychological Perspective*. Cambridge: Cambridge University Press, 1989.

Kaplan, Stephen. "The Restorative Environment: Nature and Human Experience." *The Role of Horticulture in Human Well-Being and Social Development*. Edited by Diane Relf, 134–142. Portland: Timber Press, 1992.

Kaplan, Stephen. "The Restorative Benefits of Nature: Toward an Integrative Framework." *The Journal of Environmental Psychology* 15 (1995): 169–182.

Karadada, Jack. *Uunguu Plants and Animals: Aboriginal Biological Knowledge from Wunambal Gaambera Country in the North-West Kimberley, Australia*. Wyndham, WA, Australia: Wunambal Gaambera Aboriginal Corp., 2011.

Keith, Heather, Brendan G. Mackey, and David B. Lindenmayer. "Re-Evaluation of Forest Biomass Carbon Stocks and Lessons from the World's Most Carbon-Dense Forests." *Proceedings of the National Academy of Sciences* 106, no. 28 (2009): 11635–11640.

Kellert, Stephen R. and E. O. Wilson, eds. *The Biophilia Hypothesis*. Washington, DC: Island Press, 1995.

Kew Royal Botanic Gardens. "*Ficus benghalensis* (Banyan)." Accessed March 8, 2016. http://www.kew.org/science-conservation/plants-fungi/ficus-benghalensis-banyan.

Kew Royal Botanic Gardens. "*Welwitschia mirabilis* (Tree Tumbo)." Accessed December 4, 2015. http://www.kew.org/science-conservation/plants-fungi/welwitschia-mirabilis-tree-tumbo.

Kingsbury, Noël and Andrea Jones. *The Glory of the Tree: An Illustrated History*. Richmond Hill, ONT, Canada: Firefly Books, 2014.

Klinger, Dr. Lee. *Sudden Oak Life: Observations on Oak Health, Tree Care, Organic Farming, Gardening, Forest Decline, Acid Rain, Climate Change, Gaia*. Accessed October 12, 2015. https://suddenoaklifeorg.wordpress.com/.

Kossinets, Gueorgi and Duncan J. Watts. "Origins of Homophily in an Evolving Social Network." *American Journal of Sociology* 115 (2009): 405–450. Accessed February 28, 2010. doi:10.1086/599247.

Krohn, Elise. "Western Red Cedar." *Wild Foods and Medicines*. Accessed December 4, 2014. http://wildfoodsandmedicines.com/cedar/.

Kühn, Stefan, Bernd Ullrich, and Uwe Kühn. *Deutschlands Alte Bäume: Eine Bildreise zu den sagenhaften Baumgestalten zwischen Küste und Alpen*. Munich, Germany: Buchverlag GmbH & Co. LG, 2003.

Lake, Morris. *Australian Rainforest Woods: Characteristics, Uses and Identification*. Collingwood, VIC, Australia: CSIRO Publishing, 2015.

Lanner, Donald. *Conifers of California*. Los Olivos, CA: Cachuma Press, 1999.

Lea, Ted, Wynn Miles, and Terry McIntosh." Garry Oak Ecosystems Recovery Team Research Colloqium 2006." Victoria, BC, Canada: Canadian Forestry Service, February 24, 2006. Accessed June 30, 2016. http://www.wnps.org/ecosystems/west_lowland_eco/documents/GOERTResearchColloquium2006Proceedings.pdf.

Lewington, Anna and Edward Parker. Ancient Trees: *Trees That Live for a Thousand Years*. London, England: Collins & Brown, 1999.

Lewis, Thomas, Fari Amini, and Richard Lannon. *A General Theory of Love*. New York: Vintage Books, 2000.

Lightfoot, Kent G. and Otis Parrish. *California Indians and Their Environment: An Introduction*. Berkeley: University of California Press, 2009.

Little, Elbert L. Jr. "Digital Representations of Tree Species Range Maps." *USGS Geosciences and Environmental Change and Science Center*. Accessed January 25, 2016. http://esp.cr.usgs.gov/data/little/.

Logan, William Bryant. *Oak: The Frame of Civilization*. New York: W.W. Norton & Company, 2005.

Long, Andrew. *Aboriginal Scarred Trees in New South Wales: A Field Manual*. Hurstville, NSW, Australia: Department of Environment and Conservation, 2005. Accessed January 18, 2016. http://www.environment.nsw.gov.au/resources/cultureheritage/ScarredTreeManual.pdf.

Louv, Richard. *Last Child in the Woods: Saving Our Children from Nature-Deficit Disorder*. Chapel Hill, NC: Algonquin Books, 2005.

Louv, Richard. *The Nature Principle: Reconnecting with Life in a Virtual Age*. Chapel Hill, NC: Algonquin Books, 2011.

Mandal, S. G., R. V. Shete, K. J. Kore, K. V. Ortair, B. N. Kale, A. K. Manna. "Review: Indian National Tree: (*Ficus bengalensis*)." *International Journal of Pharmacy and Life Sciences* 1, no. 5 (September, 2010): 286-273. Accessed March 20, 2016. www.ijplsjournal.com/issues%20PDF%20files/sep2010/5.pdf.

Manson, B. R. "The Life History of Silver Beech (*Nothofagus menziesii*)." *Proceedings (New Zealand Ecological Society)* 21 (1974): 27-31. http://www.jstor.org/stable/24061487.

Marazita, Elizabeth. "Reflexology." *MedFinds*. Accessed September 5, 2015. http://www.medfinds.com/therapy_definitions.html?definition=reflexology.

McMahon, Lynne, Brendan George, and Robyn Hean. "*Eucalyptus pilularis*." PrimeFacts for Profitable, Adaptive, and Sustainable Industries. New South Wales Department of Industry and Investment. September 2010. Accessed January 18, 2016. http://www.dpi.nsw.gov.au/__data/assets/pdf_file/0008/356084/Eucalyptus-pilularis.pdf.

McSpadden, Kevin. "You Now Have a Shorter Attention Span Than a Goldfish." *TIME Magazine*. Accessed April 28, 2016. http://time.com/3858309/attention-spans-goldfish/.

Menon, Subhadra, Pallava Bagla, and Aruna Ghose. *Trees of India*. New Delhi, India: Timeless Books, 2000.

"Modern Day Muir: Steve Sillett, PhD. Sequoia/Redwood Canopy Research, Humboldt State." YouTube video, 4:53. Posted by "Oakland Museum of California," October 7, 2014. https://www.youtube.com/watch?v=e-oKQBZMTvU.

Moerman, Daniel E. "Native American Ethnobotany: A Database of Foods, Drugs, Dyes and Fibers of Native American Peoples, Derived From Plants." University of Michigan, Dearborn College of Art, Sciences, and Letters. Accessed October 12, 2015. http://herb.umd.umich.edu/herb/search.pl

Moerman, Daniel E. *Native American Ethnobotany*. Portland: Timber Press, 1998.

Moerman, Daniel E. *Native American Food Plants: An Ethnobotany Dictionary*. Portland: Timber Press, 2010.

Monumental Trees. "Elevation Distribution for *Adansonia digitata*." Accessed February 7, 2016. http://www.monumentaltrees.com/en/elevation/adansoniadigitata/.

Monumental Trees. "Elevation Distribution for *Quercus robur*." Accessed April 1, 2016. http://www.monumentaltrees.com/en/elevation/quercusrobur/.

Monumental Trees. "*Small-Leaved Lime*." Accessed April 1, 2016. http://www.monumentaltrees.com/en/trees/smallleavedlime/.

Moore, Michael. *Medicinal Plants of the Pacific West*. Santa Fe: Museum of New Mexico Press, 1993.

Morton, Andrew. *Tree Heritage of Britain and Ireland: A Guide to the Famous Trees of Britain and Ireland*. Ramsbury, England: Airlife Publishing Ltd, 2004.

Moss, Laura. "You're Never Too Old to Climb a Tree." *Mother Nature Network*. Accessed October 25, 2015. http://www.mnn.com/earth-matters/wilderness-resources/stories/youre-never-too-old-climb-tree.

Muir, John. *Travels in Alaska*. New York: Houghton Mifflin Company, 1915.

Myers, Katherine J., Tedmund J. Swiecki, and Alyson E. Mitchell. "Understanding the Native Californian Diet: Identification of Condensed and Hydrolyzable Tannins in Tanoak Acorns (*Lithocarpus densiflorus*)." *Journal of Agricultural and Food Chemistry* (2006). Accessed January 19, 2012.

National Institutes of Health, Research Initiative of Scientific Enhancement. "Medicinal Plants of the Southwest, Plants Database." *New Mexico State University*. Accessed October 12, 2015. http://medplant.nmsu.edu/newplants.shtm.

National Register of Big Trees: Australia's Champion Trees. "*Ficus macrophylla*: Fig - Moreton Bay." Accessed March 25, 2016. http://www.nationalregisterofbigtrees.com.au/listing_view.php?listing_id=529.

National Register of Big Trees: Australia's Champion Trees. "*Ficus virens*: Fig - White "The Temple Fig." Accessed July 18, 2016. http://www.nationalregisterofbigtrees.com.au/listing_view.php?listing_id=1100.

Native Indian Tribes. "Birch Bark Canoes." Accessed February 3, 2016. http://www.warpaths2peacepipes.com/native-american-life/birch-bark-canoe.htm.

Newberry, Ron. "Group Aims to Save Garry Oaks." Whidbey News-Times, April 11, 2015. Accessed June 30, 2016. http://www.whidbeynewstimes.com/community/299403851.html.

New South Wales Department of Primary Industries: Parks, Water and Environment. "Paddock Plants Fact Sheet: Tallowwood." Modified October 19, 2010. Accessed March 27, 2016. http://www.dpi.nsw.gov.au/content/agriculture/resources/private-forestry/paddock-plants/Eucalyptus-microcorys-Tallowwood.pdf.

New South Wales Office of the Environment and Heritage. "Illawara Aboriginal Resource Use Coastal Plain Habitat." *Murni, Dhungang, Jirrar: Living in the Illawarra – Aboriginal People and Wild Resource Use*. Accessed March 26, 2016. http://www.environment.nsw.gov.au/resources/culture-heritage/illawarraAboriginalResourceUseCoastal.pdf.

New Zealand Department of Conservation: Te Papa Atawhai. "Beech Forest." Accessed January 23, 2016. http://www.doc.govt.nz/nature/native-plants/beech-forest/.

New Zealand Tree Register. "*Podocarpus totara*: Mills Totara." New Zealand Notable Trees Trust. Accessed April 8, 2016. http://register.notabletrees.org.nz/tree/view/1009.

Nicolle, Dean. *Native Eucalypts of South Australia*. Dean Nicolle, 2013.

Nicolle, Dean. *Smaller Eucalypts for Planting in Australia. Their Selection, Cultivation and Management*. Dean Nicolle, 2016.

Nicolle, Dean. *Taller Eucalypts for Planting in Australia. Their Selection, Cultivation and Management*. Dean Nicolle, 2016.

Nicolle, Dean. "Old Blotchy: Big and Famous Trees Gallery." Accessed March 26, 2016. http://www.dn.com.au/Big_Famous_Trees_gallery/pages/Old-Blotchy-Corymbia-maculata-spotted-gum.html.

Nilsson, Kjell, Marcus Sangster, Christos Gallis, Terry Hartig, Sjerp de de Vries, Klaus Seeland, and Jasper Schipperijn, eds. *Forests, Trees and Human Health*. New York: Springer, 2011. Kindle edition.

Notten, Alice. "*Welwitschia mirabilis*." South African National Biodiversity Institute. Accessed August 29, 2012. http://www.plantzafrica.com/plantwxyz/welwitschia.htm.

Olive Tree Museum of Vouves. "Welcome to the Olive Tree Museum of Vouves." Accessed September 6, 2015. http://www.olivemuseumvouves.com/default.aspx?lang=en.

Oregon State University. "Oregon White Oak (*Quercus garryana*)." *Oregon Wood Innovation Center: Connecting People, Ideas, Resources*. Accessed June 20, 2016. http://owic.oregonstate.edu/oregon-white-oak-quercus-garryana.

Original Gold Rush Colony. "Traditional Aboriginal Uses for Plants of the Eurobodalla." Accessed January 19, 2016. http://www.goldrushcolony.com.au/australian-gold-history-culture-info/indigenous-australian-aboriginal-and-koorie-culture/traditional.

Ortiz, Beverly. "With Respect: Mabel McKay." *News from Native California* 7, no. 3 (1993).

Orwa, C., A. Mutua, R. Kindt, R. Jamnadass, and S. Anthony. "*Eucalyptus camaldulensis*." *Agroforestree Database: A Tree Reference and Selection Guide*, version 4.0 (2009). Accessed April 1, 2016. http://www.worldagroforestry.org/treedb/AFTPDFS/Eucalyptus_camaldulensis.pdf.

Orwa, C., A. Mutua, R. Kindt, R. Jamnadass, S. Anthony. "*Toona ciliata*." *Agroforestree Database: A Tree Reference and Selection Guide*, version 4.0 (2009). Accessed January 20, 2016. http://www.worldagroforestry.org/treedb/AFTPDFS/Toona_ciliata.pdf.

Orwin, Joanna. "Story: Kauri Forest - Kauri Forest Ecology." *Te Ara: The Encyclopedia of New Zealand*. Accessed January 22, 2016. http://www.teara.govt.nz/en/kauri-forest/page-2.

Orwin, Joanna. "Story: Southern Beech Forest - Ecology." *Te Ara: The Encyclopedia of New Zealand*. Accessed January 24, 2016. http://www.teara.govt.nz/en/southern-beech-forest/page-3.

Pakenham, Thomas. In *Search of Remarkable Trees on Safari in South Africa*. New York: Walker & Company, 2007.

Pakenham, Thomas. *Meetings with Remarkable Trees*. New York: Random House, 1996.

Pakenham, Thomas. *The Remarkable Baobab*. New York: W.W. Norton & Company, 2004.

Pakenham, Thomas. *Remarkable Trees of the World*. New York: W.W. Norton & Company, 2002.

Parker, Edward. *Photographing Trees*. Richmond, England: Kew Pub, 2012.

Paterson, Jacqueline M. *Tree Wisdom: The Definitive Guidebook to the Myth, Folklore and Healing Power of Trees*. London, England: Thorsons, 1996.

Pavlik, Bruce M., Pamela Muick, Sharon Johnson, and Marjorie Popper. *Oaks of California*. Los Olivos, CA: Cachuma Press and The California Oak Foundation, 1991.

Pennacchio, Marcello, Lara V. Jefferson, and Kayri Havens. *Uses and Abuses of Plant-Derived Smoke: Its Ethnobotany as Hallucinogen, Perfume, Incense, & Medicine*. New York: Oxford University Press, 2010.

Peters, Josephine G., Cheryl Beck, and Beverly Ortiz. *After the First Full Moon in April: A Sourcebook of Herbal Medicine from a California Indian Elder*. Walnut Creek, CA: Left Coast Press, 2011.

Petrides, George A. *A Field Guide to Trees and Shrubs*. Boston: Houghton Mifflin Company, 1972.

Petrides, George A., Roger Tory Peterson, and Olivia Petrides. *A Field Guide to Western Trees (Peterson Field Guides: 44)*. New York: Houghton Mifflin Company, 1992.

Plants For a Future. "*Cupressus macrocarpa*: Monterey Cypress." Accessed January 24, 2016. http://www.pfaf.org/user/Plant.aspx?LatinName=Cupressus+macrocarpa.

Plants For a Future. "*Olea europaea*: Olive." Accessed March 24, 2016. http://www.pfaf.org/user/Plant.aspx?LatinName=Olea+europaea.

Plants For a Future. "*Salix alba*: White Willow." Accessed April 2, 2016. http://www.pfaf.org/user/Plant.aspx?LatinName=Salix+alba.

Plotkin, Bill. "The Wild Human." *Shift: At the Frontiers of Consciousness* 19 (June-August, 2008). Accessed February 2, 2012. http://library.noetic.org/library/publication-articles/wild-human.

Point Lobos Foundation. "Plant Communities." Accessed January 24, 2016. http://www.pointlobos.org/nature/plant-communities.

Points of Interest Australia. "Australian Red Cedar (*Toona ciliata* var. *australis*)." Accessed January 20, 2016. https://poi-australia.com.au/australian-red-cedar-toona-ciliata-var-australis/.

Pojar, Jim and Andy Mackinnon. *Plants Of The Pacific Northwest Coast: Washington, Oregon, British Columbia & Alaska*. Auburn, CA: Lone Pine Publishing, 2004.

Pollet, Cédric. *Bark: An Intimate Look at the World's Trees*. London: Frances Lincoln, 2010.

Poropat, Peter. *Barks and Trunks: Rainforest Trees of Eastern Australia*. Goonellabah, Australia: Dragonwick Publishing, 2009.

Power Trees. "Urban Forestry." Accessed August 12, 2013. http://www.powertrees.com/urbanforestry.htm.

Preston, Robert. *The Wild Trees: A Story of Passion and Daring*. New York: Random House, 2007.

Prideaux, Bruce, ed. *Rainforest Tourism, Conservation and Management: Challenges for Sustainable Development*. The Earthscan Forest Library. New York: Routledge, 2014.

Prindle, Tara. "Uses for Birchbark." *NativeTech: Native American Technology and Art*. Accessed August 27, 2013. http://www.nativetech.org/brchbark/brchbark.htm.

Quattrocchi, Umberto. *CRC World Dictionary of Medicinal and Poisonous Plants: Common Names, Scientific Names, Eponyms, Synonyms, and Etymology*. Boca Raton, FL: CRC Press, 2012.

Queensland Government, Department of Environment and Heritage Protection, "Wildlife of Curtain Fig National Park." *WetlandInfo*. Accessed April 1, 2016. http://wetlandinfo.ehp.qld.gov.au/wetlands/facts-maps/wildlife/?AreaID=national-park-curtain-fig.

Queensland Government, Department of National Parks, Sport and Racing. "About Curtain Fig." *Curtain Fig National Park*. Accessed April 1, 2016. http://www.nprsr.qld.gov.au/parks/curtain-fig/about.html.

Queensland Government, Department of National Parks, Sport and Racing. "Nature, Culture and History." *Curtain Fig National Park*. Accessed January 19, 2016. http://www.nprsr.qld.gov.au/parks/curtain-fig/culture.html.

Rasolofo, V. M. "Mangroves of Madagascar." In *Conservation and Sustainable Utilization of Mangrove Forests in Latin America and Africa Regions, Part II: Africa: Mangrove Ecosystems Technical Reports 3*, (1993), edited by Diop, E. S. International Society for Mangrove Ecosystems and Coastal Marine Project of UNESCO, Okanawa, Japan. ISSN 0919-2646.

Recreation Oak Bay. "Oak Bay Parks Services: Information Guide for Tree Protection (Bylaw 4326)." Accessed June 30, 2016. https://www.oakbay.ca/sites/default/files/recreation/documents/tree_prot_info_guide11.pdf.

Revolvy. "Sclerophyll." Accessed March 29, 2016. http://www.revolvy.com/main/index.php?s=Sclerophyll&item_type=topic.

Reynaud, Françoise. *The Tree in Photographs*. Los Angeles: J. Paul Getty Museum, 2010.

Robinson, Les. "Aboriginal Uses of Plants Around Sydney." *Australian Plants Online*. Accessed March 4, 2016. http://anpsa.org.au/APOL10/jun98-6.html.

Roman, Lara A. and Frederick N. Scatena. "Street Tree Survival Rates: Meta-Analysis of Previous Studies and Application to a Field Survey in Philadelphia, PA, USA." *Urban Forestry and Urban Greening* 10 (2011): 269-274. Accessed August 12, 2013. www.actrees.org/wp-content/uploads/.../roman-scatena-2011-street-tree-mortality.pdf.

Roodt, Veronica. *Trees & Shrubs of the Okavango Delta: Medicinal Uses and Nutritional Uses, The Shell Field Guide Series: Part I*. Arcadia, South Africa: Shell Oil Botswana, 1998.

Roszak, Theodore, Mary E. Gomes, and Allen D. Kanner, eds. *Ecopsychology: Restoring the Earth, Healing the Mind*. San Francisco: Sierra Club Books, 1995.

Sandved, Kjell B., Ghillean T. Prance, and Anne E. Prance. *Bark: The Formation, Characteristics, and Uses of Bark Around the World*. Portland: Timber Press, 1993.

Schellberg, Dirk and Tony Langham. *Didgeridoo: Ritual Origins and Playing Techniques*. Diever, Holland: Binkey Kok, 1994.

Selhub, Eva M. and Alan C. Logan. *Your Brain on Nature: The Science of Nature's Influence on Your Health, Happiness, and Vitality*. Mississauga, ON: John Wiley & Sons Canada, 2012.

Sewall, Laura. "Seeing Where We Are." *Shift: At the Frontiers of Consciousness* 11 (June-August, 2006). Accessed February 12, 2012. http://library.noetic.org/library/publication-articles/seeing-where-we-are.

Shigo, Alex L. *Modern Arboriculture - Touch Trees: A Systems Approach to the Care of Trees and Their Associates*. Durham, NC: Shigo and Trees, Associates, 1991.

Shigo, Alex L. *A New Tree Biology Dictionary: Terms, Topics, and Treatments for Trees and Their Problems and Proper Care*. Durham, NC: Shigo and Trees, Associates, 1986.

Sibley, David Allen. *The Sibley Guide to Trees*. New York: Knopf, 2009.

Siyabona Africa: Botswana Safari. "Trees of Tuli Block." Accessed February 10, 2016. http://www.itravelto.com/trees-tuli-block-botswana.html.

Siyabona Africa: Kruger National Park. "Baobab." Accessed April 2, 2016. http://www.krugerpark.co.za/africa_baobab.html.

Smale, Mark, David Bergin, and Greg Steward. "The New Zealand Beeches: Establishment, Growth, and Management." *New Zealand Indigenous Tree Bulletin* 6. Rotorua, NZ. Accessed March 2, 2016. http://maxa.maf.govt.nz/sff/about-projects/search/07-128/beech-bulletin.pdf.

Smith, Michael. "The Kutthung, or Kattang, Dialect was Spoken Amongst the Australian. "Aborigines Living Along the Southern Bank of the Karuah River and the South Shore of Port Stephens, New South Wales, Australia." Accessed March 26, 2016. http://users.hunterlink.net.au/~madms/kutthu.html.

Smyth, Robert B. *The Aborigines of Victoria: With Notes Relating to the Habits of the Natives of Other Parts of Australia and Tasmania. Vol. 2*. London: Adamant Media Corporation, 2005.

Society for Barefoot Living. "Free Your Feet and Your Mind Will Follow." Accessed September 5, 2015. http://www.barefooters.org/.

Spafinder: Wellness 365. "Global Spa Trends 2015 Report: Forest Bathing." Accessed January 14, 2016. http://www.spafinder.com/blog/trends/2015-report/forest-bathing/.

Starhawk. *The Spiral Dance: A Rebirth of the Ancient Religion of the Great Goddess*. San Francisco: Harper and Row, 1979.

State Library of Queensland. "South East Queensland Placenames." Accessed March 16, 2016. http://www.slq.qld.gov.au/resources/atsi/languages/queensland/southeast-queensland-placenames.

Statistic Brain Research Institute. "Footwear Industry Statistics." Accessed September 5, 2015. http://www.statisticbrain.com/footwear-industry-statistics/.

Steele, John G. *Aboriginal Pathways: In Southeast Queensland and the Richmond River*. St. Lucia, New York: University of Queensland Press, 1984.

Stein, William I. "Oregon White Oak." Accessed June 30, 2016. https://www.na.fs.fed.us/pubs/silvics_manual/volume_2/quercus/garryana.htm.

"Stephen Sillett/FINAL." from Past, Present and Future of Redwoods: A Redwood Ecology and Climate Symposium hosted by Save the Redwoods League. YouTube video, 46:39. Posted by "2001CMP," October 22, 2014. https://www.youtube.com/watch?v=iNBBcN_SCNY.

Stokes, John and Donald Rodger. *The Heritage Trees of Britain and Northern Ireland*. London, England: Constable & Robinson, Ltd., 2004.

Stritch, Larry. "*Betula papyrifera*." *The IUCN Red List of Threatened Species*. Accessed February 3, 2016. http://dx.doi.org/10.2305/IUCN.UK.2014-3.RLTS.T194502A2342659.en.

Stuart, John D., John O. Sawyer, and Andrea J. Pickart. *Trees and Shrubs of California*. Berkeley: University of California Press, 2001.

Sustainability in Prisons Project. "We Connect Prisons with Nature." Accessed October 14, 2014. http://sustainabilityinprisons.org/.

Taranaki Educational Resource: Research, Analysis and Information Network. "*Metrosideros robusta* (Northern Rata)." Accessed January 23, 2016. http://www.terrain.net.nz/friends-of-te-henui-group/trees-native-botanical-names-m-to-q/rata.html.

Tasmania Parks and Wildlife Service: Department of Primary Industries: Parks, Water and Environment. "Huon Pine." March 2011. Accessed January 21, 2016. http://www.parks.tas.gov.au/file.aspx?id=6575.

Tasmania Parks and Wildlife Service: Department of Primary Industries: Parks, Water and Environment. "Native Conifers of Tasmania." Accessed March 5, 2016. http://www.parks.tas.gov.au/?base=3240.

Thornburgh, Dale A. "*Quercus chrysolepis* Liebm: Canyon Live Oak." Accessed February 3, 2016. http://www.na.fs.fed.us/pubs/silvics_manual/volume_2/quercus/chrysolepis.htm.

TJs Garden. "Diseases of Trees - Mistletoe Growing on a Tree - Infected Mesquite Trees." Accessed February 5, 2016. https://tjsgarden.com/tag/mesquite-trees/.

Townsend, Mardie and Rona Weerasuriya. *Beyond Blue to Green: The Benefits of Contact with Nature for Mental Health and Well-Being*. Beyond Blue Limited: Melbourne, Australia, 2010.

Tsunetsugu, Yuko, Bum-Jin Park, and Yoshimfumi Miyazaki. "Trends in Research Related to 'Shinrin-Yoku' (Taking in the Forest Atmosphere or Forest Bathing) in Japan." *Environmental Health and Preventive Medicine* 15 (2010): 27-37. doi:10.1007/s12199-009-0091-z.

Tudge, Colin. *The Tree: A Natural History of What Trees Are, How They Live, and Why They Matter*. New York: Three Rivers Press, 2005.

Ulrich, Roger S., Robert F. Simons, Barbara D. Losito, Evelyn Fiorito, Mark A. Miles, and Michael Zelson. "Stress Recovery During Exposure to Natural and Urban Environments." *Journal of Environmental Psychology* 11 (1991): 201–230.

Uncharted Africa Safari Company. "Guides and Conservation." Accessed May 14, 2016. http://www.unchartedafrica.com/page.php?p_id=390.

United States Forest Service. "*Betula papyrifera*." *Index of Species Information*. Accessed August 27, 2013. http://www.fs.fed.us/database/feis/plants/tree/betpap/all.html.

United States Forest Service. "*Hesperocyparis macrocarpa*." *Index of Species Information*. Accessed January 24, 2016. http://www.fs.fed.us/database/feis/plants/tree/hesmac/all.html.

Van Noort, Simon and Jean-Yves Rasplus. "*Ficus sycomorus sycomorus* Linnaeus 1753." *FigWeb*. Iziko Museums of South Africa. Accessed February 11, 2016. http://www.figweb.org/Ficus/Subgenus_Sycomorus/Section_Sycomorus/Subsection_Sycomorus/Ficus_sycomorus_sycomorus.htm.

Van Pelt, Robert. *Forest Giants of the Pacific Coast*. Seattle: University of Washington Press, 2001.

Van Wyk, Ben-Erik. *People's Plants: A Guide to Useful Plants of South Africa*. Gauteng, South Africa: Briza Publications, 2007.

Van Wyk, Piet. *Field Guide to the Trees of the Kruger National Park*. Cape Town, South Africa: Struik, 1994.

Vennell, Robert. "Lord of the Trees: The Botany of Middle Earth." *The Meaning of Trees: The History, Use and Significance of New Zealand's Native Flora*. Accessed July 26, 2015. https://meaningoftrees.com/.

Venolia, Carol. *Get Back to Nature Without Leaving Home: 10 Simple Ways to Feel Happier, Healthier, and More Connected in Everyday Life*. Come Home to Nature. Accessed April 29, 2016. http://www.comehometonature.com/ebook-get-back-to-nature-without-leaving-home/.

Vesely, David and Gabe Tucker. "A Landowner's Guide for Restoring and Managing Oregon White Oak Habitats." Pacific Wildlife Research, October, 2004. Accessed June 30, 2016. http://www.blm.gov/or/districts/salem/files/white_oak_guide.pdf.

Wardle, P. "Biological Flora of New Zealand." *New Zealand Journal of Botany* 5, no. 2 (1967): 276-302. doi: 10.1080/0028825X.1967.10428746.

Watson, Rupert. *The African Baobab*. Cape Town, South Africa: Struik, 2007.

Weisman, Alan. *The World Without Us*. New York: St. Martin's Press, Thomas Dunne Books, 2007.

White, Gemma. "The Bush Tucker Survival Guide." *Survival, Tracking, and Awareness: The Ancient Skills for Living with the Earth*. Accessed March 4, 2016. http://www.survival.org.au/bush_tucker_survival_guide.php.

Wickens, Gerald and Pat Lowe. *The Baobabs: Pachycauls of Africa, Madagascar and Australia*. Berlin, Germany: Springer, 2008.

Wilkin, Dieter and Julie Burgher. "Oregon Oak." United States Department of Agriculture Natural Resources Conservation Service, National Plant Data Center. Accessed June 30, 2016. https://plants.usda.gov/plantguide/pdf/pg_quga4.pdf.

Willard, Dwight. *A Guide to the Sequoia Groves of California*. Berkeley: Yosemite Association/Heyday Books, 2000.

Williams, Cheryll. *Medicinal Plants in Australia. Vol. 1, Bush Pharmacy*. Kenthurst, NSW, Australia: Rosenberg Publishing, 2010.

Williams, Florence. "Take Two Hours of Pine Forest and Call Me in the Morning." *Outside Magazine*. Accessed April 28, 2016. http://www.outsideonline.com/1870381/take-two-hours-pine-forest-and-call-me-morning.

Williams, Terry Tempest. *An Unspoken Hunger*. New York: Vintage Books, 1995.

Williams, Terry Tempest. *Desert Quartet: An Erotic Landscape*. New York: Pantheon, 1995.

Wilson, E. O., *Biophilia*. 3rd ed. Boston: Harvard University Press, 1984.

Wilson, Matthew. "A Look at the Ancient Roots of the Olive Tree, From Adam to Zeus." *The Financial Times: Gardens*. Accessed March 25, 2016. http://www.ft.com/cms/s/2/61ee463e-0b3a-11e4-9e55-00144feabdc0.html.

Wohlleben, Peter. *The Hidden Life of Trees: What They Feel, How They Communicate—Discoveries from a Secret World*. Vancouver, BC, Canada: Greystone Books, 2016.

Woodford, James. "Big Tree Nearly Out of the Woods." *The Sydney Morning Herald*. Accessed March 28, 2016. http://www.smh.com.au/news/Environment/Big-tree-nearly-out-of-the-woods/2005/01/17/1105810845441.html.

Woodland Trust. "White Willow (*Salix alba*)." Accessed April 2, 2016. https://www.woodlandtrust.org.uk/visiting-woods/trees-woods-and-wildlife/british-trees/native-trees/white-willow/.

Wood Solutions. "Tallowwood - *Eucalyptus microcorys*." Accessed March 27, 2016. http://www.woodsolutions.com.au/Wood-Species/Tallowwood.

Wu, Tim. "The Slow-Photography Movement." *Slate Magazine*. Accessed April 29, 2016. http://www.slate.com/articles/life/obsessions/2011/01/the_slowphotography_movement.html.

Acknowledgements

Deep gratitude for the trees and the spirits of the lands who welcomed and guided me in my work, and deep gratitude for all the humans that made this book possible:

For inspiration– the remarkable Thomas Pakenham whose photography books sent me on a treasure hunt for my own intimate meetings with remarkable trees around the world. For seminal artistic collaboration with the wild– Caiyloirch Rupert Marques, Devin Fleurdujon, and Grahame "Platy" Fairall, who started it all. For courage and beauty, my treegirls– Nicteha Cohen, Anna Leppur, Kathy Stewart, Suzen Dyslin, and Char Horning.

For exceptional nature guiding and photographic assistance: Nico Visser at Nature Friend Safaris, Namibia; Mike Schyff at Copper Sun Tours, South Africa; Chabaimaketse Selei at Expert Africa, Botswana, Brett Greenaway at Wilderness Safaris, South Africa; Manda "Everything is Possible" Razakavonison and staff at Cactus Tours, Madagascar; Laurent Dick of Wild Alaska Tours; Ken Wu, TJ Watts and the Ancient Forest Alliance, Canada. For guiding, warm hospitality and leech and tick safety– Robert Stavro, Australia. For enthusiastic eucalypt guiding and expertise– Dean Nicolle, Australia. For making my ancient olive dreams come true– Dimitri Psarras, Haris Papagiannakis, and Eleni Chaireti of Green Tour, Crete. For exceptional guiding and photography mentoring– Justin Reznick, Kohve Tavakkol, along with the support of the Legendary Photographers of Christchurch, New Zealand: Terry Koyama, James Newkirk, Lynda Holman, Eva Bernstein, and Ron Bernstein. For invaluable photography assistance– StarMan, Martin Haigh, Lewis Regan, Chip Regan and Candace Regan. For legendary nature guiding through the wilds of Miami, Florida and photographic support– The Legend himself. For generous photographic assistance, tech support and tenacious copiloting on the karmic ride– Damien McAnany.

For unconditional support and belief in my vision– my extended family, Derrick Jensen, the New College of California community, the global European Ecopsychology Community, Tina Fields, Maria Owl Gutierrez, Liz Reuter, Melissa Patterson, Dominique Peters, Rebecca Valentine and Alison Willets. For providing me a sanctuary while I worked– Elisabeth Sherman, and home-base support while I travelled in the field– Arianna 'Aunsheen' Husband. For giving me writing refuge in their home with access to their ethnobotany library– Zoë Gardner, Ben Woodward, and Noah Norbert. For infinite life support, love and energy richness– Lynn Abraham, John Amaral, Christina Amaral, and the entire Body Centered Leadership field.

For impassioned, embodied moral and technical support– Erica Mueller. For indespensible research assistance and moral support– Sequoia Etcheverry. For superlative administrative assistance– Diane Osborne, Samara Seibel and Itan Mushik. For tireless and expert editing and cheerleading– Carol Venolia, Teja Watson, Tom Wilhite, and Diana Young. For exceptional care in proofing and fine art printing– Darren Briggs and Sara Silver at Skylark Images . For generous book review and moral support– Paul Hawken, Antoine de Hardy le Beaulieu and Fred Hageneder. For indespensible publishing consulting– Cynthia Frank and Cypress House. For superb publicity– Jane Wesman Public Relations. For enthusiastic, dedicated, innovative book design and bringing TreeGirl to life beautifully on the page– Annett Boerner.

Heartfelt gratitude for all the tree lovers who supported the funding of the publication of this book.

About the Author

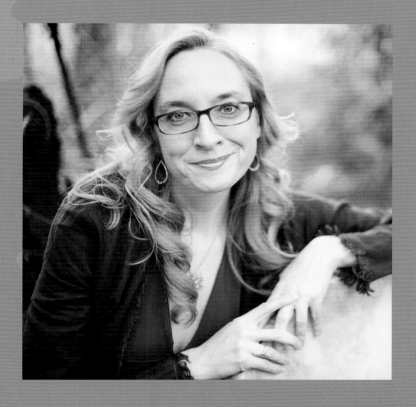

Julianne Skai Arbor, aka TreeGirl, has been on the forefront of the fields of ecotherapy and nature connection since she began creating self-portraits with trees in 1995. To date, she has intertwined with over seventy species of trees in thirteen countries. She is certified as an arborist with the International Society of Arboriculture, as a California naturalist with the University of California, and has taught interdisciplinary college-level conservation education for over ten years, including pioneering the first program in environmental arts. She holds graduate degrees in Environmental Education and Arts and Consciousness Studies. With her passion for trees she creates an experiential bridge to connect people with nature through forest ecotherapy, portrait sessions, and immersive nature retreats with trees around the world. She lives in Sonoma County, California amidst the native oak and coast redwood trees. Her gallery of images can be found at www.TreeGirl.org and www.modelsociety.com/Photographer/TreeGirl.

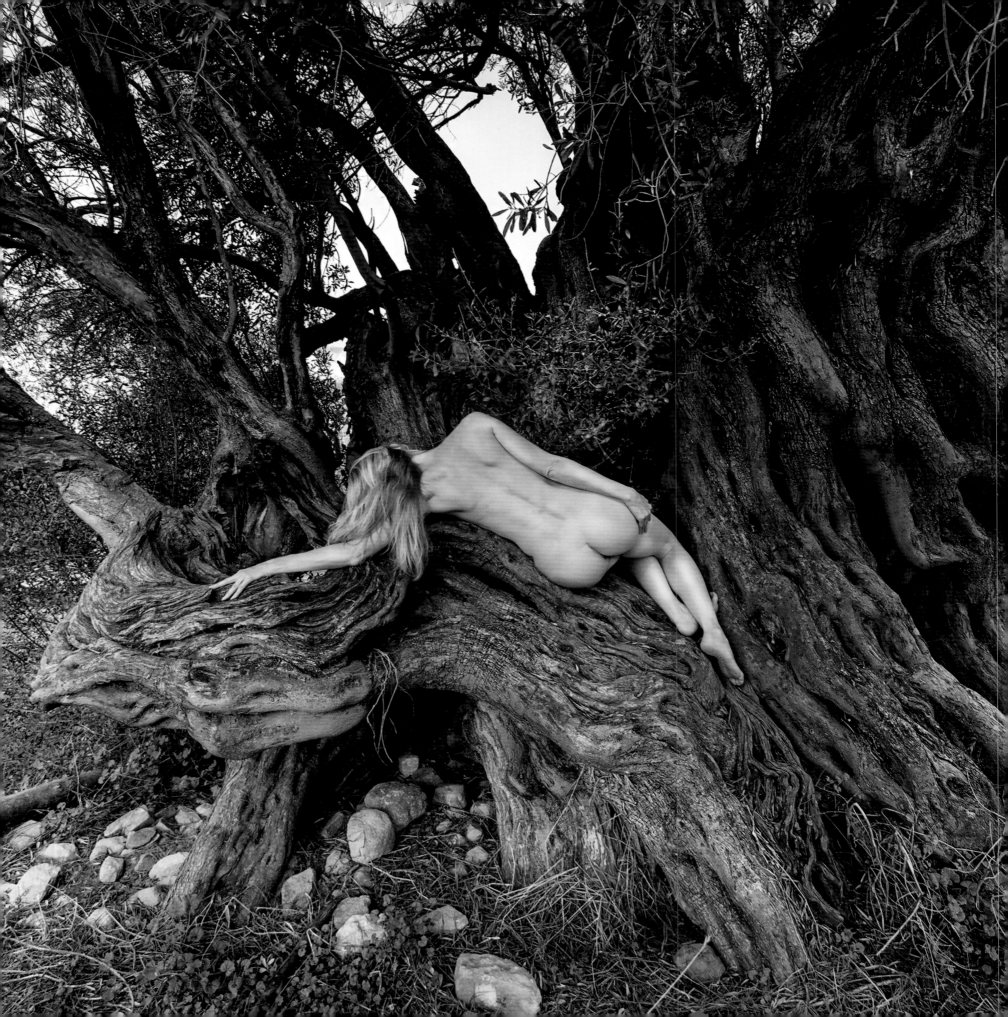

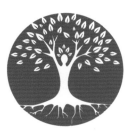

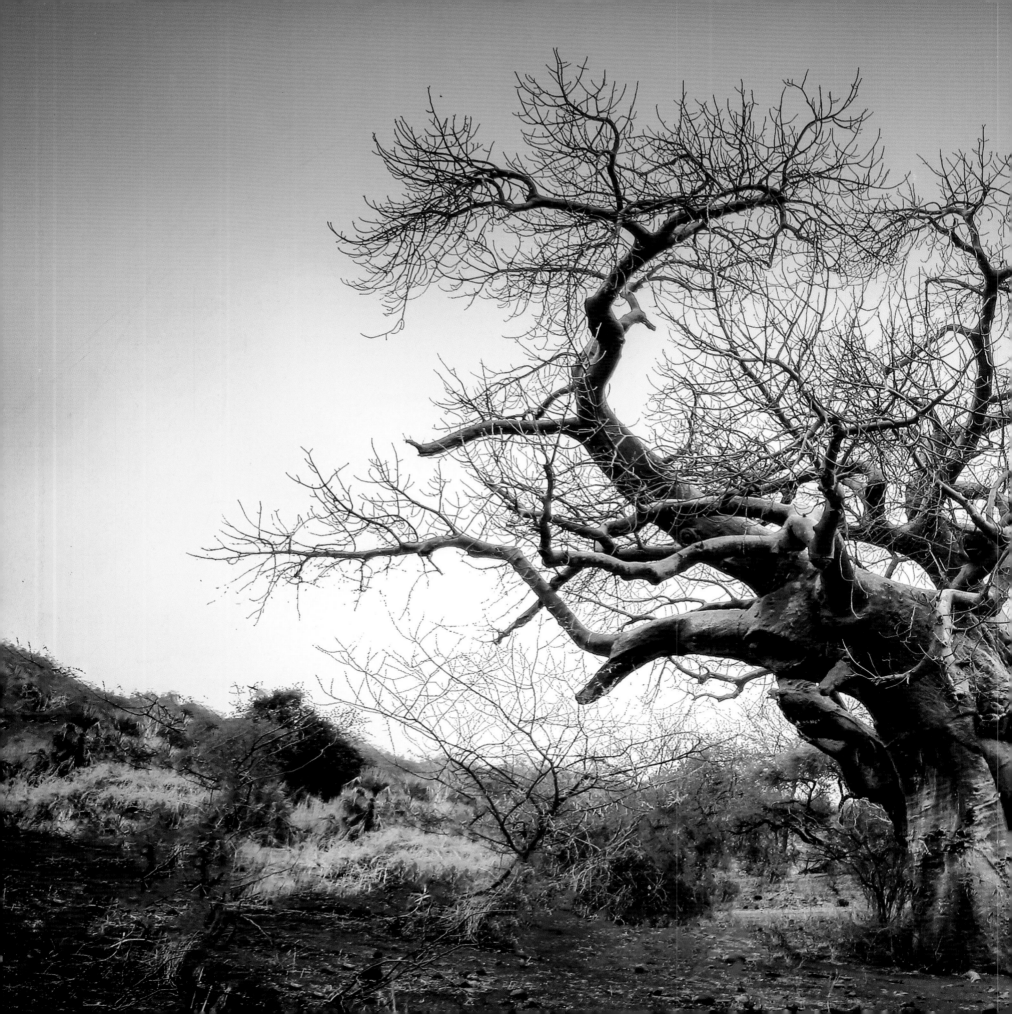